BASIC ELECTRICITY

Books by Rufus P. Turner

BASIC ELECTRICITY

BASIC ELECTRONIC TEST INSTRUMENTS

A potential difference of several million volts causes the lightning flash, the most spectacular example of electricity in nature. (*Courtesy, Bell Telephone Laboratories*)

Rufus P. Turner

Registered Professional Engineer

BASIC ELECTRICITY

Technical Division

Holt, Rinehart and Winston, New York

Third Printing, February, 1961

08567-0217

Printed in the United States of America
Library of Congress Catalog Card Number: 57-11620

To Mary

ELECTRICITY ....

carrier of light and power ....

devourer of time and space ....

bearer of human speech over land and sea ....

greatest servant of man ....

itself unknown.

Charles W. Eliot (1834–1926)

Preface

This book is intended as a text for beginning students of electricity. For this reason, the author has endeavored to make its explanations and illustrations as clear as possible and to keep the theme descriptive in nature. Mathematical derivations have been avoided except in certain instances where their illustrative value seems unique and exclusive.

The material has been assembled expressly for use by students in trade schools and technical institutes so that they may provide themselves with a theoretical background needed in their practical applications and for further study. However, more advanced students and electrical workers might find the text helpful as a quick refresher on basic topics forgotten through lack of use. It is felt that a painstaking study of this book will provide a strong background for the later study of electronics.

Illustrative examples are employed freely to demonstrate step-by-step methods of calculation. Chapter summaries as well as questions and exercises appear at the ends of chapters. It is hoped that these features will be of some aid to both students and instructors. For the home-study student, the exercises and questions offer an opportunity for testing progress.

With any subject that has grown as large as electricity, many details must of necessity be omitted when the subject is to be surveyed in a single volume. This is especially necessary when the material is addressed to the beginning student who, understandably, may become thwarted by the sheer bulk of superfluous data so often appearing in print. The author has tried to eliminate nothing that is essential, but has endeavored to give the student a broad view of basic electrical theory.

The writing of this text has not been a conscious attempt to sugar-coat electrical theory or to offer a royal road of short cuts to technical knowledge. The student needs to study each section diligently and in sequence. Indeed, the only purpose in bringing forth a new book on electricity at present is to introduce a fresh attack and to bring a new clarity to the presentation of old fundamentals, all of this liberally interspersed with modern applications of basic theory. The author hopes he has achieved these objectives.

RUFUS P. TURNER

Los Angeles, California
August, 1957

Acknowledgments

The author acknowledges with appreciation illustrations furnished by the following organizations:

Bell Telephone Laboratories, Inc.
Bogue Electric Manufacturing Co.
Central Scientific Company
Cornell-Dubilier Electric Corporation
De Jur-Amsco Corporation
Department of Water & Power, the City of Los Angeles
Electric Specialty Company
General Electric Company
International Rectifier Corporation
International Resistance Company
New York Life Insurance Co.
Nippert Electric Products Co.
Simpson Electric Company
Western Electric Company
Weston Electrical Instrument Corp.

Contents

BASIC ELECTRICITY

Chapter 1

Basic Fundamentals

1.1 HISTORICAL BACKGROUND

Man has been having some sort of experience with electricity since the time he first appeared on earth. He could not have escaped this because electricity is found throughout nature. The lightning flash is one evidence of its existence. Other primitive examples are the northern lights, the crackling and sparking of hair and fur when it is stroked or combed, and the ability of the electric eel to deliver a powerful shock. Electricity is present even in the human body as a part of the life process. But man had lived on earth countless centuries before he first recognized electricity for itself and began to build his store of knowledge concerning it.

The record begins in Greece. There, Thales, a statesman, philosopher, and traveler, had gained distinction by predicting accurately an eclipse, using methods he is believed to have learned from the Babylonians. But his lasting fame, as far as electrical scientists are concerned, was to come from what then appeared a trivial discovery. In about 600 B.C., he noticed that amber, a yellow fossil resin, when rubbed, acquired the curious ability to attract light bits of material. This is our earliest record of such an observation, but we recognize the possibility that some unnamed person may have noticed the effect in still earlier times. Thales might even have picked up his information from someone else during one of his many journeys.

We have no record that Thales gave any name to the phenomenon. But the Greek word for amber is *elektron.* Several hundred years later William Gilbert listed a number of other materials which, when rubbed, behaved like amber, and he called these *electrics*—that is, materials that behave like elektron. From this stems our word "electricity."

Although Saint Augustine distinguished between electricity and magnetism as early as A.D. 426, over a thousand years after the discovery by Thales, it was only at the beginning of the seventeenth century that important early electrical developments took place. Around 1600, William Gilbert, English physician to Queen Elizabeth, became an avid experimenter. He made an *electroscope,* an instrument for detecting the presence of electricity; and, as mentioned in the preceding paragraph, was the first person to use the word "electric." Between 1650 and 1671, Otto von Guericke invented a machine for generating electricity by means of friction. It consisted of a sulphur globe rotated on a spindle while being rubbed against the palm of the operator's hand resting upon it. In 1675, Sir Isaac Newton in England discovered electrostatic induction.

The next century witnessed several basic discoveries of great significance. Gray showed the difference between conductors and insulators in 1729, and Dufay distinguished between positive and negative electricity in 1733. Between 1745 and 1746, Winckler, von Kleist, and Musschenbroek independently discovered a device, the *Leyden jar,* for storing electricity. In 1746, Gralath built an *electrometer,* the first instrument for measuring electricity. In America in 1752, Benjamin Franklin performed his famous kite experiment to demonstrate the identity of lightning and electricity; and in France in 1785, Charles Augustin Coulomb proved the law of inverse squares in connection with electric and magnetic attraction and repulsion. The production of electricity by chemical action and in the form of a current was observed for the first time, in 1786, when Aloisio Galvani noticed that frog legs contracted when touched by two connected dissimilar metals. This was the modest beginning of the electric battery. Alessandro Volta, an Italian professor, in 1799 crystallized Galvani's experiment by inventing the *voltaic pile,* the first primary battery. Volta earlier had invented an electroscope and the *electrophorus,* a device for generating electricity by means of friction.

The nineteenth century was the golden age of early electrical invention and discovery. Starting in 1800 with William Nicholson (England) and his colleague Carlisle, who employed an electric current to decompose water, new observations came in rapid succession. Hans Christian Oersted (Denmark, 1820) discovered that an electric current deflects a magnet. André Marie Ampère (France, 1820) demonstrated electromagnetic reaction. Michael Faraday (England, 1821) caused a current-carrying conductor to rotate about a magnetic field, and this laid the foundation for the electric motor. In 1831, he constructed a disc dynamo, the first mechanical "generator" of electric current. In 1825, William Sturgeon (England) made the first electromagnet.

In Germany in 1825, Georg Simon Ohm discovered the law that

bears his name. In 1831, Joseph Henry (United States) and Faraday (England) independently discovered mutual induction; and during the following year, Henry discovered self-induction. Karl Friedrich Gauss and Wilhelm Edward Weber (Germany, 1833–1838) built and operated a rudimentary telegraph. Samuel F. B. Morse (United States, 1837) invented the practical telegraph. The arc light was first used commercially in 1846. Mirand invented the electric bell in 1850. Cyrus Field (United States, 1858) laid the first Atlantic cable. James Clerk Maxwell (England, 1864) developed his celebrated electromagnetic theory of light, the mathematical foundation of radio.

Henry A. Rowland (United States, 1873) explained the magnetic circuit. Alexander Graham Bell (United States, 1876) invented the telephone. Thomas A. Edison (United States, 1879) invented the practical incandescent lamp. Ernest W. von Siemens (Germany, 1879) successfully demonstrated a three-car electric train utilizing a third rail and carrying up to 30 passengers at a speed of 8 miles per hour. Heinrich Rudolph Hertz (Germany, 1888) verified experimentally the predictions of Maxwell by demonstrating electromagnetic waves and thus opening the field of practical radio. In Germany in 1895, William Conrad Roentgen discovered X rays. This was followed by the discovery of radioactivity by Antoine Henri Becquerel (France, 1896). In Italy in 1897, Guglielmo Marconi sent radiotelegraph messages over a distance of 9 miles. Between 1897 and 1899, the electron was identified as a corpuscle by Joseph J. Thomson (England) and was given its name (again recalling the Greek name for amber) by G. Johnston Stoney (England, 1891).

Twentieth-century electrical development has been distinguished by the widespread use of radio, development of picture transmission and reception, introduction and commercialization of television, rise of electronics, development of radar, invention of electronic computers, development of semiconductor devices, and utilization of atomic energy. Progress also has been made in the use of electricity in the diagnosis and treatment of disease.

The crystal detector for radio reception was introduced in 1906 and succeeded, although J. A. Fleming had introduced the two-element vacuum tube two years earlier in England. Lee de Forest (United States, 1906) invented the three-element vacuum tube, the device which laid the foundation of modern electronics. Transatlantic radiotelegraphy was accomplished for the first time from the United States in 1915.

The first long-distance broadcast and reception of television in the United States took place in 1927. Robert J. Van de Graff (United States, 1929) designed an electric machine capable of generating several million volts.

Ernest O. Lawrence invented the cyclotron "atom smasher" in the United States in 1931. Although fragmentary observations of basic radar phenomena date back to the 1922 experiments of A. Hoyt Taylor (United States), radar reached its state of practical development during World War II (1941–1945). During this same period, improved modern counterparts of the early crystal detector, silicon and germanium diodes, were developed. The germanium diode led to the invention of the transistor (John Bardeen and W. H. Brattain, United States, 1948), a device which promises to take over many of the functions formerly performed exclusively by vacuum tubes.

1.2 WHAT IS ELECTRICITY?

Electricity is a form of energy and is another manifestation of the deep, underlying forces of the universe. Although we know a great deal about its effects, only a few electrical workers attempt to understand the exact nature of electricity. Our large store of electrical knowledge concerns mainly what electricity *does* and how we can control this energy; nevertheless, scientists have pried deeply into the nature of electricity and its relationship to other forms of energy.

The complex explanations offered by advanced physicists and mathematicians are beyond the scope of this book. However, the first chapters will present physical pictures of electricity at rest and in motion, and the student must make these concepts a part of his thinking.

1.3 STATIC ELECTRICITY

One of the commonest effects of electricity is the one that caught Thales' eye. When some materials are rubbed with certain others, they become electrified, or charged and will attract bits of paper and other light chaff just as a magnet attracts iron. A clean, dry, glass rod, for example, may be charged by rubbing it with a silk cloth. So may a stick of hard rubber when rubbed with flannel, or a stick of sealing wax rubbed with cat's fur. The energy with which these materials are charged is termed *static* electricity because it is motionless. It also is called *frictional* electricity from the fact that the charge is generated by rubbing.

An electrified body will retain its charge for a certain length of time, but eventually the electrification will leak off. The body also may be discharged quickly, when desired, by connecting it to the earth. A plastic comb, after passing briskly through the hair on a cold, dry day, is found to be strongly charged and will attract objects from a good distance. After a few minutes, however, the charge will have dissipated to such an extent that only nearby objects will be drawn to the comb, and soon the charge disappears completely. If the fully charged comb is touched to a cold-water pipe, or similar metallic object which is connected to the earth, it will lose its charge almost immediately.

1.4 ELECTROSTATIC FIELD

Extending in all directions around a charged body, there is a region inside of which objects are attracted to the body. The strength of this attraction diminishes rapidly as the space between the body and object increases.

This region of influence is called the *electrostatic field* or simply *electric* field. The field is imagined to contain many *electrostatic lines of force* which extend outward in all directions from the charged body. Collectively, these lines of force are termed *flux*. The number of lines per unit area is the *flux density* (example, so many lines per square centimeter).

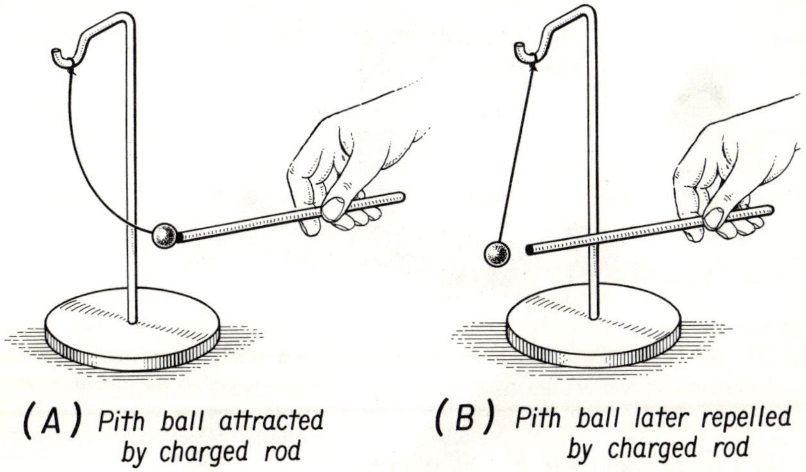

Fig. 1–1 Demonstration of Charge Polarity

1.5 POSITIVE AND NEGATIVE ELECTRIFICATION

Two kinds of electrification, *positive* and *negative*, are recognized. Their difference is easily demonstrated by the experiment shown in Figure 1–1.

A pith ball is suspended on a silk thread. Rub a glass rod with a silk cloth and bring the end of the rod near the ball. Under the influence of the electrostatic field about the rod, the ball will be attracted. [See Figure 1–1(A).] The ball sticks to the rod for a short time and acquires some of the charge of the rod. (This is the process of *electrostatic induction.*) Then it falls away.

Once again, bring the tip of the charged rod close to the ball. This time, the ball is pushed away, as in Figure 1–1(B). The charged rod now no longer attracts the ball but *repels* it.

This experiment shows that when the ball is neutral (uncharged), it is attracted by the charged rod. But when the ball has acquired some of the same sort of charge as the rod, it is repelled.

Now, with the ball charged from contact with the charged glass rod, rub a stick of sealing wax with cat's fur and bring the end of the stick near the ball. This time the charged ball is attracted. This indicates that the sealing wax is not charged in the same way as the glass.

Plate 1–1 An Electric Machine. This *Wimshurst* machine generates intense charges of static electricity. It consists basically of two glass discs each containing spaced, metal-foil sectors. As the crank is turned, the discs rotate in opposite directions. The foil sectors pass under metal brushes which rub against them. Potentials as high as 60,000 volts are obtainable with this type of machine.

From these experiments, we deduce that (1) *like charges repel,* and (2) *unlike charges attract* each other. This is a basic phenomenon which is encountered in all applications of electricity.

These two opposite kinds of electricity are distinguished as *positive* (the kind possessed by the glass rod) and *negative* (the kind possessed by the sealing wax). This nomenclature was introduced by Benjamin Franklin. The condition of positiveness or negativeness is termed *polarity.* Positive polarity is indicated by the plus sign (+) and negative by the minus sign (—). Positive and negative polarity are encountered in all applications of electricity.

1.6 CHARGE ATTRACTION AND REPULSION

It has already been noted that like charges (+ and +, or — and —) repel each other and that unlike charges (+ and —, or — and +) attract. Figure 1–2(A) shows lines of force between two spheres having like charges of electricity. Figure 1–2(B) shows lines of force between

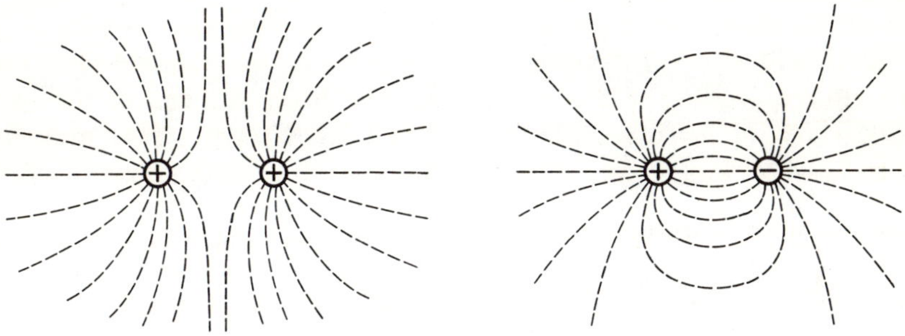

Fig. 1–2 Charges and Fields

similar spheres having opposite charges. These illustrations show only those lines lying in the plane of the page. Remember that a similar arrangement of lines exists in all planes up to the vertical and above and below the page, since the field extends in all directions around the charges.

Note that the lines of force between the two positive charges in Figure 1–2(A) oppose each other and do not meet. (Two negative charges, also being alike, would give the same effect.) But in Figure 1–2(B), many of the lines of force extend from one to the other of the unlike charges.

Coulomb's law expresses the force between two charged bodies in air or vacuum:

$$F = (Q_1 Q_2)/d^2 \quad (1\text{–}1)$$

where F = the force (in dynes) acting between the charged bodies
Q_1 = charge on one body (in statcoulombs)
Q_2 = charge on other body (in statcoulombs)
d = separation between bodies (in centimeters)

The *statcoulomb* (also called the *electrostatic unit* or esu) is the electrostatic unit of charge and is of such amount that if a charge of this value is placed 1 cm from an identical charge (in free space), it is repelled with a force of 1 dyne. (1 dyne = 2 millionths of a pound.) Although the coulomb is the practical unit of quantity of electricity, it is so large that the statcoulomb, a much smaller unit, commonly is used. (1 coulomb = 3 billion statcoulombs.) The coulomb is named in honor of Charles Augustin Coulomb (1736–1806).

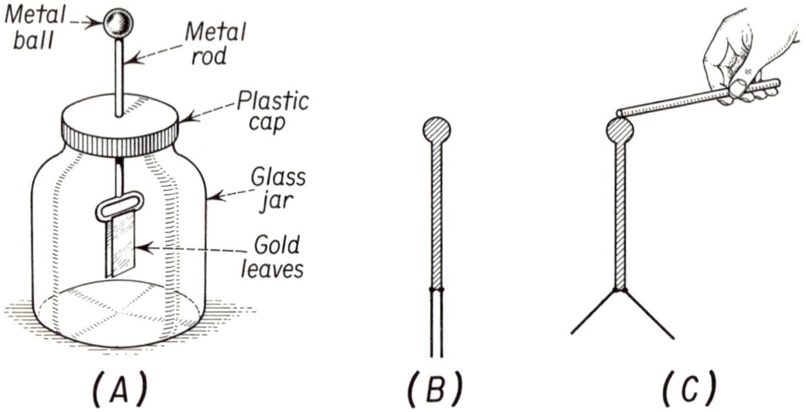

Fig. 1–3 How the Electroscope Detects a Charge. (A) The electroscope is a simple device for detecting an electric charge. It has the simple construction shown above. Two gold leaves are hung in a jar on the end of a rod or wire connected externally to a smooth, metal ball. (B) Normally, the gold leaves hang vertically. (C) But when a charged object is touched to the ball (or sometimes just brought near the ball), the leaves spread apart. The reason for this is that, being connected together, both leaves receive the same charge (+ or −) and accordingly repel each other.

1.7 ELECTROSTATIC POTENTIAL AND POTENTIAL DIFFERENCE

If a heavy box is lifted from the floor and placed on a high shelf, *potential energy* is stored in the box as a result of the work done against gravity to lift it to the shelf.

Similarly, when a charge is moved in an electrostatic field against the opposing action of the field, potential energy is stored in the charge by the work done to move the latter. The amount of potential energy in this instance depends upon the position of the charge in the field. Electrical potential thus is "electrical level." A unit electrostatic field has such strength that when a unit charge is placed in the field, the force acting on the charge is 1 dyne.

It is easy to see that between two given points in an electrostatic field there is a *potential difference*. This is equal to the work which must be done to move a unit positive charge from the point of lower potential to the point of higher potential. The unit of potential difference is the *volt*, named in honor of Alessandro Volta (1745–1827).

Potential differences in electrostatic fields are expressed in practical work as so many *volts per inch* or *volts per centimeter.*

Two points, *A* and *B*, in an electric field have a potential difference of 1 volt (v) when an amount of work equal to 1 joule (j) per coulomb is done against the field in moving a positive charge from Point *A* (lower potential) to Point *B* (higher potential). The following equation shows the relationship between work, potential difference, and charge.

$$W = QE \tag{1-2}$$

where W = work (in joules) (1 j = 0.738 ft-lb)
Q = electric charge (in coulombs)
E = potential difference (in volts)

Similarly

$$E = W/Q \tag{1-3}$$

and

$$Q = W/E \tag{1-4}$$

1.8 CONSTITUTION OF MATTER

In order to understand more clearly some of the electrical facts which follow, it is necessary to consider, at this point, the structure of matter.

All matter may be grouped into three categories: solid, liquid, and gaseous. Solids may be classified further as metals and nonmetals.

If a particular kind of matter cannot be broken down finally into any other kind but retains its unique features in spite of being ever so finely divided, it is called an *element.* There are 98 known *natural* elements and a few others that have been produced artificially in atom smashers. Copper, sulphur, aluminum, iron, and iodine are examples of elements. When a particular kind of matter is a chemical union (not merely a mixture) of two or more elements, it is called a *compound.* A compound is a sort of chemical marriage between several elements. Sodium chloride (common table salt) is an example of a compound. It is a chemical union of the metal, sodium, and the gas, chlorine. Other well-known compounds are hydrogen peroxide (hydrogen and oxygen), carbon dioxide (carbon and oxygen), water (hydrogen and oxygen), and carbon tetrachloride (carbon and chlorine).

All matter is made up of extremely tiny building blocks. The building block of an element is the *atom.* The atoms of any given element are identical particles. Thus, every copper atom "looks" like every other copper atom but differs from, say, a silver atom. (We place the word "looks" in quotation marks because the atom is too small to be seen even with the aid of the most powerful microscope.) The building block of the compound is the *molecule.* Each molecule consists of one or more atoms of each of the elements that combine to form the compound. The same number of atoms may be found in each molecule of a given compound. For example, each sodium chloride (NaCl) molecule contains 1 sodium (Na) atom and 1 chlorine (Cl) atom.

At one time, the atom was thought to be the smallest subdivision of matter, but scientific investigations, first carried on during the latter part of the nineteenth century, disclosed still smaller particles inside the atom. The atom is believed to consist of a heavy central core, called the *nucleus,* and of one to many lighter particles, called *electrons,* which travel around the nucleus in fixed paths or orbits. The atom thus is a submicroscopic replica of our solar system with its planets traveling around the sun.

It has been determined theoretically and verified experimentally that the nucleus has a net positive electric charge, and that each of the electrons has a negative charge. The sum of the negative electron charges equals the positive charge of the nucleus, and the two charges therefore cancel each other with the result that the atom normally is neutral; that is, it exhibits no net electrification. The nucleus consists of a certain number of electrons and *protons* (positive particles) bound tightly together.

The nature of an element is determined by the characteristic number of electrons and protons in its atoms. This fact accounts for the strong individuality of the atoms. The electrons and protons in one atom, however, are the same kind that are found in *all other* atoms. What makes the atoms, and the elements they form, different is the *number* of these tiny particles in each atom.

The hydrogen atom is thought to be the simplest in nature, consisting of 1 electron rotating about 1 proton. The heavier an element, the more electrons and protons its atoms contain. The nucleus of the oxygen atom contains 16 protons and 8 electrons bound tightly together (the positive charge thus is dominant, since there are more protons than electrons) and 8 rotating (*orbital* or *planetary*) electrons each traveling in its own path around the nucleus. The nucleus thus has a charge of $+16$ and $-8 = +8$. This is balanced by the eight negative (-1) charges of the orbital electrons $= -8$. The oxygen atom accordingly is neutral, since $+8 - 8 = 0$.

Another particle found inside the nuclei of atoms is the *neutron.* The neutron has no charge. It is found bound like a cementing agent with the protons and electrons in the nucleus.

It is interesting to consider some of the important facts concerning atoms, and we can do this without getting too deeply involved in chemistry. One fact is that the atom is nearly all empty space, therefore it is not a "solid" but only seems so. This is apparent from Figure 1–4(A) which illustrates the hydrogen atom. The diameter of the electron in this atom is 4×10^{-13} cm. That of the nucleus is only 1/1,834 of this figure. This means that if the nucleus were the size of a regulation baseball and located at New York, the electron could be 300 ft in diameter, and its orbit would pass through Kansas.

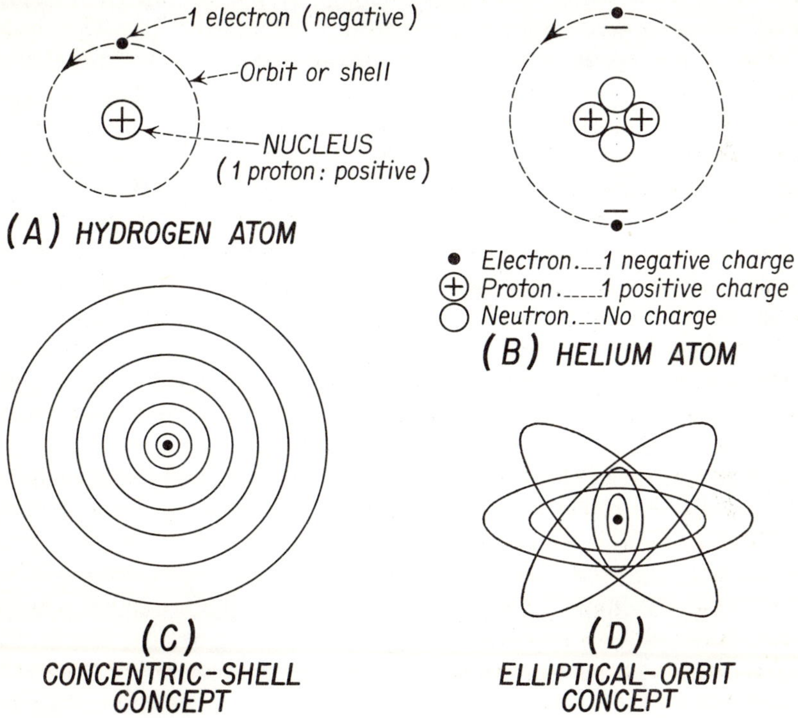

Fig. 1–4 Atomic Details

Figure 1–4(B) shows the structure of an atom of the gas, *helium*. Here, there are two orbital electrons (total negative charge = −2) and two protons (total positive charge = +2). The protons are cemented into the nucleus with the aid of 2 neutrons (total charge = 0). The net charge of the helium atom thus = +2 − 2 = 0.

Nature's heaviest atom is that of *uranium* 238, which has 92 orbital electrons. Its nucleus consists of 92 protons and 146 neutrons.

No person knows exactly how an atom looks inside. It is convenient to imagine the electrons traveling in well-ordered circular orbits around the nucleus. One concept is that the electrons occupy concentric shells, around the nucleus, as depicted by Figure 1–4(C), the separation of these shells increasing as we move out from the nucleus. A later concept is that the electron orbits are nonconcentric ellipses, as shown in Figure 1–4(D).

The electron completes its trip around the nucleus 10 thousand trillion times each second at a velocity of 4,000 miles per second. But it does not fly off into space because the equal and opposite positive charge of the proton nucleus is sufficient to hold it, like a rope, to its circular orbit. It is no wonder, then, that although atoms are filled with empty space, matter appears solid to our senses.

The charge on the electron and also on the proton has been found to be 4.774×10^{-10} statcoulomb (Robert A. Millikan, United States, 1916). The symbol for this charge is ϵ, the Greek letter *epsilon.*

$$\epsilon = 4.774 \times 10^{10} \text{ esu} = 16 \times 10^{-20} \text{ coulomb} \quad (1\text{–}5)$$

This is an exceedingly tiny amount of electricity.

So far, we have considered only planetary electrons traveling in orderly paths. Actually, these are not the only ones to be found outside of the nucleus in the atom. In some atoms, there are found also a few electrons in the outermost shells which are not confined to fixed orbits and may move about with very little constraint, sometimes even being drawn out of the atom. These are called *free electrons.* They are somewhat similar to the comets among the planets. Metallic elements, such as copper, silver, gold, and aluminum, are rich in free electrons. Some metals have more of them than others.

Scientists have discovered particles other than electrons, protons, and neutrons in atoms. They have also come to think that the electron is not a particle at all, as we ordinarily think of particles, but a wave phenomenon. These matters are beyond the scope of this book.

1.9 IONIZATION

It was mentioned in the preceding section that atoms normally are neutral; that is, they exhibit no net electric charge. However, if one or more electrons are removed from the atom, there will be insufficient negative charge to neutralize the positive charge of the nucleus, and the atom will exhibit a positive charge. Similarly, if an atom gains an electron from outside, it then will have more negative than positive charges and will exhibit a net negative charge.

Atoms which are in an unbalanced state of this kind and thus show electrification are called *ions.* The process of ionization consists of adding electrons to, or removing them from, the atomic structure. A positively charged atom is a positive ion, or *cation;* a negatively charged atom is a negative ion, or *anion.*

When the glass rod was rubbed with silk in Section 1.3, it became positively charged (as explained in Section 1.5) because some of the free electrons in the outer orbits of the glass atoms were removed by the rubbing process. These electrons then were added to the silk atoms, so the silk acquired a negative charge at the same time. Likewise, rubbing the sealing wax with cat's fur transferred electrons from the fur to the wax, and the latter became negatively charged. Thus, ionization is responsible for the generation of static electricity in some substances.

Any mechanism which knocks electrons from an atom or injects electrons into an atom causes ionization. Such media include friction, heat, intense electric fields, X rays, radioactivity, light rays, and high pressure. Ionization occurs in metals, nonmetals, gases, and liquids.

1.10 ELECTRON DRIFT

Since an electron exhibits a negative charge, a free electron can be repelled by a sufficiently strong negative charge or attracted by a positive charge. If a positive charge is located close outside the atomic structure, a free electron may move out of the atom toward the charge.

In a long, slender wire, atoms may be visualized as arranged in line. If a positively charged body is placed in contact with the right end of the wire, and a negatively charged body in contact with the left end, a free electron may be drawn from the atom in the right end of the wire by the positive charge. That atom then becomes a positive ion, since it is missing an electron. Since its net charge is positive, it is able to attract a free electron from its neighboring atom on the left which, in turn, also becomes a positive ion and attracts a free electron from its neighbor. This action continues down the line, with electrons *drifting successively* from atom to atom until finally the leftmost atom of the wire becomes a positive ion and attracts an electron from the external negatively charged body. All atoms in the wire will have become neutralized, and the drift will cease at this point if the two external charges are removed.

This electron drift is the basis of electric current flow. Here we have electric charges in motion, in contrast to the stationary charges of static electricity considered earlier in this chapter. Electron drift is most pronounced in metals, since these materials have the largest numbers of free electrons to contribute to the flow. Electrons, being negative, move toward the positive charge. Current flow therefore is conceived as taking place from negative to positive.

It is evident also that a negative ion can drift toward a positively charged body, and a positive ion toward a negatively charged body. This can happen, however, only in materials such as liquids and gases where atoms are not bound as tightly together as in a metallic wire and can enter into a drift movement. Certain liquids, as we shall see later, are composed of mixtures of positive and negative ions; if charged bodies are immersed into such liquids, the positive ions will migrate to the negative body, and the negative ions will move to the positive body. This simultaneous movement constitutes a current. This is designated an *ionic* current, to distinguish it from the electronic current.

1.11 ELECTRIC CURRENT

Current electricity due to the electron drift described in Section 1.10 is electricity in motion. Current (I) is measured in terms of quantity of electricity (Q) and time (t). It usually is expressed basically in coulombs per second. The unit of current is the *ampere* (amp) named in honor of André Marie Ampère (1775–1836). The relationship between current, quantity, and time is shown by the following equation:

(1–6) $$I = Q/t$$

where I = current (in amperes)
Q = quantity of charge (in coulombs)
t = time interval (in seconds)

1 ampere = 1 coulomb per second. This means that a quantity of electricity equal to 1 coulomb flows past any point in the system in 1 sec. When the current strength is 1 amp, 6,300 quadrillion electrons flow past a given point each second.

Current flows only between points exhibiting a difference of potential. Electrons flow toward a positively charged body. The latter, therefore, is the point of higher potential. The potential of the earth is zero. When a negatively charged body is connected to the earth, electrons flow from it to the earth, because a negative quantity is lower than zero. Conversely, when a positively charged body is connected to the earth, electrons flow from the earth to it.

A potential difference must be maintained in order for a current flow to be supported. It is the purpose of all generators of electricity so to maintain a difference of potential.

1.12 CONDUCTORS, INSULATORS, SEMICONDUCTORS

A conductor is a material that supports current flow. Since the atoms of metallic elements contain the most free electrons, the best practical conductors are metals. Some metals are better conductors than others. Carbon, a nonmetal, is an inferior conductor. Ionic conduction takes place in gases and in *electrolytes* (acid, alkali, or salt solutions).

An insulator (also called *dielectric* or *nonconductor*) is a material that does not support current flow. Insulators have only an occasional free electron. Since there is no completely perfect insulator, an exceedingly tiny current will flow through even a good dielectric if the potential difference is made quite high. This current, owing to the occasional free electrons, is so minute as to be ignored in most instances. Examples of insulators are glass, mica, rubber, dry wood, dry paper, most plastics, pure water, shellac, and dry fabrics.

Insulators are used for structures which must come in contact with current-carrying bodies but must not carry away any of the current. Thus, supports for wires are made of insulating material, such as ceramics or plastics, and wires themselves are covered with rubber, enamel, cotton, or silk.

Semiconductors are a whole class of materials which lie near the borderline of conductors and insulators, sharing some of the properties of each. When two pieces of properly chosen semiconductor material are placed together in intimate contact, as we shall see later, the junction between the two acts like a good conductor when current passes over it in one direction, but almost like an insulator when current tries to pass in the opposite direction. Examples of semiconductors are silicon, germanium, copper oxide, and copper sulphide.

1.13 SOURCES OF ELECTRICITY

There are six principal sources of electricity.

1. Friction. Static electricity is generated by the action of friction between two materials. This has been discussed in Section 1.3.

2. Chemical Action. A great deal of the world's electricity is produced by batteries. These devices generate a difference of potential by means of chemical action.

3. Heat Action. Two dissimilar metals bonded intimately together in a junction will, when heated, exhibit a difference of potential. Such a bond is called a *thermocouple*. The tip of an iron wire, for example, may be welded to that of a copper wire. When this junction is heated, the iron wire shows a positive charge, and the copper wire has a negative charge. Electricity generated by heat action is termed *thermoelectricity*.

4. Light Action. Photocells are semiconductor devices which convert light energy directly into electrical energy. Either sunlight or artificial illumination may be employed. This action is due to the ability of light energy to knock free electrons out of atoms of the semiconductor material. This process is termed *photoelectricity*.

5. Pressure. A difference of potential appears across the faces of certain crystals, such as quartz, when they are squeezed or stretched. This action is termed *piezoelectricity*.

6. Mechanical Action. All electricity in large, useful amounts important commercially is at present produced by rotating machines making use of magnetism. These machines, known as generators, are turned by water power, wind, gas engines, or steam engines, and sometimes by electric motors. Small generators are found in automobiles. In some portable applications, small generators are driven by hand cranks or foot pedals.

Only small amounts of electricity ordinarily are produced by action of heat, light, and pressure. An additional source, not yet of commercial importance, is the direct action of radioactivity.

SUMMARY OF CHAPTER 1

Historical. Thales observed static electricity in about 600 B.C. Electrical discovery and development reached a peak in the nineteenth and twentieth centuries. Modern developments include communications, electronics, and atomic energy.

Static Electricity. Electricity may be generated by the action of friction between two suitable materials. A glass rod rubbed with silk acquires a *positive* charge. A stick of sealing wax rubbed with cat's fur acquires a *negative* charge. Like charges repel each other; unlike charges attract. Coulomb's law, $F = (Q_1Q_2)/d^2$, expresses the force acting between two charged bodies.

An *electrostatic field* consists of *lines of force* about a charge. *Flux density* is the number of lines per square centimeter.

A charge equal to 1 statcoulomb (1 esu) in free space placed 1 cm from an identical charge will repel the latter with a force of

1 dyne. A unit charge in a unit field is acted upon by the field with a force of 1 dyne. 1 coulomb is equal to 3×10^9 esu.

Potential difference is expressed in *volts.* The potential difference between two points in an electric field is 1 v when work equal to 1 j per coulomb must be performed to move a positive charge from one of the points to the other.

Constitution of Matter. The building block of a compound is the *molecule,* while the building block of an element is the *atom.* Each atom consists of a dense, positive, central nucleus (composed of protons and electrons) around which revolve one or more planetary or orbital electrons. Electrically, the atom normally is neutral, since the total negative charge due to the electrons in the orbits equals the net positive charge of the nucleus. Loss of an electron gives an atom a net positive charge. Acquisition of an electron gives the atom a net negative charge. A positively charged atom is a positive *ion* (cation); a negatively charged atom is a negative ion (anion). Adding electrons to, or subtracting them from, atoms is a process called *ionization.*

The charge of the electron $\epsilon = 16 \times 10^{-20}$ coulomb. *Free electrons* can drift from atom to atom under the influence of a charge and may even be drawn from or injected into an atom. Metals are rich in free electrons.

Electric Current. The drift of electrons constitutes the electric current. Electronic current flows from negative to positive. Ionic current, which is the drift of ions, flows in both directions simultaneously, anions to the positive and cations to the negative body.

Current strength is measured in amperes. 1 ampere = 1 coulomb per second. $I = Q/t$.

Conductors, Insulators. A conductor is a material supporting current flow. Such a material has many free electrons. The best practical conductors are the metals. Some metals are superior to others as current conductors.

Insulators, or nonconductors, are materials which are very deficient in free (conduction) electrons and thus do not support current flow, except in infinitesimal amounts.

Sources of Electricity. Electricity may be produced by (1) friction, (2) chemical action, (3) heat action, (4) light action, (5) pressure, and (6) mechanical action.

Chapter 2

Basic Circuits and Direct Currents

2.1 BASIC ELECTRIC CIRCUITS

In Sections 1.10 and 1.11, Chapter 1, the electric current was described as a drift of electrons. In order to utilize this current for useful work, it must be passed through some electrically operated device. This is accomplished by connecting the device to the electric generator in such a way that electrons leave the generator, pass through the device and operate it, and return to the generator.

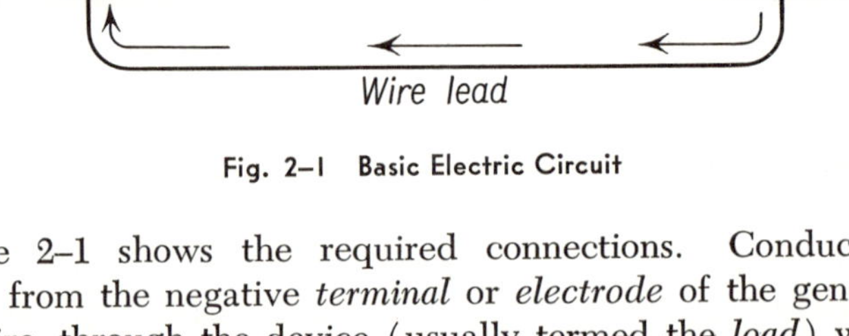

Fig. 2–1 Basic Electric Circuit

Figure 2–1 shows the required connections. Conduction electrons flow from the negative *terminal* or *electrode* of the generator, via the top wire, through the device (usually termed the *load*) which they actuate, and back to the positive terminal of the generator via the bottom wire. The electrons continue to circulate in this manner around this *circuit,* the current flowing as long as the connections are intact and the generator is able to maintain the necessary difference of potential. The load in the circuit may be a motor, lamp, bell, heater, or other suitable device.

A circuit is any route which current must travel to and from the generator and through the load. There are many different types of circuits. Some are combinations of the simpler ones. The one shown in Figure 2–1 is the simplest. If the electrons always travel in the same direction around the circuit, as shown in Figure 2–1, they constitute a *direct current* (dc). If operation of the generator alternately changes the polarity of its terminals from positive to negative and back to positive, current will flow completely around the circuit in one direction, then reverse and flow completely around the other way, repeating this *cycle* in a rhythmic manner. We then have an *alternating current* (ac).

In the electric circuit, the purpose of the generator is to maintain the electric pressure needed to force a current of electricity through the wires to operate the load device. This electric pressure is termed *electromotive force* (emf). Emf is measured in volts and designated by the letter E.

Figure 2–2 shows some of the more complicated circuits. In Figure 2–2(A), several loads are connected together, end-to-end, and to the generator so that the current must pass through each load in succession. This is termed a *series circuit.* It is important to remember that in a series circuit the same amount of current flows through each series element. Thus, if current I_1 flows through Load 1, I_2 through Load 2, and I_3 through Load 3, and if the total current from the generator is I_t, then

$$I_t = I_1 = I_2 = I_3 \tag{2–1}$$

In Figure 2–2(B), each of the three loads is connected across the generator. Because of the shape of this arrangement, this is called a *parallel circuit.* It sometimes is known also as a *shunt circuit.* In any parallel circuit, the total current (I_t) from and to the generator divides. One portion (I_1) flows through Load 1, another portion (I_2) through Load 2, and another portion (I_3) through Load 3. The total current is the sum of these branch currents. Thus

$$I_t = I_1 + I_2 + I_3 \tag{2–2}$$

Figure 2–2(C) is a *parallel-series circuit,* so called because it consists of two parallel branches connected in series. The total current (I_t) in this circuit divides the first time into two branches, I_1 flowing through Load 1 and I_2 through Load 2. After passing through the loads, these currents recombine as I_3 which, in turn, divides into I_4 which flows through Load 3, and I_5 which flows through Load 4. Equation (2–2) shows that total current is the sum of the currents in each parallel branch. Thus $I_t = I_1 + I_2$, and $I_3 = I_4 + I_5$. But since

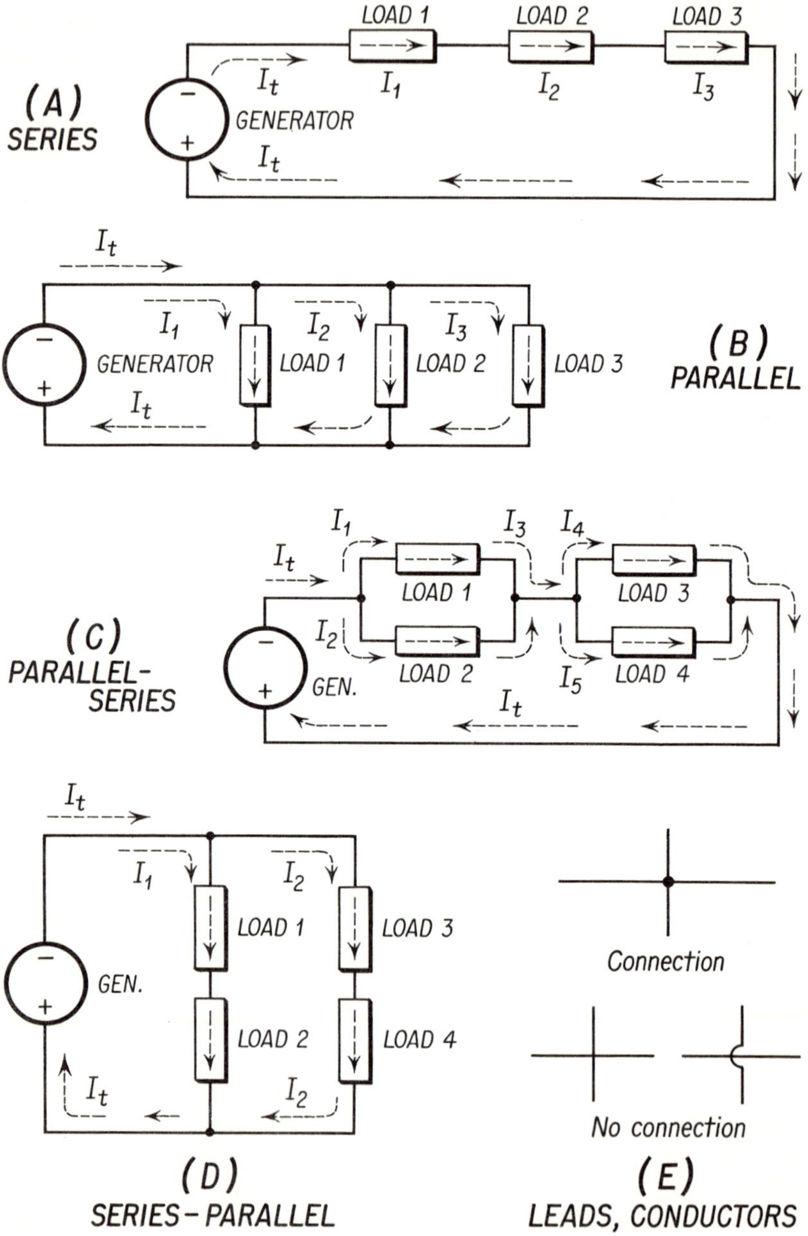

Fig. 2–2 Typical Circuits

the two parallel branches (Load 1, Load 2; and Load 3, Load 4) are in series, $I_t = I_3$. So $I_t = I_1 + I_2 = I_4 + I_5 = I_3$.

Figure 2–2(D) is a *series-parallel circuit,* so called because it consists of two series circuits connected in parallel across the generator. Since the total current in a parallel circuit is the sum of the currents in the parallel branches [Equation (2–2)], $I_t = I_1 + I_2$.

Figure 2–2(E) shows how connections are indicated between conductors (wires, leads) in an electrical circuit drawing. In the upper figure, the dot indicates that the crossing wires are connected together at the point of intersection. In the lower figure, the arc ("cross-over") in one lead, or simply the lack of a dot, indicates that no connection is made.

Generators of electricity also may be connected in series or in parallel. Figure 2–3 (A) shows the series connection. Here, the positive

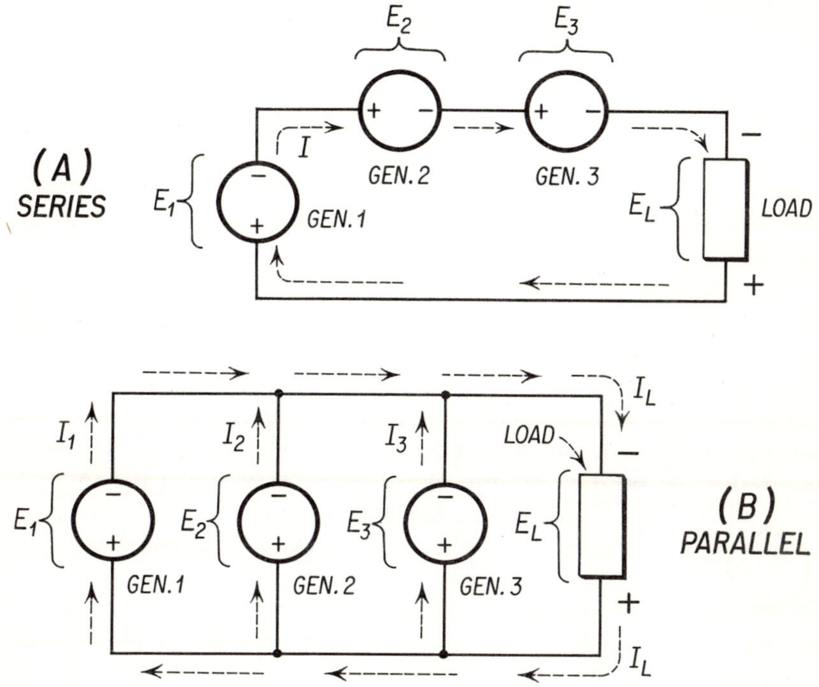

Fig. 2–3 Generator Connections

terminal of one generator is connected to the negative terminal of the other. If *similar* terminals (that is, positive and positive, or negative and negative) were connected together, the generators would oppose each other, since like charges repel. As in the series-load circuit [Figure 2–2(A)], the same current flows in all portions of this circuit, so if the generators are identical, current I flowing through the load has the same value as the current output of any one of the generators. The emf's (E_1, E_2, and E_3) add, however, and the voltage E_L across the load is

$$E_L = E_1 + E_2 + E_3 \tag{2-3}$$

Figure 2–3(B) shows the parallel connection of generators. Similar to the parallel-load circuit [Figure 2–2(B)], the total current flowing through the load is the sum of the individual generator currents (I_1, I_2, I_3). Thus

$$I_L = I_1 + I_2 + I_3 \tag{2-4}$$

If each generator delivers the same voltage, the voltage E_L across the load will be equal to the voltage of any one of the generators. Thus

$$E_L = E_1 = E_2 = E_3 \tag{2-5}$$

2.2 CELL AND BATTERY OPERATION

The simplest practical device for producing d-c electricity in useful quantities is the *cell.* This device employs chemical action. Two or more cells connected together form a *battery.* A single cell often is incorrectly called a battery.

Primary cells and batteries utilize chemical action directly to produce electricity and ordinarily must be discarded when the chemical action is exhausted. An example is the familiar dry cell. *Secondary* or *storage* cells and batteries are unable to generate electricity until a suitable *charging* current from some other source of electricity has been passed through them for several hours. When secondary cells become exhausted (that is, their supply of electricity has been used up), they may be recharged by passing current through them again. An example of the secondary battery is the automobile storage battery.

Primary Cell. The simplest primary cell consists of two plates or electrodes of dissimilar metals immersed in a suitable chemical solution and separated from each other. Copper and zinc, for example, may be immersed in a dilute solution of sulphuric acid, as shown in Figure 2–4. An emf appears between the plates, and current will flow through an external conductor connected between them.

The following principles underlie operation of the primary cell: The molecules of certain chemical compounds, such as acids, bases, and salts, become *dissociated* when dissolved in water; that is, their molecules break down into positive and negative ions. Chemicals which

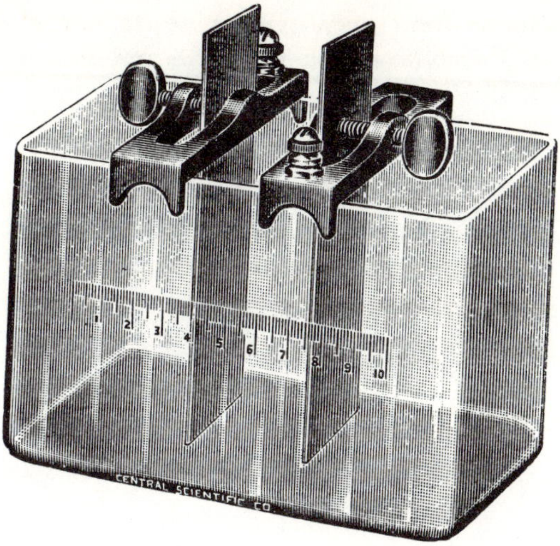

Plate 2–1 An Experimental Cell. A rectangular glass jar holds the electrolyte. The two electrode plates are suspended into the electrolyte from holders riding along the top of the jar. The holders may be moved to vary the separation between the plates. This separation is read from the scale etched in the front of the jar.

become ionized in solution are known as *electrolytes.* The electrolyte used in the simple cell in Figure 2–4 is a water solution of sulphuric acid. Each of the molecules of sulphuric acid (H_2SO_4) contains

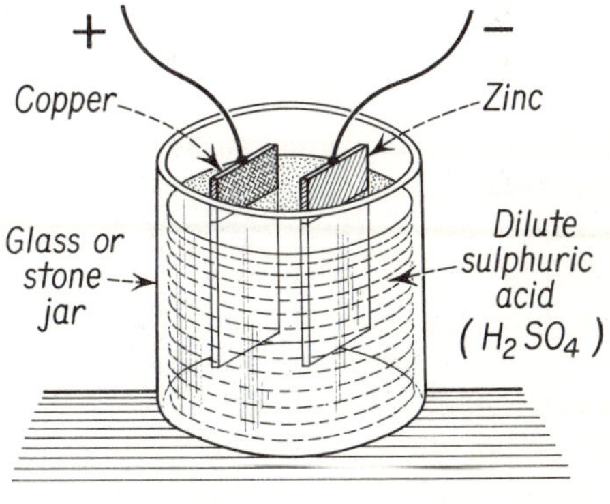

Fig. 2–4 Basic Primary Cell

1 hydrogen (H_2) and 1 sulphate group (SO_4). In water solution, the molecule dissociates into 1 hydrogen ion (H_2^+) having 1 positive charge and 1 sulphate ion (SO_4^-) with 1 negative charge. This chemical action is represented by the equation

$$H_2SO_4 \rightleftarrows H_2^+ + SO_4^- \tag{2–6}$$

Reading from left to right, this expression shows that sulphuric acid dissociates into positive H_2 and negative SO_4 ions. From right to left, it shows that these ions may recombine to form sulphuric acid. Figure 2–5 shows the ionic action when the circuit is completed; positive H_2 ions migrate to the copper electrode and, on contact with it,

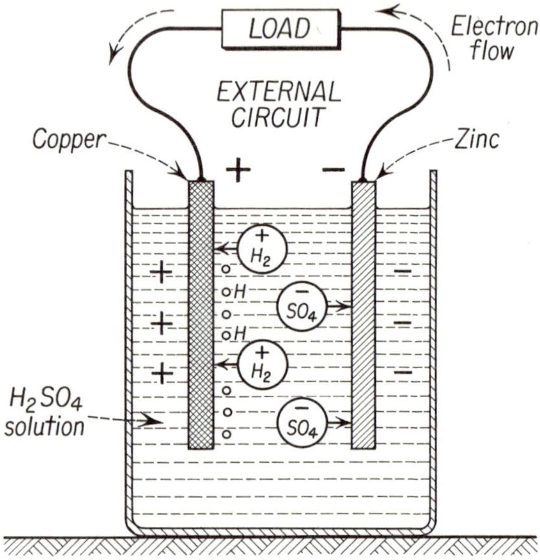

Fig. 2–5 Action of Primary Cell

give up their positive charges to this electrode and become neutral hydrogen atoms. The copper thus becomes positively charged. The hydrogen atoms combine in pairs to form hydrogen gas (H_2) which bubbles out of the electrolyte in the vicinity of the copper electrode. At the same time, the negative SO_4 ions migrate to the zinc electrode and, on contact with it, give up their negative charges to this electrode. The resulting neutral SO_4 atoms then combine with the zinc to form zinc sulphate ($ZnSO_4$) which dissolves in the water of the electrolyte. The zinc thus becomes negatively charged. Electrons will flow in the external circuit through the load as a result of the difference of potential between the copper and zinc electrodes.

This explanation shows how the plates become charged as the result of chemical action. Note that current flows as an electron drift in the external circuit and as a drift of ions inside the cell, through the electrolyte. The difference of potential between the plates is approximately 1 v. In this cell, the zinc electrode is consumed slowly as current is drawn from the cell, being converted into zinc sulphate, as explained previously. The cell is said to be "burned out" when the zinc has wasted away and the electrolyte has become depleted to such an extent that a useful emf no longer is generated. Much of the hydrogen gas that is generated clings to the copper electrode, eventually insulating it, reducing the efficiency of the cell, and contributing to the final exhaustion of the device as a source of emf. This coating of the copper electrode with an insulating jacket is called *polarization.*

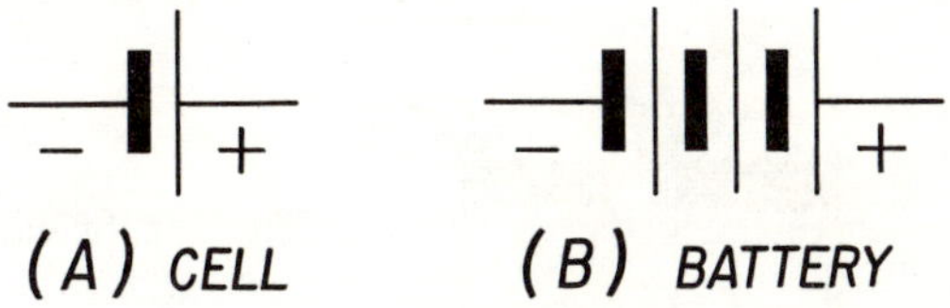

Fig. 2–6 Circuit Symbols for Cell and Battery

While copper and zinc are the electrode metals in the simple cell illustrated here, other metals and carbon, a nonmetal, are used in cells. The scheme for selecting these metals is the *electrochemical series* illustrated by Table 2–1. When two materials are chosen from this series, the one nearest the top of the list will be the negative electrode.

Table 2–1 Electrochemical Series

Zinc	METALS
Iron	
Tin	
Lead	
Copper	
Silver	
Platinum	
Carbon	NONMETAL

Cells may be connected in series to obtain a higher voltage (at the same current) or in parallel to obtain a higher current (at the same voltage).

The type of cell just described commonly is called a "wet" cell. Several different types of wet primary cells were in use during the early 1900's. A disadvantage of the wet cell is its sloppiness. There also is a tendency to lose electrolyte by evaporation, and installation is limited to the upright position. In modern electricity, the wet cell has been supplanted by the *dry cell* which, we shall see, is not completely dry but certainly is nonspilling and has other advantages over the wet type.

Figure 2–7 shows principal details of the dry cell. This cell is enclosed in a zinc can (usually cylindrical in shape) which is the negative electrode. The interior of the can is lined with gauze. The

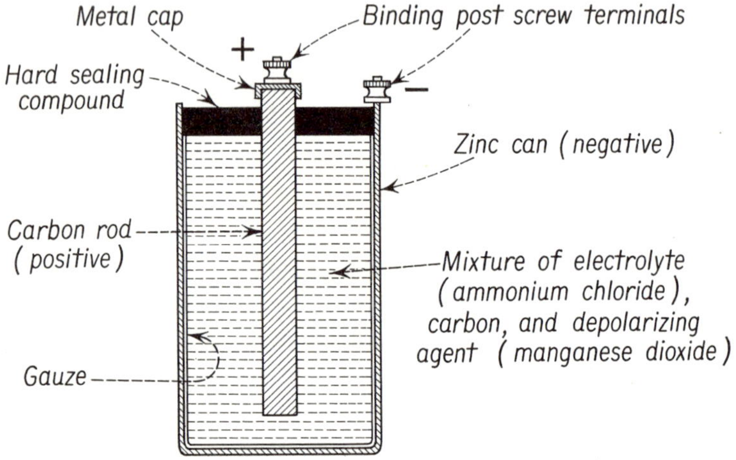

Fig. 2–7 Details of Dry Cell

positive electrode is an axially mounted carbon rod. The can is filled with an active chemical mixture and the top is sealed tightly to prevent evaporation. The active mixture contains the electrolyte, ammonium chloride ("sal ammoniac"), powdered carbon, and manganese dioxide. The latter compound serves as a depolarizing agent and acts by preventing the formation of gas bubbles on the carbon rod.

The emf of a dry cell is approximately $1\frac{1}{2}$ v. Dry cells are made in a variety of sizes from the smallest pen-size flashlight type to the familiar large, round "No. 6" size. Radio "B" batteries contain a number of small cells connected in series for the desired voltage. Figure 2–6 shows the symbols used to represent the cell and battery in electrical circuit diagrams.

Another type of primary cell of comparatively recent invention, which finds use in portable electronic devices, is the *mercury cell.* This type is manufactured in small sizes, which suits it to installations in miniaturized equipment, and has the advantage that it maintains its emf closely at 1.34 v throughout its life.

Secondary Cell. Current is delivered by a secondary cell as the result of chemical action, and in this respect it resembles the primary cell. The secondary cell is exhausted when it no longer can maintain a difference of potential in a circuit. Unlike the primary cell, however, it may be replenished by passing a current through it in the reverse direction. This reverses the chemical action and restores the cell to its original potency.

The secondary or *storage cell,* although of the wet type, is important in modern electricity. The main reasons for its survival are that it is not destroyed by normal discharge but can be recharged, it can supply

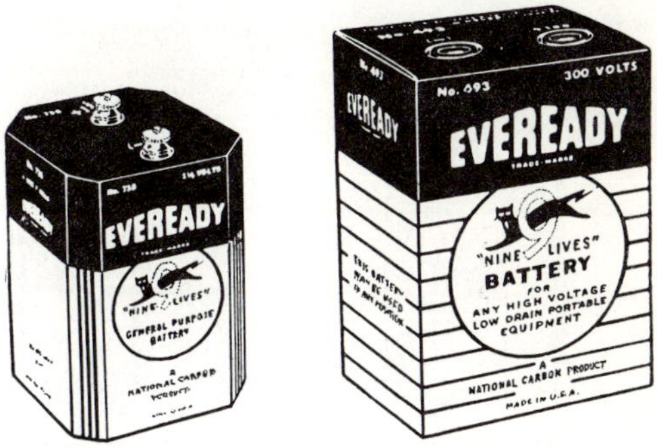

Fig. 2–8 Typical Dry Cell Battery

larger currents and a somewhat higher voltage than can the dry cell, and it is capable of longer service periods than is the dry cell. Storage batteries are applied widely in starting, lighting, and ignition service in automobiles, aircraft, and marine installations. They are well suited to use as stand-by sources of emergency direct current.

The most familiar storage cell is the *lead-acid* type. A simple demonstration type can be made by immersing two separated lead plates into dilute sulphuric acid. If these plates are connected to an external source of emf and current is allowed to pass through the cell for some time, the surface of the plate connected to the positive terminal of the voltage source will be converted to lead peroxide (PbO_2). The surface of the other plate will be converted to spongy lead (Pb). The cell then is charged and will deliver a small amount of current to an external circuit. The emf developed by the cell will be approximately 2.1 v.

As the cell is discharged by drawing current from it, the negative sulphate (SO_4) ions of the electrolyte combine with the lead (Pb) of the negative plate, imparting a negative charge to the latter and forming lead sulphate ($PbSO_4$). The positive hydrogen (H) ions migrate to the lead peroxide-coated plate and, on arrival, give up their positive charges to that plate and become atoms of hydrogen. They then react with the lead peroxide to form water. Lead sulphate also is formed at the positive plate by SO_4 ions from the surrounding sulphuric acid. The remaining hydrogen ions react with the oxygen (O_2) of the lead peroxide to form more water. During the discharge, the acid is being weakened by this water dilution, and the plates are being coated with lead sulphate. These two factors eventually cause the cell to become completely discharged.

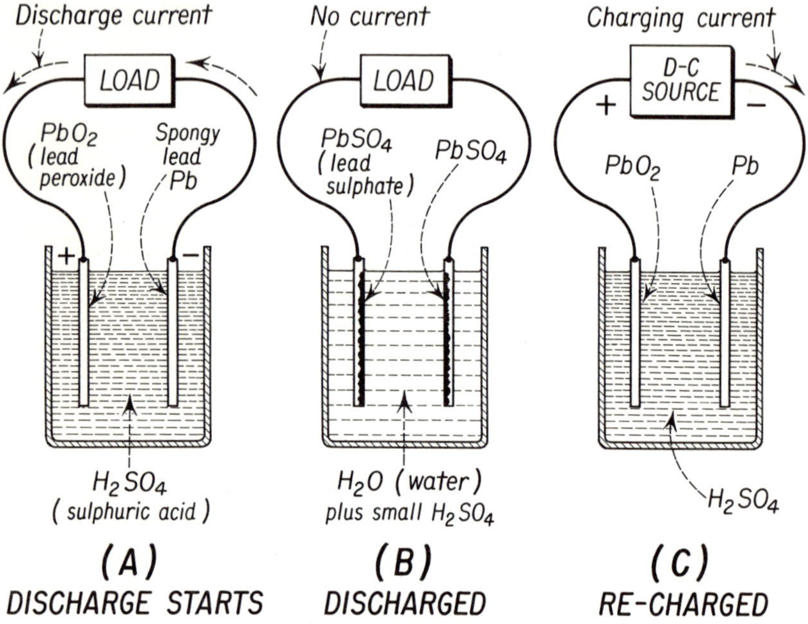

Fig. 2–9 Lead-Acid Storage Cell Action

When the cell is charged, the lead peroxide plate is connected to the positive terminal of the external d-c source, and the spongy lead plate is connected to the negative terminal. As current passes through the cell, a chemical action takes place which is the reverse of that occurring during discharge. The water is decomposed into positive hydrogen ions and negative oxygen ions. Some of the hydrogen ions migrate to the negative plate and react with the undesired lead sulphate on its surface to form sulphuric acid. This action replenishes the acid lost by water dilution during discharge and restores the negative plate to spongy lead by destroying the lead sulphate. The negative oxygen ions migrate to the positive plate to combine with the lead of the undesired lead sulphate there and form lead peroxide, the proper positive-plate coating. The remainder of the hydrogen ions from the ionization of the water combine with the sulphate portion of the lead sulphate to form sulphuric acid. Thus, additional destruction of lead sulphate takes place, and the sulphuric acid is replenished further. From this explanation, it is seen that charging forms lead peroxide (positive plate), spongy lead (negative plate), and sulphuric acid; while discharging forms lead sulphate (which insulates both plates) and water (which dilutes the acid). This action is shown by the reversible chemical equation:

$$PbO_2 + Pb + 2H_2SO_4 \underset{\text{charge}}{\overset{\text{discharge}}{\rightleftarrows}} 2PbSO_4 + 2H_2O \qquad \textbf{(2-7)}$$

(PbO_2: pos. plate; Pb: neg. plate; H_2SO_4: electrolyte)

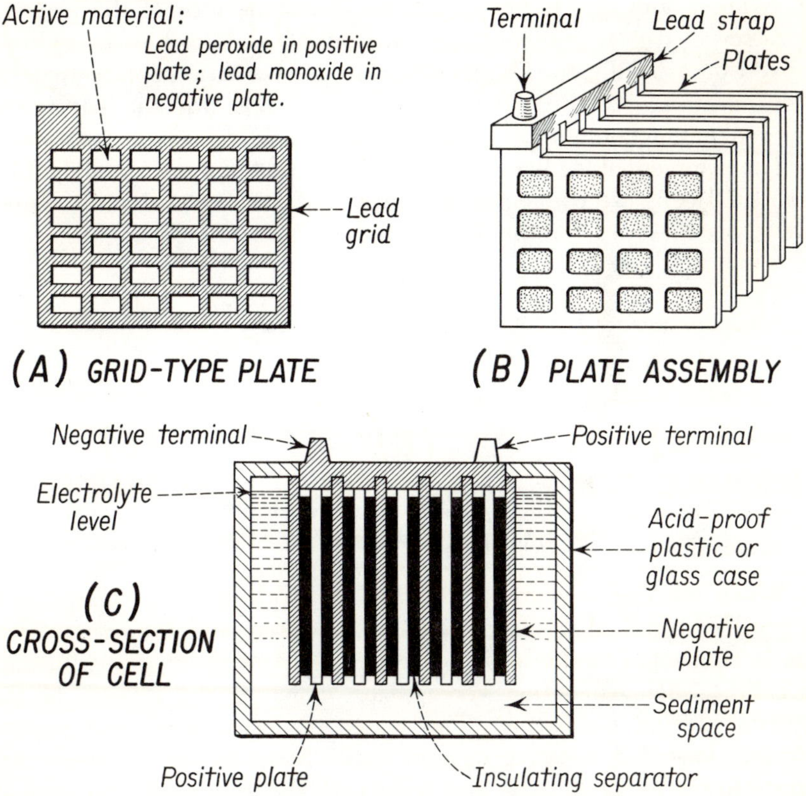

Fig. 2–10 Details of Commercial Lead-Acid Storage Cell

Figure 2–9 shows diagrammatically storage-cell action during charge and discharge.

Commercial lead-acid storage cells are not as simple in their construction as the two-plate illustrative cell just described, although the operating principle is the same for both.

Grid-type lead plates, similar to Figure 2–10(A), are used in commercial cells. Paste-type active material is pressed into the open spaces of the grids. Lead peroxide is the active material used to fill the positive plates, and lead monoxide (litharge) is used to fill the negative plates.

Positive and negative plate assemblies each consist of several plates held in place and spaced by a lead strap, or bar, on top of which the output terminal is mounted. [See Figure 2–10(B).]

The plate assemblies are suspended in an acidproof case with positive and negative plates interleaved and with wood or plastic separators between them to prevent contact. [See Figure 2–10(C).] The case is filled with electrolyte to a level above the tops of the plates.

A screw cap is provided on the top of each cell case. This is removed occasionally for filling with distilled water to replace that lost by evaporation. The cap has a small vent hole for the passage of hydrogen gas generated by the chemical action.

The commercial lead-acid cell delivers an emf of approximately 2.1 v. Three such cells are connected in series in the common 6-v automobile battery; six in series for the 12-v battery.

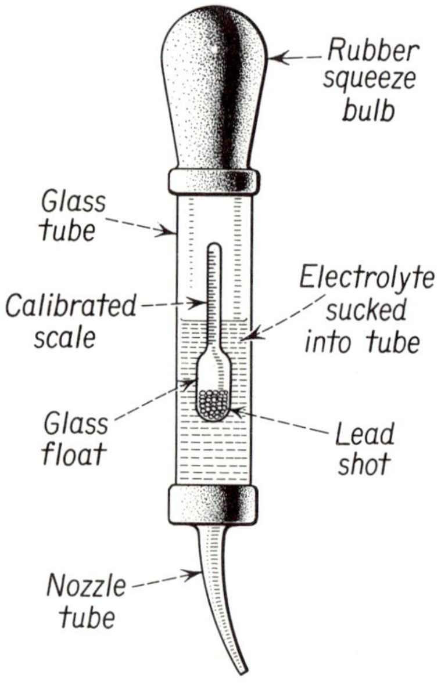

Fig. 2–11 Hydrometer for Testing Lead-Acid Battery

The voltage is not a reliable check in determining when the lead-acid cell is fully charged. A more satisfactory method is to measure the specific gravity of the electrolyte. The reason for this is that the sulphuric acid of the electrolyte is restored when the cell is charged, and the specific gravity will be higher than that of the dilute solution present at discharge. Specific gravity is tested with a *hydrometer*. (See Figure 2–11.) The nozzle of this device is inserted through the uncapped hole of the cell and into the electrolyte, and a quantity of electrolyte is drawn up into the hydrometer tube by squeezing and releasing the rubber bulb. The glass float is weighted by lead shot and floats in the electrolyte to a level determined by the specific gravity. The scale on the neck of this float reads the specific gravity at the level of the liquid. The cell is considered completely charged when the specific gravity is 1.285 to 1.300, and discharged at 1.150 to 1.175.

A second type of storage cell is the *Edison* or *nickel-iron-alkali* cell. This type is not as widely used as the lead-acid cell. The Edison cell employs a positive plate consisting of nickel hydroxide, $Ni(OH)_2$, in steel tubes built into a steel grid. The negative plate has pockets which hold iron oxide (FeO). The electrolyte is a 21 per cent solution of potassium hydroxide (KOH) in water, with a small amount of lithium hydroxide ($LiOH$). The fully charged voltage of an Edison cell is 1.2 v. The voltage measurement is satisfactory for indicating charged and discharged condition of the Edison cell, since the specific gravity of its electrolyte does not change appreciably in these two conditions.

2.3 VOLTMETER AND AMMETER

Electromotive force is measured with a *voltmeter* connected across the generator or the load in a circuit. Current is measured with an *ammeter* connected in series with the generator and load. The connections of these instruments are shown in Figure 2–12. The voltmeter and ammeter will be described in greater detail in a later chapter.

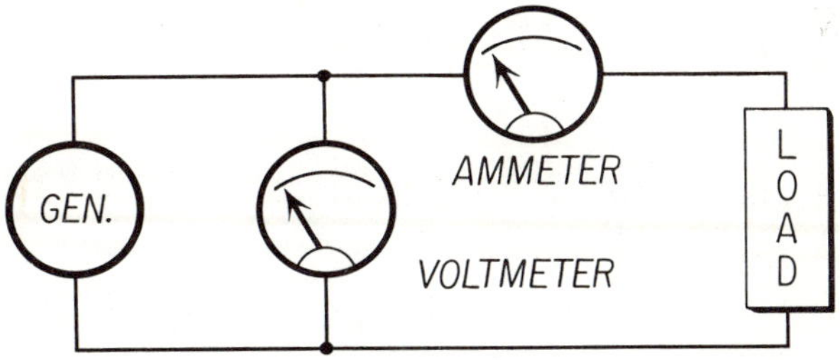

Fig. 2–12 Basic Electric Meter Connections, Showing Voltmeter and Ammeter Connections

Potential and emf are measured not only in volts but also in kilovolts (thousands of volts), millivolts (thousandths of a volt), and microvolts (millionths of a volt). Similarly, current is measured in amperes, milliamperes (thousandths of an ampere), and microamperes (millionths of an ampere).

With the aid of meters, the voltage of a battery or other electrical source may be measured, as may also the current forced through a circuit by the battery.

2.4 RESISTANCE

Conduction electrons do not pass from atom to atom entirely without impediment. Opposition to their flow is offered by collisions with other electrons and with atoms. The result is that the current often is reduced, and the conductor is heated by the energy lost in the collisions. All conductor materials offer some opposition to current flow, some more than others, since there is no entirely perfect conductor. This opposition, which is somewhat similar to mechanical friction, is called *resistance.*

Georg Simon Ohm (1787–1854) discovered that resistance (R) varies directly as the voltage (E) applied to a circuit and inversely as the current (I) through the circuit. The unit of resistance is the *ohm,* named in honor of G. S. Ohm. The Greek capital letter omega, Ω, is the symbol for the ohm; while the lower-case omega, ω, is the symbol for megohms (millions of ohms).

Ohm's law for direct current is expressed by the equation

(2–8) $$R = E/I$$

where R = resistance (in ohms)
I = current (in amperes)
E = emf (in volts)

Also

(2–9) $$E = IR$$

and

(2–10) $$I = E/R$$

Example: In a certain circuit, the emf is measured as 6.3 v, the current as 1.5 amp. What is the resistance?

By Equation (2–8),

$$R = E/I = 6.3/1.5 = 4.2\Omega$$

Resistance varies directly as the length of a conductor and the *specific resistance,* or *resistivity,* of the conductor material and inversely as its cross-sectional area. Resistivity is designated by the Greek letter rho, ρ. For a wire conductor

(2–11) $$R = \frac{\rho l}{d^2}$$

where ρ = resistivity of the material (in ohms)
l = length (in feet)
d = diameter (in mils)

The property of resistance may be exploited when we wish to reduce the strength of current flowing in a circuit. This is done by passing the current through a special high-resistance "conductor." Nichrome, German silver, and iron are metals which may be used for wires that offer resistance. Carbon is a nonmetal having appreciable resistance. From Equation (2–10) we see that the current (I) from a 6-v battery (E) can be limited to 0.6 amp by passing it through a 10-ohm conductor (R); that is, $I = E/R = 6/10 = 0.6$ amp.

How is a 10-ohm conductor obtained? One way would be to use a length of copper wire having that value of resistance. Standard wire tables show, for example, that 9.34 ft of No. 40 copper wire has a resistance of 10 ohms. But 9 ft is cumbersome to handle. For convenience and compactness, that length of insulated wire might be wound on a nonconducting stick or tube. This makes one type of 10-ohm unit—properly called a *resistor*. (A resistor is a component for introducing a desired amount of resistance at a required point in a circuit.)

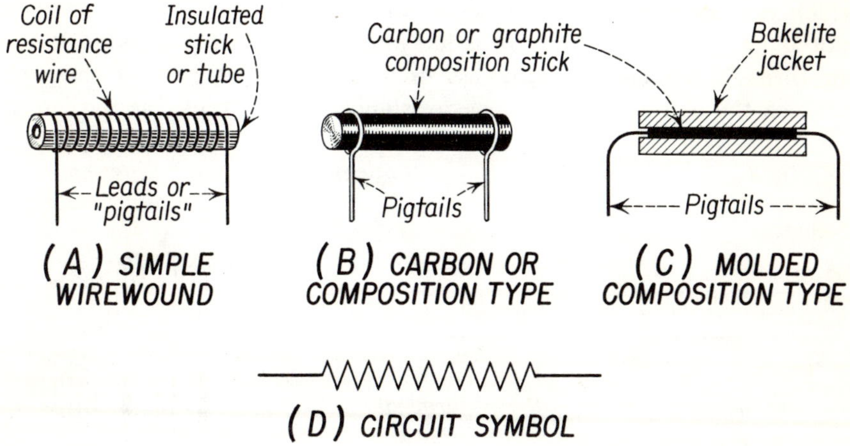

Fig. 2–13 Typical Fixed Resistors

Commercial wire-wound resistors employ resistance wire (iron, German silver, Nichrome, manganin, etc.) instead of copper, since a shorter length of such wire is needed for a given resistance. Figure 2–13(A) shows a simple wire-wound resistor. Another type of resistor is made from a stick or rod of carbon, or graphite mixed with a binder, as shown in Figures 2–13(B) and (C). A lead ("pigtail") or terminal at each end of the stick makes contact with the resistance material.

A resistor of the type just described is known as a *fixed* resistor, since, at a given temperature, its ohmic value is constant. Figure 2–13(D) shows the circuit symbol for a fixed resistor.

There also are *variable* resistors, so called because their resistance values may be adjusted to a desired value. The simplest variable resistor consists of a straight length of resistance wire with a sliding metal clip

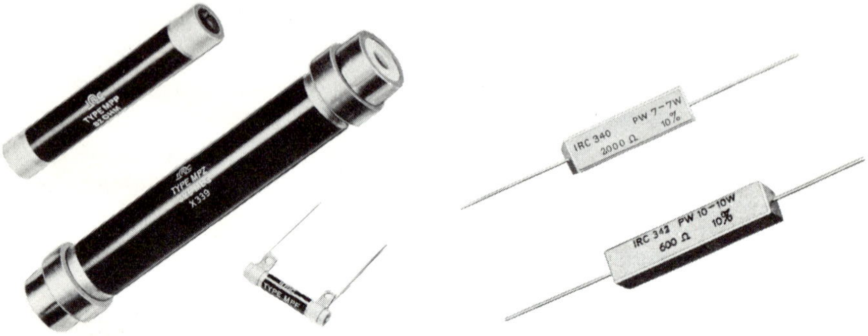

Fig. 2–14 Typical Fixed Resistors (Courtesy, International Resistance Co.)

which the operator can move along to make contact with any desired spot along the wire. If current is fed into one end of the wire and taken out through the clip, it will pass only through that part of the wire between the end and clip. Since resistance varies directly as the length

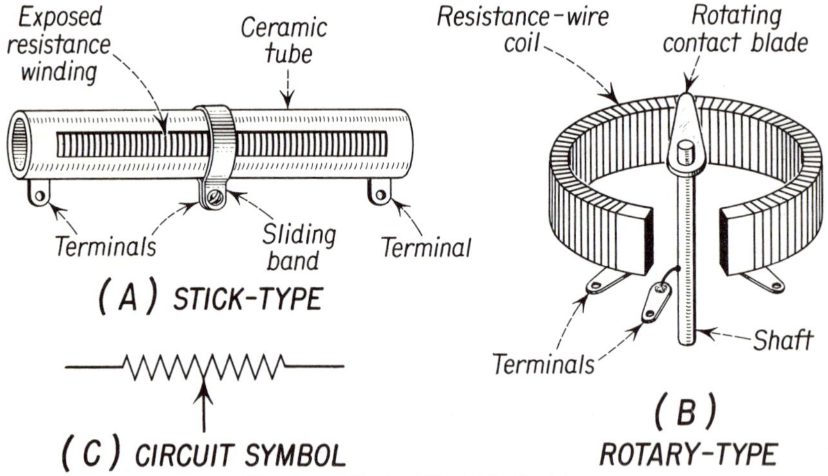

Fig. 2–15 Typical Variable Resistors

of a conductor [Equation (2–11)], the resistance here will be highest and the current lowest when the distance between the end and clip is greatest, and vice versa.

Obviously, a straight wire must be quite long and cumbersome to handle when a high resistance value is desired. But, like the wire-wound fixed resistor, a long piece of resistance wire may be wound around an insulating tube and a variable resistor thus obtained if some means is provided for contacting the individual turns of this coil of wire. Figure 2–15(A) shows a stick-type variable resistor of this sort. The

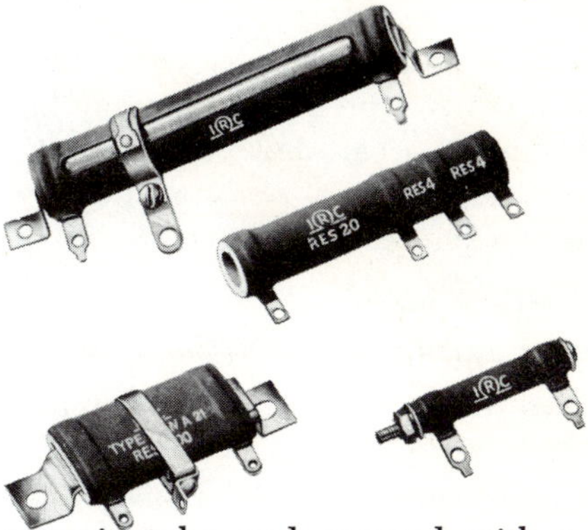

Plate 2–2 Wirewound Resistors. The upper and lower-left units are of the variable stick type. See Fig. 2–15 (A). (Courtesy, International Resistance Co.)

resistance wire is wound around a ceramic tube and covered with a baked porcelain enamel jacket except for a straight, lengthwise slot through which the turns of wire are exposed. A sliding metal ring, fitted around the stick, makes contact with the wire turns. This

Plate 2–3 Variable Resistors of Rotary Type (Courtesy, International Resistance Co.)

ring forms one of the resistor terminals, and its position from one end of the stick determines how much of the available resistance is used. When the ring is halfway along the stick, for example, only half of the total resistance of the coil is used.

Figure 2–15(B) shows another type of variable resistor. Here, the coil of resistance wire is wound on a card of insulating material and bent into an arc. A contact blade is rotated around this arc to make contact with the turns of the coil and govern the length of wire used and thus the resistance. This type of variable resistor is known sometimes as a *rheostat.* Some rheostats use an arc of carbon composition material instead of a wire coil.

Like fixed resistors, variable resistors are manufactured in a wide variety of resistance values and physical sizes. Figure 2–15(C) shows the circuit symbol for the variable resistor.

2.5 CONDUCTANCE

Resistance expresses the difficulty a current has in flowing through a conductor or circuit. It may be likened to friction which impedes the rolling of a ball down a chute. *Conductance,* the opposite of resistance, expresses the ease of current flow. Conductance (G) is the reciprocal of resistance; that is

(2–12) $$G = 1/R$$

The unit of conductance is the *mho,* which is the word "ohm" spelled backward. A good conductor has low resistance and high conductance.

Example: A certain electrical device has a resistance of 125 ohms. What is its conductance?

From Equation (2–12),

$$G = 1/R = 1/125 = 0.008 \text{ mho}$$

The Ohm's-law fractions [Equations (2–8), (2–9), and (2–10)] may be inverted when resistance is substituted for conductance. Thus

(2–13) $$G = I/E$$

(2–14) $$I = EG$$

(2–15) $$E = I/G$$

Just as resistivity is the specific resistance of a conducting material [Equation (2–11)], *conductivity* is the specific conductance of a material. Conductivity, designated by the Greek letter gamma, γ, and measured in mhos per centimeter cube of the material, is the reciprocal of resistivity (ρ). Thus

(2–16) $$\gamma = 1/\rho$$

2.6 RESISTANCE CIRCUITS

Resistors may be connected together to form various circuits which are similar to the basic electric circuits shown in Figure 2–2 or are combinations of these basic arrangements. Many electrical devices resemble resistors in their behavior (examples: incandescent lamps, heaters, fuses), and circuits containing such devices may, for analysis, be viewed as a combination of resistances.

Resistances in Series. [See Figure 2–16(A).] Since each conduction electron must travel completely around the circuit, the same value of current (I) flows through each resistor (R_1, R_2, R_3). By Ohm's law [Equation (2–9)] this current flow causes a voltage $e_1 = IR_1$ to appear across resistor R_1, $e_2 = IR_2$ across R_2, and $e_3 = IR_3$ across R_3. The sum of these voltage *drops* $e_1 + e_2 + e_3$ equals the battery voltage E. Thus: $E = IR_1 + IR_2 + IR_3 = I(R_1 + R_2 + R_3) = IR_{\text{total}}$. From this, we

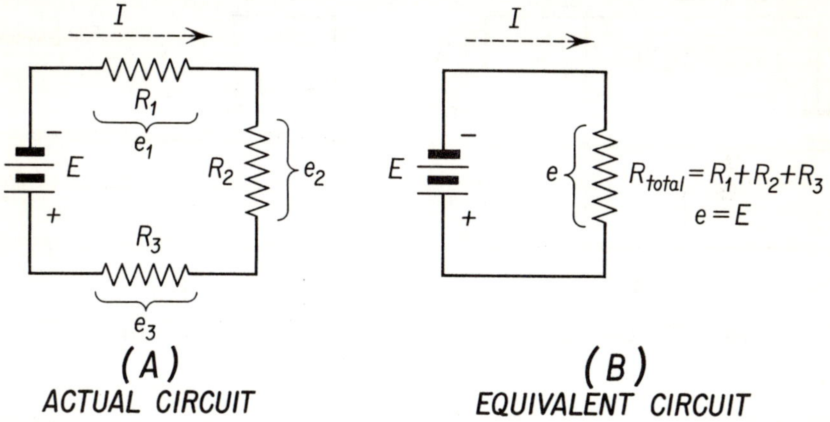

Fig. 2–16 Resistances in Series

see that the total resistance for the series circuit is

(2–17) $$R_t = R_1 + R_2 + R_3$$

If there are more than three resistances, their values are added to the sum in Equation (2–17).

Example: 100-, 6,000-, 300-, and 50-ohm resistors are connected in series. What is the total resistance of the circuit?

From Equation (2–17),

$$R_t = R_1 + R_2 + R_3 + R_4 = 100 + 6{,}000 + 300 + 50 = 6{,}450\Omega$$

Resistances in Parallel. [See Figure 2–17(A).] The total current (I_t) flowing from the battery divides into three parts; I_1 flowing through R_1, I_2 through R_2, and I_3 through R_3. These branch currents produce voltage drops $e_1 = I_1R_1$ across R_1, $e_2 = I_2R_2$ across R_2, and $e_3 = I_3R_3$ across R_3. Since the total current is the sum of the branch currents, $I_t = I_1 + I_2 + I_3 = E/R_1 + E/R_2 + E/R_3$. This is the same as the expression $E\left(\frac{1}{R_1} + \frac{1}{R_2} + \frac{1}{R_3}\right)$, or E/R_t, where R_t is the total resistance in the circuit. From this, $1/R_t = 1/R_1 + 1/R_2 + 1/R_3$, or the total resistance

(2–18)
$$R_t = \frac{1}{\dfrac{1}{R_1} + \dfrac{1}{R_2} + \dfrac{1}{R_3}}$$

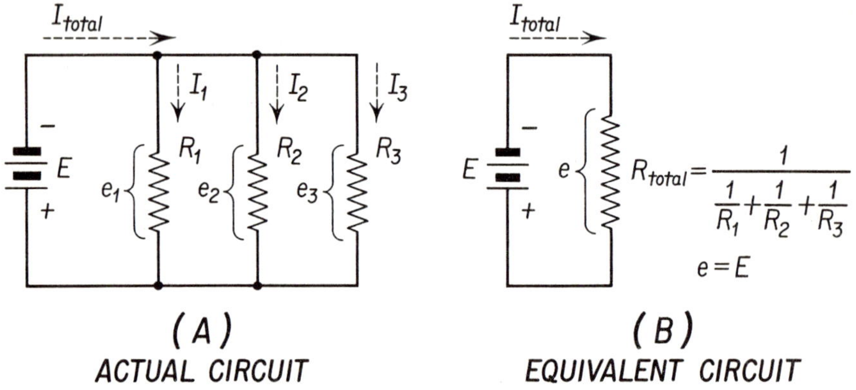

Fig. 2–17 Resistances in Parallel

If there are more than three resistors, their reciprocals are added to the denominator in Equation (2–18).

Example: 2,500-, 10,000-, 50,000-, and 100,000-ohm resistors are connected in parallel. What is the total resistance of the circuit?

From Equation (2–18),

$$R_t = \frac{1}{\dfrac{1}{2500} + \dfrac{1}{10{,}000} + \dfrac{1}{50{,}000} + \dfrac{1}{100{,}000}}$$

$$= \frac{1}{0.0004 + 0.0001 + 0.00002 + 0.00001}$$

$$= 1/0.00053 = 1886\Omega$$

Resistances in Series-Parallel. The facts just learned about series and parallel circuits may be applied together to combined circuits of resistances. Figure 2–18(A) shows a series-parallel resistance circuit. Here, each parallel leg contains two resistances (R_1 and R_2, and R_3 and R_4) in series. From Equation (2–17), the total resistance (R_a) of Leg 1 is $R_1 + R_2$, and the total resistance (R_b) of Leg 2 is $R_3 + R_4$. The total resistance of these two legs in parallel then is $1/(1/R_a + 1/R_b)$, or the total resistance (R_t) of the circuit is

$$R_t = \frac{1}{\dfrac{1}{R_1 + R_2} + \dfrac{1}{R_3 + R_4}} \qquad (2\text{–}19)$$

Less or more series resistances may be present in each leg of the circuit. Likewise, more legs may be added in parallel.

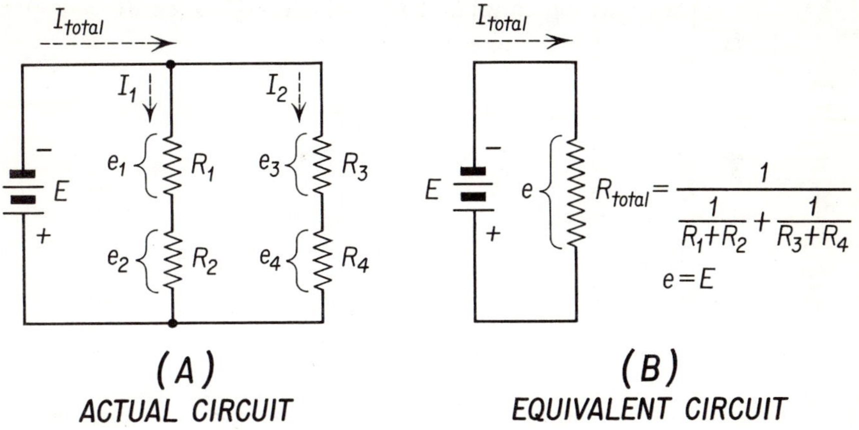

Fig. 2–18 Resistances in Series-Parallel

Example: In a practical case, $R_1 = 3\Omega$, $R_2 = 5\Omega$, $R_3 = 10\Omega$, and $R_4 = 50\Omega$. What is the total resistance of this circuit?

$$R_t = \frac{1}{\dfrac{1}{3+5} + \dfrac{1}{10+50}} = \frac{1}{\dfrac{1}{8} + \dfrac{1}{60}}$$

$$= \frac{1}{0.125 + 0.0167} = 1/0.1417 = 7.06\Omega$$

Resistances in Parallel-Series. [See Figure 2–19(A).] Here, the resistance (R_a) of Leg 1 $= 1/(1/R_1 + 1/R_2)$, and the resistance (R_b) of Leg 2 $= 1/(1/R_3 + 1/R_4)$. These two legs are connected in series, so the total resistance (R_t) of the circuit is the sum of R_a and R_b. Thus

$$R_t = 1/(1/R_1 + 1/R_2) + 1/(1/R_3 + 1/R_4) \qquad (2\text{–}20)$$

More resistances may be connected in parallel in each leg. Likewise, more resistance legs may be added in series.

Example: In a practical case, R_1 and R_3 each is a 15-ohm lamp, while R_2 and R_4 are 10-ohm shunting resistors. What is the total resistance of this circuit?

$$\begin{aligned} R_t &= 1/(1/15 + 1/10) + 1/(1/15 + 1/10) \\ &= 1/(0.066 + 0.10) + 1/(0.066 + 0.10) \\ &= 1/0.166 + 1/0.166 = 6.024 + 6.024 \\ &= 12.05\Omega \end{aligned}$$

Equivalent Circuits. Note that in each example (Figures 2–16 through 2–19), the circuit may be simplified to a single resistance, R_t, equal to

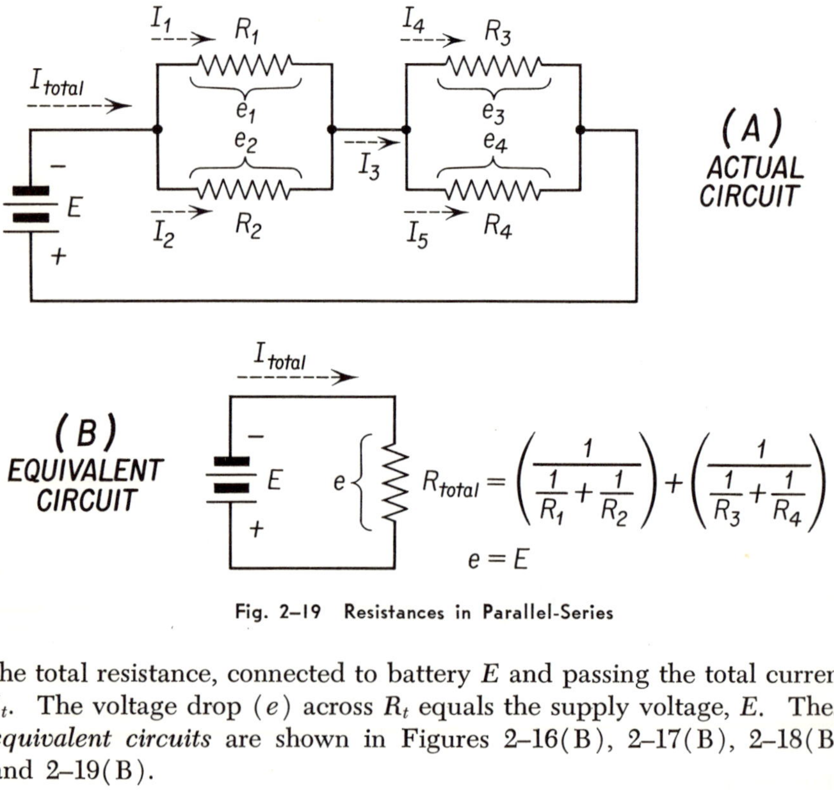

Fig. 2–19 Resistances in Parallel-Series

the total resistance, connected to battery E and passing the total current, I_t. The voltage drop (e) across R_t equals the supply voltage, E. These *equivalent circuits* are shown in Figures 2–16(B), 2–17(B), 2–18(B), and 2–19(B).

2.7 DROPPING RESISTOR

In electrical equipment, a series resistor is employed occasionally to reduce the supply voltage to a desired value across a load device.

In Figure 2–20, the dropping resistor (R_1) is connected between the battery (supply voltage E) and the load device which has resistance R_2. Current I_t flows through both resistors in series. This current sets up a voltage drop $e_1 = I_tR_1$ across dropping resistor R_1. The voltage across the *load resistance* (R_2) then is $e_2 = I_tR_2$. Load voltage e_2 thus is equal to $E - e_1 = E - (I_tR_1)$. Thus e_2 has been dropped from the supply value E by an amount equal to I_tR_1 volts.

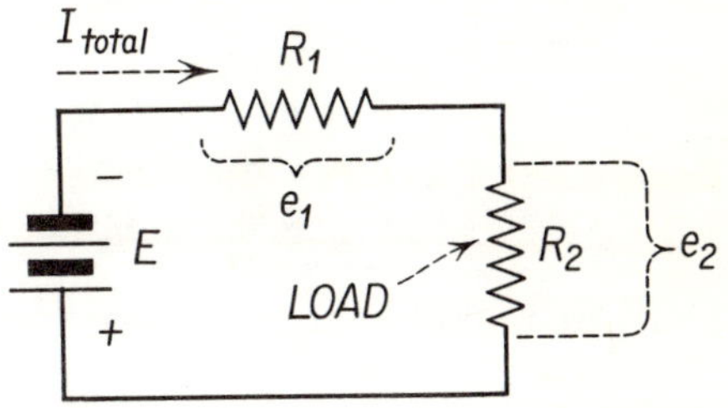

Fig. 2–20 Principle of Dropping Resistor

When the supply voltage (E), desired voltage (e), and current (I) are known, the value of the required dropping resistor may be calculated with the aid of the following equation:

$$R = (E - e)/I \qquad (2\text{–}21)$$

where E and e are in volts, and I is in amperes.

Example: A 28-v lamp draws 1.35 amp. What value of series dropping resistor will be required to operate this lamp from the 115-v power line?

From Equation (2–21),

$$R = (E - e)/I = (115 - 28)/1.35 = 87/1.35 = 64.4\Omega$$

2.8 D-C POWER

Power is the rate at which electrical energy is generated or used. The unit of power is the *watt* (w), named in honor of James Watt (1736–1819). Power is expressed also in fractions of a watt: milliwatts (thousandths of a watt) and microwatts (millionths of a watt) and in multiples of a watt: kilowatts (thousands of watts) and megawatts (millions of watts). There are 746 watts in 1 horsepower. The abbreviation for power is P, and for horsepower, *hp*.

In a d-c circuit, the power (P) may be computed by multiplying the voltage (E) by the current (I), where E is in volts, I in amperes, and P in watts.

$$P = EI \qquad (2\text{–}22)$$

If the voltage is not known, but the current and resistance (in ohms) are,

$$P = I^2R \qquad (2\text{–}23)$$

Similarly, if the voltage and resistance are known, but current is not,

(2–24) $$P = E^2/R$$

Example: Using all of the foregoing methods of calculation, find the power in a circuit in which the voltage = 110 v, current = 5 amp, and resistance = 22Ω

$$P = EI = 110(5) = 550 \text{ w}$$
$$P = I^2R = 5^2(22) = 25(22) = 550 \text{ w}$$
$$P = E^2/R = 110^2/22 = 12{,}100/22 = 550 \text{ w}$$

An important consideration in electrical equipment is power *loss.* Power is lost in the form of heat associated with circuit resistance. In the dropping resistor (Figure 2–20), power equal to $(e_1)^2/R_1$ is lost in this resistor. The power used by the load device (resistor R_2) is $(e_2)^2/R_2$.

There is a difference between power (P) and work (W). Work is done *by* electric power in operating apparatus. Work is power generated or consumed per unit of time. Thus, practical units are the *watt-second, watt-hour, kilowatt-hour,* and *horsepower-hour.* In each case, the power is multiplied by the time ($W = Pt$). Power companies sell electricity by the kilowatt-hour.

Example: A 150-w lamp is left burning 72 hr. How many kilowatt-hours (kwhr) are consumed?

$$W = Pt = 150(72) = 10{,}800 \text{ watt-hours (whr)}$$
$$= 10.8 \text{ kwhr}$$

2.9 NORMAL AND ABNORMAL CIRCUITS

Figure 2–21(A) shows a conventional circuit in which generator voltage E forces a current, I, through load resistance R. This is called a *closed circuit* because all of its portions are closed in an unbroken path for the flow of electrons.

In Figure 2–21(B), the circuit has been broken open between Points A and B, hence the term *open circuit.* Since electrons cannot jump the gap from A to B, no current can flow from the generator (E) through the load resistor (R), and the circuit is disabled. Closing a circuit [Figure 2–21(A)] places it in operation, while opening a circuit [Figure 2–21(B)] disables it.

In Figure 2–21(C), a good conductor has been connected across the circuit from A to B. Because of the better conductivity of the Path AB, virtually all of the current (I) flows through this path, robbing resistor *R*. This condition often is spoken of by laymen as "electricity taking the path of the least resistance." We know that *most,* not all, of the current flows through this path. Because of the shorter route taken by the current in this case, Path *AB* is called a *short circuit.* Normally, short circuits are unintentional, being caused either by accident or by mistakes in wiring. A wire or piece of metal falling into electrical equipment can cause a short circuit. Short circuits are avoided, since

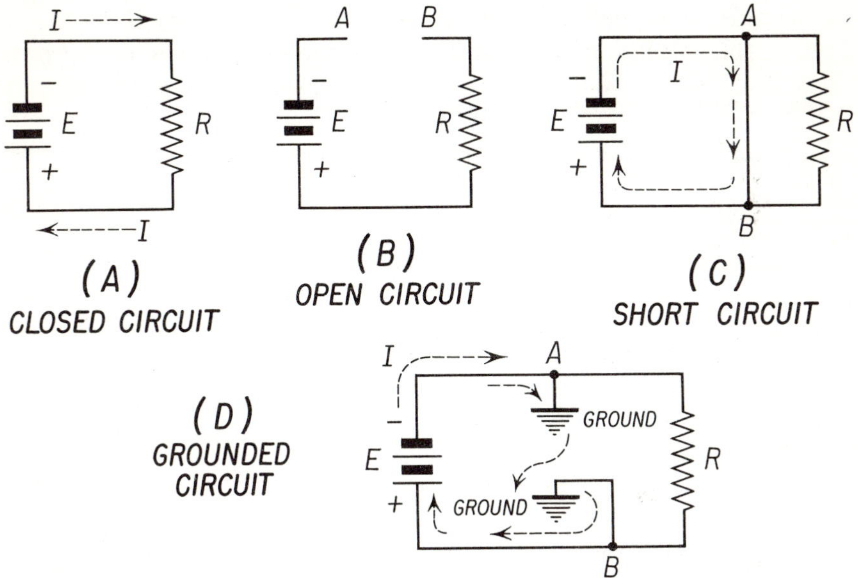

Fig. 2–21 Normal and Abnormal Circuits

they can cause damage to generators and circuit components because of the large currents they draw. A portion of a circuit as well as a complete circuit may become short-circuited.

A special kind of short circuit is illustrated by Figure 2–21(D). This is called a *grounded circuit,* or simply a *ground.* It too is a fault. The name "ground" arose from the instance of two, long, current-carrying, outdoor wires falling to the ground accidentally and making contact with it. More often, it refers to two conductors accidentally making contact with a conducting-material base, wall, or other such structure. The dashed triangle is the circuit symbol for a ground connection. Current *I* from battery or generator *E* flows from Point *A* to ground, then through the ground to Point *B* where it enters the circuit again and flows back to the battery. Load resistor *R* accordingly receives virtually no current, since the ground path (*A* to *B*) short-circuits *R*.

2.10 VOLTAGE DIVIDER

In Sections 2.6 and 2.7, it was explained how a current (I) flowing through a resistance (R) will set up a voltage drop (E) across the resistance. This is always true, since $E = IR$.

If several resistances of different value are connected in series and a current is passed through the circuit, a different voltage drop will be produced across each resistance in proportion to the resistance value. For example, a current of 3 amp passing through resistors of 10 ohms, 50 ohms, and 75 ohms in series will produce the following voltage drops. Across the first resistor, $3 \times 10 = 30$ v; across the second resistor, $3 \times 50 = 150$ v; and across the third resistor, $3 \times 75 = 225$ v. The sum of these voltage drops equals the voltage applied to the circuit (total resistance = 135Ω) to force the current of 3 amp through it = 405 v.

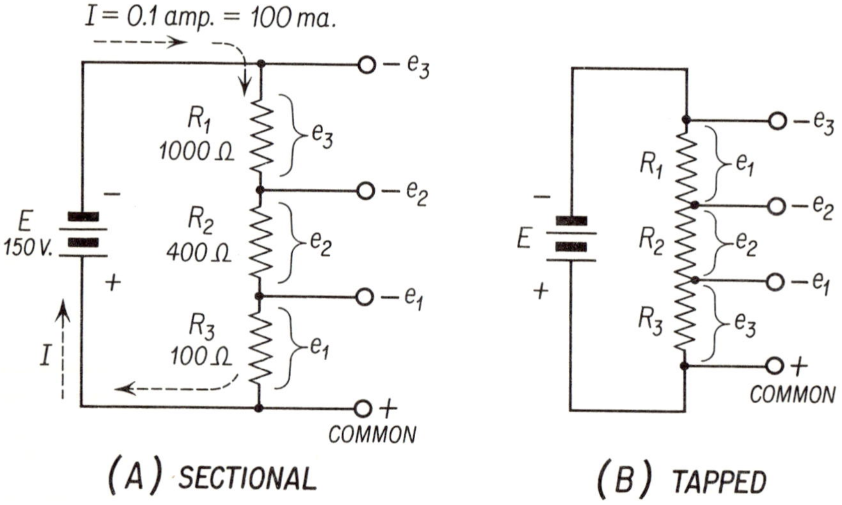

Fig. 2–22 Voltage Dividers

This principle of proportional voltage drops may be utilized to divide an available voltage by any required amount. For this purpose, several resistors are connected in series across a voltage source, as shown in Figure 2–22. This resistor arrangement is termed a *voltage divider.* In Figure 2–22(A), the total resistance $= R_1 + R_2 + R_3 = 1{,}000 + 400 + 100 = 1{,}500\Omega$. The current, $I = E/R = 150/1{,}500 = 0.1$ amp = 100 milliamperes (ma). The voltage drop e_3 across $R_1 = IR_1 = 0.1(1{,}000) = 100$ v, the voltage drop e_2 across $R_2 = IR_2 = 0.1(400) = 40$ v, and the voltage drop e_1 across $R_3 = 0.1(100) = 10$ v. Between the terminals labeled $-e_3$ and $-e_2$, 100 v accordingly are available, 40 v between $-e_2$ and $-e_1$, and 10 v between $-e_1$ and + COMMON. By making the lower terminal common, as indicated, and measuring all voltages with respect to it, we find 10 v between + and $-e_1$, $e_1 + e_2$ or 50 v from + to $-e_2$, and 150 v between + and $-e_3$

The arrangement shown in Figure 2–22(A) is termed a *sectional voltage divider* because it is made up of several sections, each consisting of a separate resistor. The same results may be obtained by making connections to appropriate portions of a *single* resistor, as shown in Figure 2–22(B), to obtain the desired amount of resistance, R_1, R_2, and R_3, between connections. These connections are called *taps,* and the entire device is known as a *tapped* voltage divider. A tapped voltage divider may be made with a resistor of the type shown in Figure 2–15(A) by providing a separate band, or ring, for each desired tap.

A study of the voltage divider shows its divisions of voltage to be proportional to its resistance sections. Thus voltage $-e_1$ = the supply voltage, E, divided by $(R_1 + R_2 + R_3)/R_3$. Similarly, $-e_2 = E$ divided by $(R_1 + R_2 + R_3)/R_2 + R_3$, and $-e_3 = E$ divided by $(R_1 + R_2 + R_3)/(R_1 + R_2 + R_3) = E/1$. Actually, this simple voltage division is possible only when no current (or only negligible current) is drawn from the voltage-divider taps. When a significant amount of current is drawn, the voltage drops across the voltage-divider sections will change from the simple calculated values, and the resistance values of these sections must be altered to yield the desired voltages.

2.11 KIRCHHOFF'S LAWS

Gustav Robert Kirchhoff (1824–1887) made two simply stated but important observations regarding currents and emf's in a circuit. The facts of these observations are expressed as laws which have been named in honor of Kirchhoff.

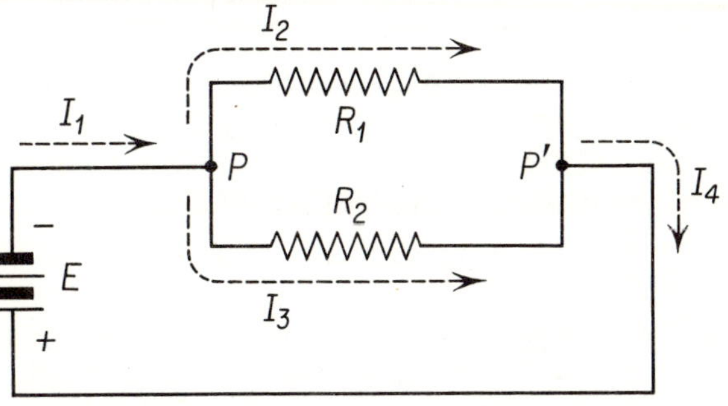

Fig. 2–23 Illustration of Kirchhoff's First Law

First Law. Kirchhoff's first law states that the sum of the currents flowing from a point in a circuit equals the current flowing to that point.

This is illustrated by Figure 2–23. Current I_1 flows from Battery E toward Point P. At P, however, the current divides into two paths, I_2 flowing through resistor R_1, and I_3 through resistor R_2. We know from the rules for parallel resistors (Figure 2–17) that $I_2 + I_3 = I_1$. Kirchhoff's first law is demonstrated again at Point P' where currents $I_2 + I_3$ flowing to this point equal current I_4 flowing away from it.

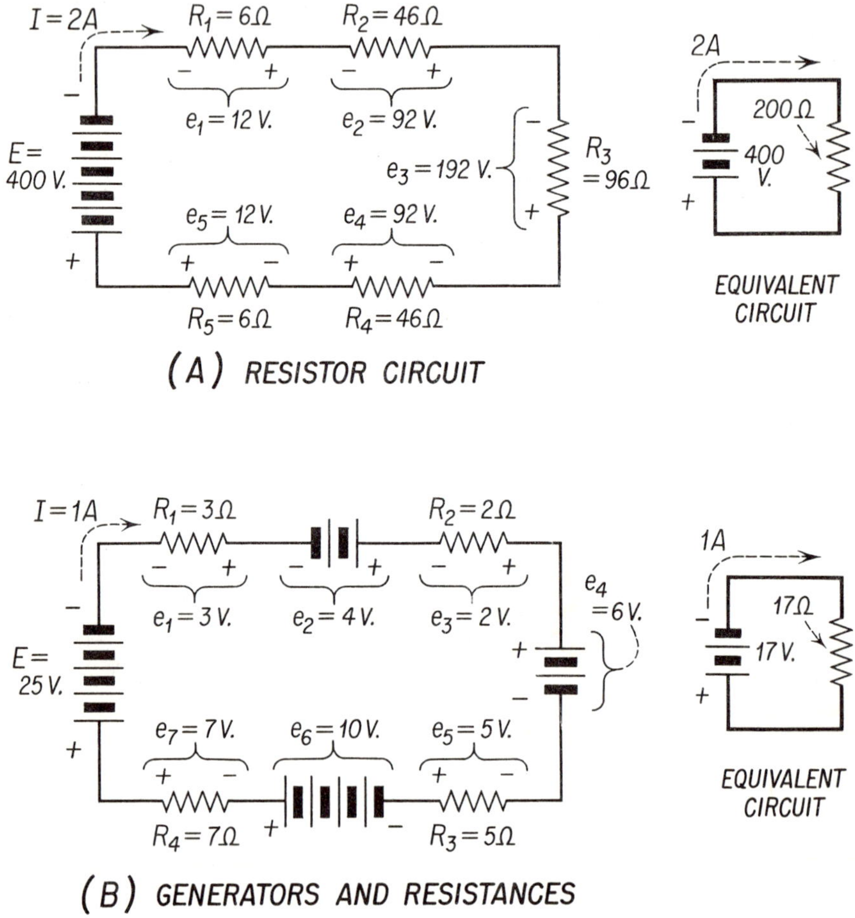

Fig. 2–24 Illustration of Kirchhoff's Second Law

Second Law. Kirchhoff's second law shows that the sum of the voltage drops around a circuit equals the emf of the supply (generator) voltage. Another way of stating this law is that the algebraic sum of all the voltage drops around the circuit, including the generator voltage, equals zero.

Figure 2–24 gives two illustrations of Kirchhoff's second law. In Figure 2–24(A), the circuit consists of five series resistors connected to a 400-v generator, E. Current I flowing through the circuit produces voltage drops e_1, e_2, e_3, e_4, and e_5 across resistors R_1, R_2, R_3, R_4, and R_5 with the polarities shown. Since the total resistance of a series circuit equals the sum of the component resistances, here $R_t = 6 + 46 + 96 + 46 + 6 = 200\Omega$. Current $I = E/R = 400/200 = 2$ amp. The voltage across each resistor equals IR. For the first resistor, this is $IR_1 = 2(6) = 12$ v. The sum of the voltage drops around the circuit $= 12 + 92 + 192 + 92 + 12 = 400$ v $= E$.

Figure 2–24(B) shows an example in which the circuit contains several additional generators (e_2, e_4, and e_6) as well as resistors. The three generators and four resistors form a series circuit connected to the 25-v generator, E. Reading from left to right around the circuit and calling out the first polarity encountered, the algebraic sum of the generator voltages is found to be $E - e_2 + e_4 - e_6$, or $25 - 4 + 6 - 10 = 17$ v. The total resistance $= R_1 + R_2 + R_3 + R_4 = 3 + 2 + 5 + 7 = 17\Omega$. The current, I, which 17 v will force through 17 ohms is 1 amp ($I = E/R$). From Ohm's law ($E = IR$), the voltage drops across the resistors accordingly are e_1, e_3, e_5, and e_7, with the polarities shown, and are 3, 2, 5, and 7 v, respectively. The algebraic sum of the voltage drops around the circuit therefore is $25 - 3 - 4 - 2 + 6 - 5 - 10 - 7 = 0$.

There are many ways of illustrating Kirchhoff's laws. While the two examples just given are simple, many complicated circuits may be reduced to these simple equivalent circuits in order to facilitate analysis.

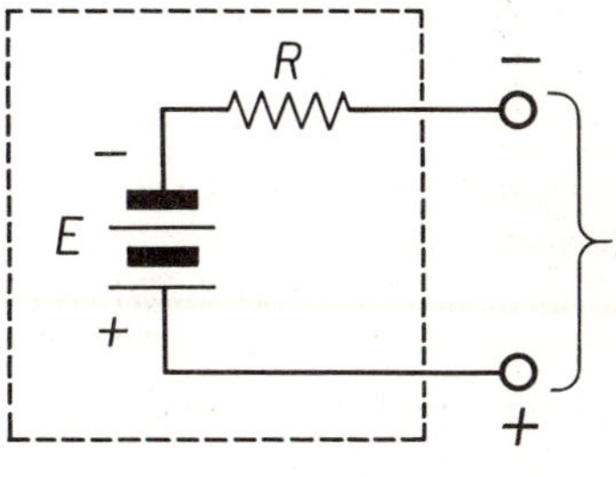

Fig. 2–25 D-C Source with Internal Resistance

2.12 INTERNAL RESISTANCE OF GENERATOR

Every device used to produce an electric current, whether a battery or some other type of generator, has by nature a certain amount of internal resistance. Current flowing through this resistance sets up a voltage drop which reduces the voltage which might otherwise be available from the generator if it had no resistance.

Figure 2–25 shows how a battery (E) is accurately represented with its internal resistance, R. (This internal resistance is largely that of the electrolyte.) When no current is drawn from the battery, there is no current flow through R and therefore no voltage drop. The full value of voltage E accordingly appears at the battery terminals. This is called the *terminal voltage, open-circuit voltage,* or *no-load voltage* of the battery. When a current, I, is drawn from the battery, a voltage drop, IR, appears across the internal resistance, R. Consequently, the output voltage drops to a value equal to $E - IR$. This is called the *full-load voltage* of the battery.

The internal resistance of a cell or battery may be measured in the following manner: (1) Measure the open-circuit voltage, E_1. (2) Connect an accurately known resistance to the battery and check (a) the current (I) flowing through the resistance and (b) the full-load voltage, E_2. (3) Calculate the internal resistance from the relationship $R = (E_1 - E_2)/I$. Resistance R is expressed in ohms, E_1 and E_2 are measured in volts, and I is in amperes.

SUMMARY OF CHAPTER 2

An **Electric Circuit** consists of a source, load, and connecting leads (conductors) in its common form. Some circuits do not contain a generator, as such.

A **Direct Current** flows always in the same direction around a circuit. The direction of electron current flow is from negative to positive.

A **Series Circuit** consists of elements connected one after another so that a current flows through every element of a series circuit.

A **Parallel, or Shunt, Circuit** consists of branched elements connected across each other. In a parallel circuit, current divides between the various branches in proportion to their electrical size.

Series-Parallel and Parallel-Series Circuits are combinations of the simpler series and parallel circuits.

A **Cell** generates electricity through chemical action resulting from ionization of an electrolyte. Two or more cells connected together form a **Battery.** The **Primary Cell,** such as the common dry cell, generates electricity *directly.* A **Secondary,** or **Storage, Cell** must first be charged by passing through it a current from some other source of d-c electricity. This cell then will generate a current. Primary cells must be discarded when they are exhausted. Secondary cells, on the other hand, may be recharged.

The purpose of a cell or battery is to maintain a difference of potential in a circuit.

In a primary cell, the breakdown of sulphuric acid into positive hydrogen ions and negative sulphate ions is represented by the equation

$$H_2SO_4 \rightleftarrows H_2^+ + SO_4^-$$

Chemical action in a lead-acid storage cell is represented by the equation

$$PbO_2 + Pb + 2H_2SO_4 \underset{\text{charge}}{\overset{\text{discharge}}{\rightleftarrows}} 2PbSO_4 + 2H_2O$$

The *specific gravity* of the electrolyte of a lead-acid storage cell, as indicated by a hydrometer, is 1.285 to 1.300 when fully charged, and 1.150 to 1.175 when discharged.

The **Voltmeter** is used to measure electromotive force. The **Ammeter** is used to measure current flow. 1 v = 1,000 millivolts (mv) = 1 million microvolts (μv). 1 kilovolt (kv) = 1,000 v. 1 amp = 1,000 milliamperes (ma) = 1 million microamperes (μa).

Resistance (R) is the opposition offered to current flow. The unit of resistance is the ohm. From Ohm's law,

$$R = E/I \qquad E = IR \qquad I = E/R$$

Resistivity (ρ) is the specific resistance of a material (resistance per unit area, or per unit volume). Resistivity is expressed in ohms.

$$\rho = (RA)/l$$

Resistors are devices for introducing a desired amount of resistance into a circuit. The resistance of a fixed resistor is constant, while that of a variable resistor is adjustable.

Conductance (G) is the opposite of resistance. $G = 1/R$. The unit of conductance is the mho.

Total Resistance. In a circuit containing resistances in series, $R_t = R_1 + R_2 + R_3 + \cdots R_n$. In a circuit containing resistances in parallel,

$$R_t = \frac{1}{\frac{1}{R_1} + \frac{1}{R_2} + \frac{1}{R_3} + \cdots \frac{1}{R_n}}$$

The **Voltage Drop** across a resistor (in volts) is equal to the product of the resistance and the current (ohms and amperes, respectively): $E = IR$. The **Dropping Resistor** and the **Voltage Divider** employ this principle to reduce voltage to a desired value.

Power (P) is measured in watts. In a d-c circuit, $P = EI = I^2R = E^2/R$. 746 w = 1 hp. 1 w = 1,000 milliwatts = 1 million microwatts (μw). 1 kilowatt (kw) = 1,000 w. 1 megawatt = 1 million w.

Kirchhoff's First Law states that the sum of currents flowing away from a point in a circuit equals the current flowing to that point. **Kirchhoff's Second Law** shows that the sum of the voltage drops around a circuit must equal the voltage of the generator, and that the algebraic sum of *all* the voltage drops around a circuit, including the generator voltage, must equal zero.

Internal Resistance is inherent in all generators of electricity. The effect of internal resistance is to reduce the terminal voltage of a generator when current is taken from the latter.

QUESTIONS AND EXERCISES

1. What is the principal difference between a primary and a secondary cell?
2. Name the electrolyte employed in the lead-acid storage battery.
3. What is the value of the specific gravity of the electrolyte in a discharged lead-acid storage battery?
4. Name the electrolyte employed in the Edison storage battery.
5. What is the unit of resistance?
6. What is the resistance of a circuit in which 3,150 v force a current of 10 ma?
7. A certain electrical device has a resistance of 273 ohms. What is its conductance in mhos?
8. One each 33-, 330-, 4,700-, and 6,800-ohm resistors are connected in series. What is the total resistance of this combination?
9. One each 3,370-, 5,280-, 2,000-, and 1,500-ohm resistors are connected in parallel. What is the total resistance of this combination?
10. A 6-v, 1-amp lamp is to be operated from the 117-v power line. What value of dropping resistor will be required?
11. The voltage across a heater is 117 v, and the current through it is 5 amp. What power is used by this heater?
12. How many kilowatt-hours will this heater use in a 48-hr period?

Chapter 3

Switches and Controls

3.1 BASIC PRINCIPLE OF SWITCH

Figures 2–21(A) and (B), Chapter 2, show the difference between a closed circuit and an open circuit, respectively. In the closed circuit, the conductors all are intact, so that a current flows. In the open circuit, one conductor has been cut. The flow of current therefore ceases, since the electrons cannot pass across the open gap.

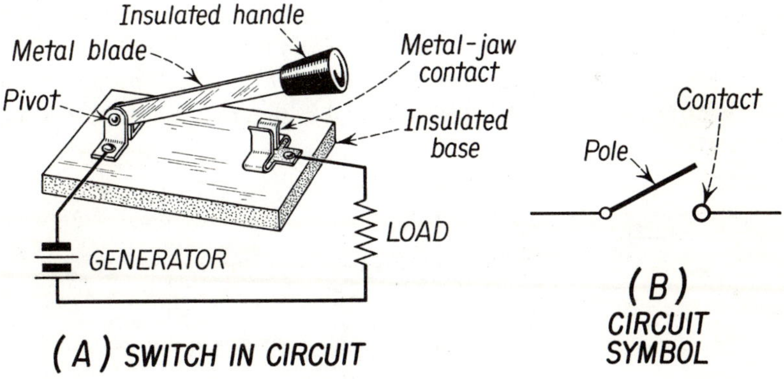

Fig. 3–1 Simple Switch

From this, it is clear that current may be started and stopped simply by closing and opening at least one path in a circuit. This is called *making* and *breaking* the circuit. The simplest way to accomplish this would be to cut a lead to interrupt current flow and to tie it back together to restore flow. But this would be neither efficient nor safe.

A *switch* is a device that performs this task of making and breaking a circuit. Figure 3–1(A) shows a simple switch. This consists of a metal blade, pivoted at its bottom end so that it can be moved in an arc, and a spring metal contact between the jaws of which the blade may be inserted. The blade is provided with an insulated handle, and both blade and contact are mounted on a base of insulating material.

When the blade is pushed all of the way down, it is gripped by the jaws of the contact and the circuit is closed. Current then flows from the generator, through the load, and back to the generator. When the blade is lifted free of the contact, the circuit is opened and the load receives no current. Because of the appearance of the blade, this particular type of switch is called a *knife switch.*

In switch parlance, the movable member (whether it is a blade or some other shape) is called a *pole* and the stationary member a *contact.* Any switch consisting of one pole and one contact, as in Figure 3–1(A), is called a *single-pole, single-throw* switch (abbreviated spst), since it has only one pole and can be thrown in only one direction to close the circuit. Figure 3–1(B) shows the circuit symbol for a spst switch.

All switches consist of poles and contacts. By combining poles and contacts in various ways, switches may be made to perform many useful functions. Besides single-throw switches, there also are double-throw, triple-throw, etc. Likewise, there are double-pole, triple-pole, four-pole, etc.

Figure 3–2 shows several of the simplest of these combinations. Figure 3–2(A) shows how an extra pole and contact might be added to the simple knife switch to obtain a double-pole, single-throw (dpst)

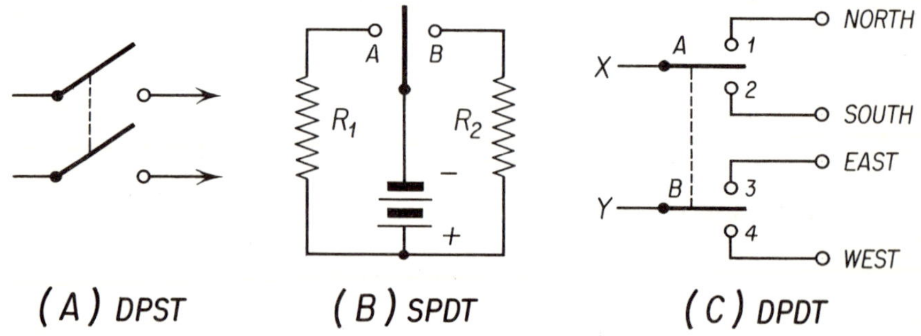

Fig. 3–2 Simple Multifunction Switches

switch. The dotted line indicates that the poles are *ganged* (attached together mechanically so as to be moved simultaneously) with no electrical connection between them. The dpst type permits both leads from the generator to be made and broken. Figure 3–2(B) shows a single-pole, double-throw (spdt) type connected in a circuit. When the pole is thrown to Contact *A*, current flows from the battery through Load R_1.

Load R_2 is dead. When the pole is thrown to Contact *B*, current then flows through R_2, and R_1 is dead. R_1 and R_2 do not have to be simple resistors. They may be lamps, bells, motors, or any other electrical devices which are to be operated from the battery.

Figure 3–2(C) shows a double-pole, double-throw (dpdt) type. When this switch is thrown to its upper position, Pole *A* engages Contact 1, and incoming line *X* is connected to the outgoing NORTH line. At the same time, Pole *B* engages Contact 3, and incoming line *Y* is connected to the outgoing EAST line. The SOUTH and WEST lines are disengaged. When the switch is thrown to its lower position, Pole *A* engages Contact 2, and *X* is connected to SOUTH, while Pole *B* engages Contact 4, and *Y* is connected to WEST. The outgoing NORTH and EAST lines then are disengaged.

3.2 MECHANICAL ARRANGEMENTS

In applying the basic switch principle, there are many possible mechanical arrangements other than the simple knife switch shown in Figure 3–1. In one kind of switch, for example, the pole is a sheet-metal spring which is pressed against the contact by a manually operated push button or lever when the switch is closed. In another type, the pole is a metallic disc sector which is rotated in or out of the gripping jaws of a stationary contact.

The *toggle* switch is a snap-action type. Familiar examples are the ON–OFF wall switches for electric lights, and the small snap switch employed in some radios, amplifiers, and television receivers. The snap action is secured through the use of a special spring assembly.

Spring-action switches are provided which remain closed (or opened) only as long as pressure is applied to a push button or lever, and return to their normal position (open or closed) when the pressure is removed. An example is the doorbell push button.

Switch-type sockets for electric lamps are provided with either *push button, key,* or *chain* switches. In the push-button type, one button closes the switch, and the other opens it. In the key type, turning the key or knob rotates a cam which presses two metal contacts together, closing the switch against the spring action of the contacts. Turning the key farther in the same direction allows the contacts to spring open. In the chain type, a cam is rotated by pulling the chain to open and close the circuit.

Some switches are designed expressly for manual or foot operation, others for operation by electricity, clockwork, vibration, expansion, or any quantity which will change sufficiently to provide the mechanical movement needed to operate the switch.

3.3 ROTARY SWITCH

The side-to-side movement of the pole in the basic switch is not the only motion that can be utilized in switching operations. The side-to-side switch is limited to one contact per pole in each position of the switch.

In the rotary switch (Figure 3–3), a pole travels around an arc or circle and is able to close the circuit successively through a number of contacts arranged around the circumference. There can be as many

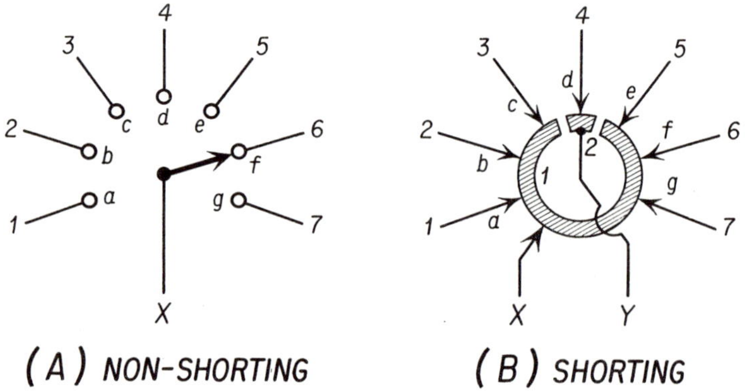

Fig. 3–3 Rotary Switches

contacts as can be spaced around the arc. This arrangement is illustrated by Figure 3–3(A). Seven spaced contacts are shown. When the pole makes contact with *a,* Line *X* is switched to Line 1. With the pole contacting *f,* as shown in the figure, Line *X* is connected to Line 6.

The switch shown in Figure 3–3(A) would be described as a single-gang, single-pole, seven-position type. It is also called a *nonshorting* type because adjacent contacts are not short-circuited by the pole. Another name is *selector switch,* which arises from the action of the rotating pole in selecting any one of a number of contacts.

A second type of rotary switch is shown in Figure 3–3(B). Here, the rotating member is a split metal ring, consisting of Sections 1 and 2 which are insulated from each other but are fastened together to turn in unison. The central piece, 2, functions as the switch pole and is shown closing the circuit from Line *Y,* through Contact *d,* to Line 4. The ring, labeled 1, short-circuits all of the unused contacts. A separate wiping contact, *X,* connects to this ring and sometimes is grounded externally.

Not all rotary switches have a large number of contact positions. A single pole and contact, for example, are found in the rotary spst type. One pole and two contacts constitute a rotary spdt type.

Rotary switches, like other types, may be ganged mechanically so that one movement will operate several switches. Multiple assemblies of this kind, available commercially in many sizes, can perform complicated switching operations. In these assemblies, each separate switch is referred to as a *section,* and the entire assembly is called a *switch.*

3.4 JACKS AND PLUGS

In certain types of circuits, usually where electricity is not present in dangerous amounts, switching is performed by the single-motion operation of clipping together or pulling apart two special contacts, each on the end of a wire lead. One of these contact devices is a *jack;* the other a *plug.* The circuit is closed by inserting a plug into the jack. The plug then is held firmly in place by the spring action of the jack. A telephone switchboard contains several plugs and many jacks.

Pin Type. Figure 3–4 shows the simplest sort of jack and plug. This is the *pin type,* so called because the metal plug usually is so thin that it resembles a pin. The spring-type jack is shown in cross section in Figure 3–4(A). Its jaws normally are pressed tightly together. When

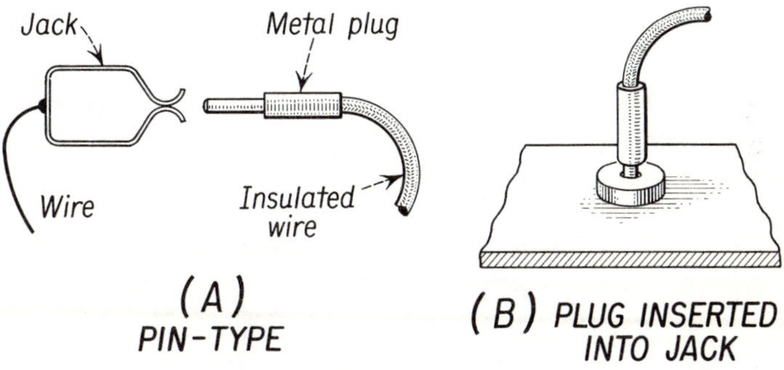

Fig. 3–4 Pin-Type Jack and Plug

the plug is inserted between them, they spread open only enough to grip the shaft of the plug and hold it in contact. The jack spring often is enclosed in a metallic or plastic housing for protection, and the housing sometimes is threaded for mounting the jack on a panel. Figure 3–4(B) shows the plug inserted into a mounted jack.

Telephone Type. The simple pin-type jack-and-plug system has the disadvantage that it handles only one lead. Only one circuit path can be made and broken with such a system. The telephone jack and plug accommodate two or more leads.

Plate 3–1 A Multicontact Pin-Type Jack and Plug. The rear block contains 21 jacks. The front bar contains 21 matching pin plugs. Leads are soldered to the terminal posts on the rear of each block. Pressing these two blocks together makes 21 circuits simultaneously; pulling them apart breaks all these circuits. (Courtesy, DeJur-Amsco Corp.)

Figure 3–5(A) shows details of the telephone-type jack and plug. The jack consists of a metal frame with an end hole for entry of the plug, and is provided with a flat, metal spring blade curved on its free end to make contact with the tip of the plug. This blade is insulated from the frame. The frame is one circuit contact, and the blade is the

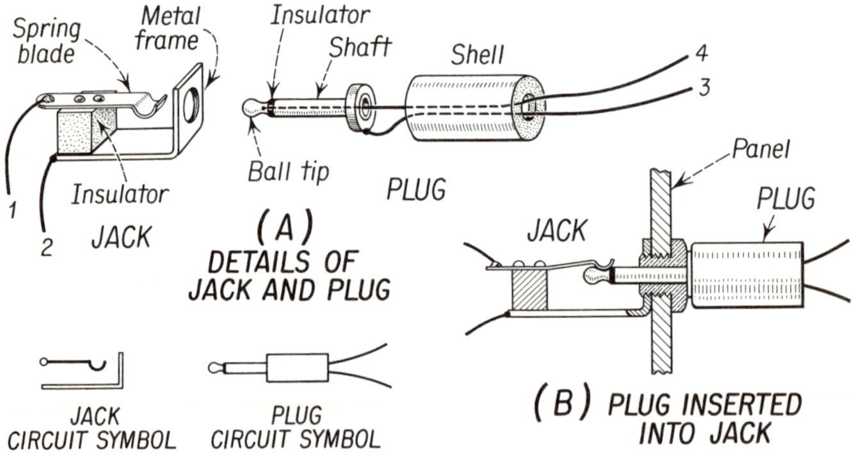

Fig. 3–5 Principle of Telephone-Type Jack and Plug

other. The end of the plug is a metal ball which is insulated from the metal shaft to which it is attached. The ball is wired to one lead (No. 4) and the shaft to the other (No. 3). A cylindrical shell, usually of insulating material, is screwed to the end of the plug.

When the plug is inserted into the jack, the spring contacts the ball, and the frame contacts the shaft. The curved part of the spring locks in behind the ball, in the groove especially prepared for it. [See Figure 3–5(B).] This completes the circuit from jack lead 1 to plug lead 4, and from jack lead 2 to plug lead 3. Inserting the plug quickly makes the circuits, and pulling it out quickly breaks them.

Complex switching operations may be accomplished with a jack and plug if additional springs and contacts are placed in proper positions in the jack assembly. Many special-purpose jacks are manufactured. Figure 3–6 shows several of the common types.

In Figure 3–6(A), the additional blade (*C*) has a lower contact (*D*). When the plug is inserted, it moves Blade *B* upward, closing an additional circuit through *B* and *C*. The plug circuit is closed through *A* and *B*. In Figure 3–6(B), the plug circuit is closed through *A* and *B*,

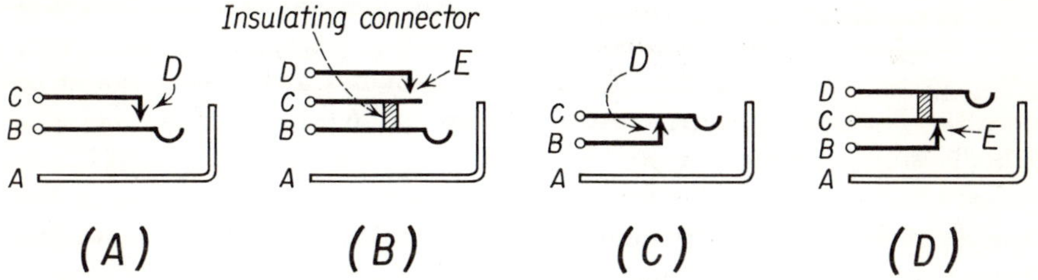

Fig. 3–6 Various Types of Telephone Jacks

but the plug moves Blade *B* upward and the insulating connector transmits this movement to Blade *C* which meets Contact *E* and closes a separate circuit through *C* and *D*. In Figure 3–6(C), a separate circuit-*opening* function is performed. Here, Blade *C* normally rests against Contact *D* of Blade *B*. This normally closes a circuit through *B* and *C*. Insertion of the plug lifts *C*, opening the *BC* circuit, while the plug circuit is closed through *A* and *C*. In Figure 3–6(D), a separate, isolated circuit is opened. Here, the plug lifts Blade *D*, and the insulating connector transmits the motion, lifting Blade *C* from Contact *E* and breaking the circuit through *B* and *C*. The plug circuit, meanwhile, is closed through *A* and *D*.

Other types of plugs are available commercially. For example, one type has three connections instead of the common two. In this plug, the shaft is divided into two sections insulated from each other, and the ball on the end of one section is insulated from both sections. In the matching jack, the frame contacts the back section of the plug shaft, while one spring contacts the other section, and a second spring contacts the ball.

Some authorities carry the classification of jacks and plugs ahead to include any type of quick-disconnect type of make-and-break device attached to the end of conducting leads. Such a wide classification would include even the plug and outlet used to connect appliances to the power line.

3.5 VARIABLE-RESISTOR CONTROLS

Switches afford only a two-way control of electricity. This is because they are make-and-break devices having only two states—*on* and *off*. No in-between states are provided. Thus, a switch will start a motor or stop it but will not make it go faster or slower. Likewise, a switch will turn a lamp on or off but will not dim it or brighten it.

Devices which permit a continuous variation of current are known by the general name *controls*. There are two general classes of controls —*step-type* and *smooth*. The step-type takes it name from the fact that it has a number of separate positions, settings, or steps [like the rotary switch in Figure 3–3(A)], each of which provides a definite amount of control. The smooth-type, instead, is so named because it allows a virtually stepless range of control from zero to maximum.

Electric controls take several forms. One of the simplest is the variable resistor. (See Section 2.4, Chapter 2, for a discussion of the variable resistor.) Resistance-type controls are classified as *rheostats* and *potentiometers*. The rheostat is a variable dropping resistor (Section 2.7), while the potentiometer is a variable voltage divider (Section 2.10). Both rheostats and potentiometers may be of either the step type or smooth type.

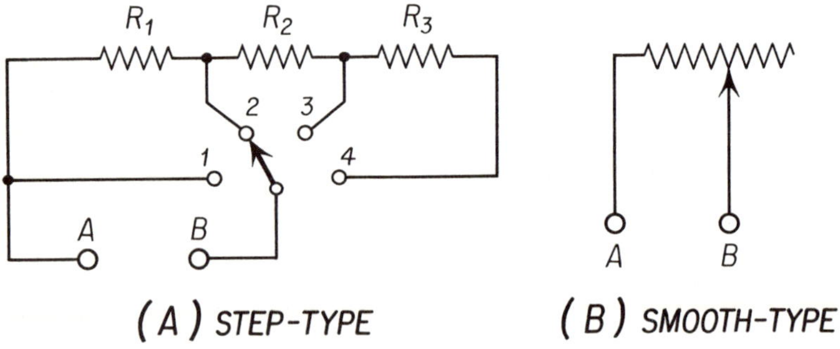

Fig. 3–7 Types of Rheostats

Rheostats. Figure 3–7(A) shows the step-type rheostat. Three resistors (R_1, R_2, and R_3) are shown connected to a single-pole, four-position switch, although a larger number of resistors can be employed if a switch with more contacts is used.

When the switch pole is in Position 1, no resistance is in the circuit, and current flows without opposition from terminal *A* to terminal *B*. With the switch set to 2, only resistor R_1 is in the circuit, and the current is reduced in proportion to the resistance of R_1. With the switch set to 3, the total resistance is $R_1 + R_2$, since the current must flow from *A*, through R_1, then through R_2, and out from Contact 3 and terminal *B*. When the switch is at 4, the total resistance is $R_1 + R_2 + R_3$. Thus, this rheostat provides a resistance range from zero to maximum $(R_1 + R_2 + R_3)$ in four steps.

A disadvantage of the step-type unit is that there is no way to obtain a resistance value between the fixed steps. In some applications, this makes no difference in the final control, but in others, such as motor-speed control, it results in "jumpy" operation of the controlled device. Smooth control is obtained by employing a variable resistor of the type shown in Figure 2–15(B), Chapter 2. This allows a smooth variation of the resistance as the contact blade is moved over the turns of wire in the resistance coil. Thus, in Figure 3–7(B), when the wiper (contact blade) is at the extreme left end, no resistance is included between terminals *A* and *B*, but when the wiper is at the extreme right end, the maximum resistance appears between *A* and *B*.

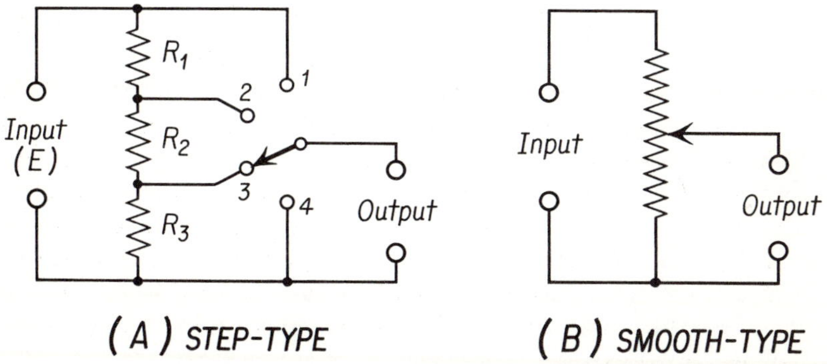

Fig. 3–8 Types of Potentiometers

Potentiometer. Figure 3–8(A) illustrates a step-type potentiometer. Only three resistors (R_1, R_2, and R_3) are shown, but a larger number may be employed by providing more contacts on the rotary switch. When the switch is set to Position 1, there is no voltage-divider action and the entire input voltage appears at the output terminals. If the resistors are equal in value, $2/3E$ appears at the output terminals when the switch is at 2, and $1/3E$ when the switch is at 3. The output voltage is zero when the switch is in Position 4, since the two output terminals then are connected to the same input terminal. The step-type potentiometer shown thus provides an output-voltage variation of zero to *E* in three steps.

Smooth potentiometer action is obtained by substituting a variable resistor of the type shown in Figure 2–15(B), Chapter 2, for the switched resistors. This modification is shown in Figure 3–8(B). When the wiper is at its topmost setting, the output voltage has the full value of the input voltage, but when it is at its lowest setting, the output voltage is zero.

SUMMARY OF CHAPTER 3

A **Switch** is a device for turning current on and off. It operates on the simple principle of interrupting (*breaking*) and closing (*making*) the circuit. A switch has at least one *pole* and one *contact.* Several poles and contacts may be included in a single switch.

Single-Throw switches may be closed in one direction only.

Double-Throw switches may be closed in either of two directions.

The **Rotary Switch** allows a single pole to close any one of a number of circuits through contacts arranged in a circular path over which the pole is rotated.

Jacks and **Plugs** in combination may be regarded as a special class of switch permitting quick make or break of a circuit. These units are equivalent to snap fasteners on the ends of conducting leads.

The **Variable Resistor** offers a simple method for varying current over a desired range. In a **Step-Type** control of this sort, various fixed values of resistance are cut into the circuit in *steps* by a selector switch. In the **Smooth-Type,** a contact blade (wiper) travels over the turns of a coil of resistance wire or a carbon resistance element, varying the resistance in an almost stepless manner. Resistor-type controls are classed as rheostats when they are used as variable dropping resistors, and as potentiometers when employed as variable voltage dividers.

Chapter **4**

Current Flow in Liquids and Gases

4.1 IONIZATION

We learned in Section 1.9, Chapter 1, that an atom which has lost one or more electrons becomes a positive *ion,* and that an atom which has gained one or more electrons becomes a negative ion. This is the process of ionization.

Section 1.11 explained that electron drift is not the only constituent of an electric current, but that a flow of ions also can be the basis of a current flow. Ions, however, being heavier than electrons, move more slowly.

Electrons flow in one direction at a time through a current-carrying conductor, while ions flow in both directions simultaneously when the current is due to ion movement.

4.2 ELECTROLYTE

Electrolytes are substances which become ionized when they are dissolved in water. These include acids, bases, and salts. Common table salt (sodium chloride, NaCl), for example, when in solution breaks down into positive Na ions and negative Cl ions. If electrodes are connected to the positive and negative terminals of a battery, as in Figure 4–1, and immersed in the solution, an electric field will appear between the electrodes when the switch is closed. This field will move the positive ions toward the negative electrode, and negative ions toward the positive electrode. The positive electrode is termed the *anode,* and the negative electrode the *cathode.*

The positive sodium ion lacks an electron and takes one from the negative electrode to which it has been attracted. The negative chloride ion possesses an extra electron which it gives up to the positive electrode to which it has been attracted. Thus, electrons enter the solution via the negative electrode and leave it via the positive electrode, and this constitutes a current flow through the solution, although the electrons

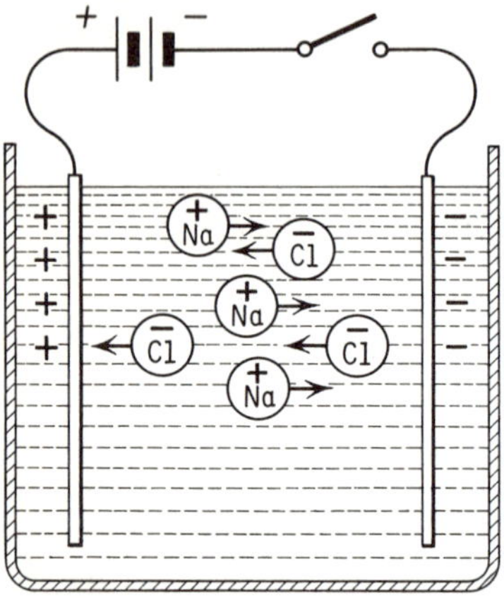

Fig. 4–1 Electrolysis in a Salt (NaCl) Solution

are transported by the movement of ions in opposite directions through the solution. At the positive electrode, the neutral chloride atoms combine in pairs (Cl_2) to form atoms of chlorine gas which bubbles out of the solution.

This process, known as *electrolysis,* accounts for the flow of electricity through liquids. A current will flow readily, however, only through those solutions which are electrolytes, because only these solutions provide the necessary ions. A sugar solution, for example, does not ionize and therefore is not a good conductor. Oils and *pure* water are insulators.

Water may be electrolyzed if it first is made conducting by the addition of a trace of acid or salt (some tap water contains enough minerals to provide conduction without this addition) and platinum electrodes are immersed into it. When a current is passed through it, the water (H_2O) is *dissociated* into positive hydrogen ions and negative oxygen ions. These ions migrate in the usual manner to the oppositely charged electrodes. Oxygen atoms combine in pairs at the positive electrode and form oxygen gas which bubbles out of the solution in that neighborhood, while hydrogen atoms combine in pairs at the negative electrode to form hydrogen gas which bubbles out of the solution around this electrode. As long as the current is sustained, the process continues until the water is used up (*decomposed*).

4.3 ELECTROPLATING

The phenomenon of electrolysis is utilized in the practical process of *electroplating.* The ability to plate one metal upon another is based upon the fact that ions in an electrolyte will transport atoms from the anode and deposit them upon the surface of the cathode. Thus, a metallic object, such as a spoon, ring, or case, may be plated with a desired metal by making the object the cathode in an electrolytic tank and making the desired metal the anode.

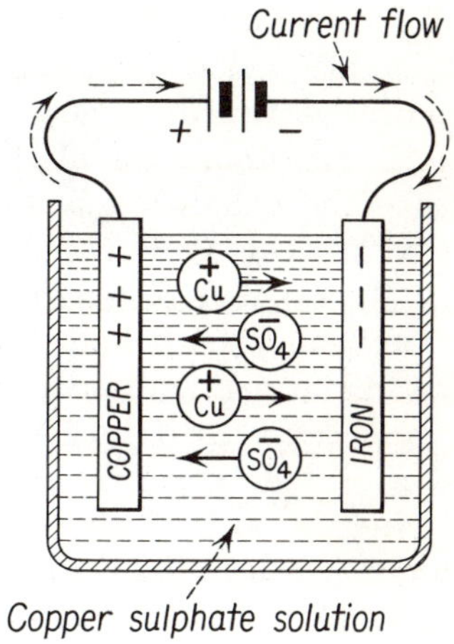

Fig. 4–2 Illustration of Electroplating

Figure 4–2 illustrates the process of electroplating copper on iron. Here, the iron to be plated is made the cathode of an *electrolytic cell,* and the copper is made the anode. The electrolyte is an acid solution of copper sulphate ($CuSO_4$). This solution dissociates into positive copper (Cu) ions and negative sulphate (SO_4) ions. When current flows through the electrolyte, positive copper ions migrate to the iron cathode where they acquire electrons, become neutral copper atoms, and are deposited on the surface of the iron. The negative sulphate ions migrate to the copper anode where they give up electrons and react with the copper to form more copper sulphate. The copper electrode thus is gradually eaten away and the iron electrode becomes coated with copper. The longer the process is continued, or the stronger the current, the heavier will be the plated coat.

While the principle of electroplating is seen to be comparatively simple, the business of electroplating is a large industry. Some of the metals regularly plated by electricity include gold, silver, nickel, chromium, and cadmium.

4.4 IONIZATION IN GASES

Gases, including air, also may become ionized and in this condition will conduct an electric current in a manner similar to that of electrolytes.

When two electrodes are immersed in a gas, negative ions will migrate to the positive electrode and positive ions to the negative electrode. Electrons freed from some of the previously neutral gas atoms likewise travel to the anode. The ions are enormous in size, compared to the electrons, and so travel more slowly. This ion movement simultaneously in both directions constitutes a flow of current through the gas.

If the voltage is made high enough, the electrons will acquire considerable velocity, and some will collide with neutral gas atoms with enough impact to dislodge more electrons from the latter, thus making ions of these atoms. Similarly, some electrons and ions strike the electrodes with such force as to knock electrons from the atoms of the electrodes. This *ionization by collision* adds to the current strength.

If the number of gas molecules is reduced (as by evacuating the tube containing the gas and electrodes), less and less voltage is required to ionize the gas and force an appreciable current through the tube. The reason for this is reduction of the gas "crowding"; that is, the number of molecules is so reduced that electrons can travel farther and thus gain higher velocity before colliding with neutral atoms. The reduction can be carried so far, however, that the chance of a molecule being struck becomes very slight, and virtually no ionization takes place.

Energy changes take place when an atom becomes an ion, and vice versa. The electrons involved in the change radiate this energy as light waves. Ionized gas glows for this reason. The *glow discharge* appears between the electrodes of a gas tube conducting current.

As the voltage applied to a gas tube is increased, a small current first flows through the tube, and this increases slowly to a point called the *ignition potential* (the voltage at which the slow current increase stops). At this voltage, the gas "breaks down"; that is, the glow discharge begins, and the current jumps suddenly to a relatively high value. At a still higher current level, a second jump occurs (the *arc discharge*), and this is characterized by a much brighter luminosity than the glow discharge.

4.5 CATHODE RAYS

If the exhaustion of a tube, such as is shown in Figure 4–3, is advanced (that is, the vacuum is high), positive ions may move the full distance between electrodes and acquire such velocity that they strike the cathode with enough force to knock out many electrons. The latter are attracted in a stream to the anode. Such streams are termed *cathode rays*. If a small hole is punched in the anode, a portion of this invisible ray will pass through the hole and strike the end of the tube behind the anode. If the tube is glass, that end may be coated with a phosphorescent material which will glow at a bright spot indicating the point at which the cathode rays strike. If the coating is Willemite, the spot will be green in color. If the entire end of the tube is coated behind the anode, the cathode rays will cast a shadow of the anode

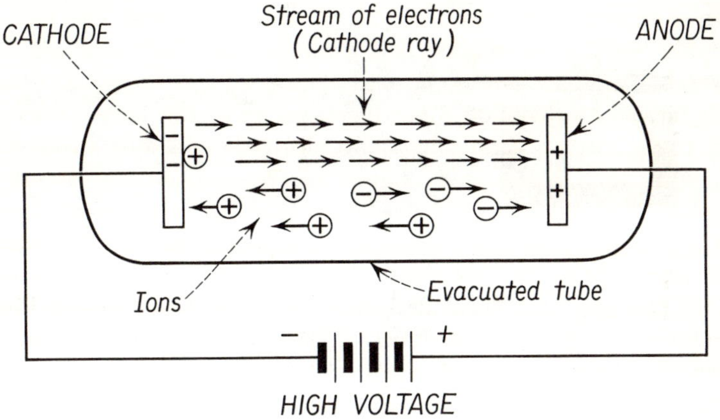

Fig. 4–3 Gas Tube Action

on the glass. This is shown in Figure 4–4 where the anode has been cut in the shape of a Maltese cross. This simple principle of phosphorescence under influence of cathode rays is utilized in the television picture tube and in the cathode-ray oscilloscope, an electronic test instrument. The principle of the emission of invisible rays from an electrode bombarded by electrons is utilized also in the X-ray tube.

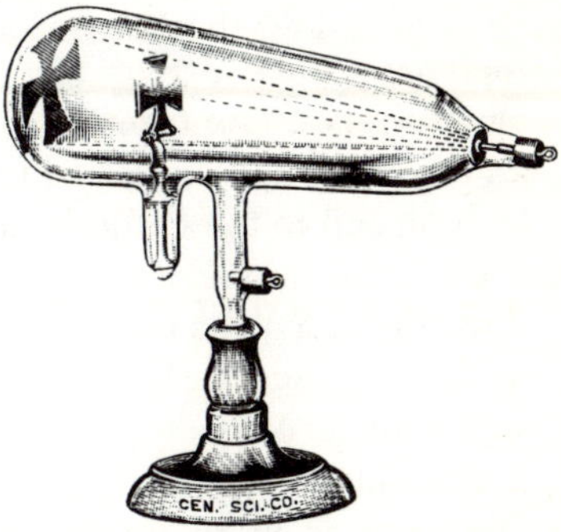

Fig. 4–4 Cathode Ray Tube with Coated Phosphorescent End to Show How Cathode Rays Cast Shadow of Cross-Shaped Anode

When a small, intercepting electrode is hung in the tube between anode and cathode, it will exhibit a negative charge when the tube is conducting. This shows that the electrons constituting the cathode rays are negative. If a charged electrode is held near the outside of the tube, the cathode-ray beam will be deflected by a negative charge and attracted by a positive one, indicating again the negativity of the electrons.

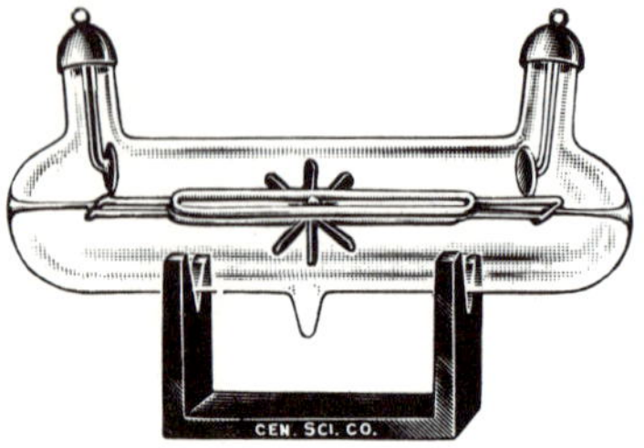

Fig. 4–5 Paddle-Wheel Cathode Ray Tube

An interesting *cathode-ray tube,* often used in lecture demonstrations, contains a lightweight paddle wheel mounted between the anode and cathode. (See Figure 4–5.) When the tube is conducting, the cathode rays strike the paddles and cause the wheel to spin.

SUMMARY OF CHAPTER 4

Ionization occurs when a previously neutral atom acquires an electric charge. If the atom gains an extra electron, its net charge is negative, and it becomes a *negative ion.* If the atom loses an electron, its net charge becomes positive, and it then is a *positive ion.*

Electrolytes are substances which become ionized in water solution. They are said to become *dissociated* into positive and negative ions. Acids, bases, and salts form electrolytes. Organic liquids, such as certain oils, sugar, and alcohols, are not electrolytes. Neither is *pure* water. Ionization permits the flow of current through an electrolyte.

Electroplating. The art of electroplating utilizes the simple process of electrolysis. The transfer of ions by electric current passing through an electrolyte is employed to carry a metal from the anode of an electrolytic cell to the cathode and to deposit it upon the cathode surface.

Gases also exhibit ionization as their atoms lose or gain electrons. Current flow through a gas is increased by *ionization by collision,* which refers to the striking of gas molecules by fast-moving electrons or ions with sufficient force to knock loose free electrons and thus creating additional ions.

Gas Discharge is luminous because electrons involved in the process of conversions between atoms and ions radiate, in the form of light waves, energy lost in the process.

Cathode Rays are streams of electrons knocked out of the cathode by positive ions bombarding that electrode. The cathode rays move to the anode and may be detected by their ability to fluoresce certain materials known as *phosphors,* to be bent by nearby electric charges, and to rotate a paddle wheel erected in their path. The negativity of the electron seems to be demonstrated by actions of the cathode-ray stream.

QUESTIONS AND EXERCISES

1. Name two substances which become electrolytes when dissolved in water.
2. Is pure water a good electrolyte?
3. Explain why ions travel more slowly than electrons.
4. Describe the process by means of which copper is electroplated on iron.
5. In gases, what is ionization by collision?
6. Why does a "harder" vacuum reduce the voltage needed to ionize the gas enclosed in a tube?
7. Describe the generation of cathode rays.
8. Why are cathode rays so called?
9. Describe an experiment which demonstrates that cathode rays are composed of *negative* charges.
10. How may the presence of cathode rays be determined visually?

Chapter **5**

Magnetism and Electromagnetism

5.1 PERMANENT MAGNET

The phenomenon of magnetism was known in ancient times. Euripides mentioned as early as 425 B.C. that lodestone attracts iron. Lodestone is a black oxide of iron which was found in volcanic rock in the town of Magnesia in Asia Minor, hence the name *magnetism.* The scientific name for lodestone is *magnetite.*

Lodestone is a *natural* magnet; that is, it needs no processing by man to make it magnetic, for nature has endowed it with this property. A piece of iron stroked with lodestone becomes magnetized to some extent and subsequently will attract other pieces of iron or steel. However, this iron is not a natural magnet, since it normally is not found magnetic in nature but must be worked on to make it so.

Lodestone and pieces of iron or steel magnetized by contact with it or with other magnets are called *permanent magnets.* This term arises from the fact that such magnets retain their magnetism for some time. In contrast, magnets that lose their magnetism soon after the magnetizing source is removed are called *temporary magnets.*

5.2 MAGNETIC POLES

If a bar-shaped magnet is explored by passing a small piece of iron throughout its immediate vicinity, one finds that the iron is attracted most forcibly by the ends of the magnet and very little, if at all, by the center. The ends are called the *poles* of the magnet. If the magnet is dipped into iron filings, it emerges with most of the filings sticking to the poles. (See Figure 5–1.)

When suspended horizontally at its center, the magnet aligns itself along a north-south line. If it is rotated out of this position, the magnet will swing back so that its *same* pole always points in the general direction of the earth's north pole. (Actually, a magnet does not point to true north but, in the United States, to a point varying from 20°E to

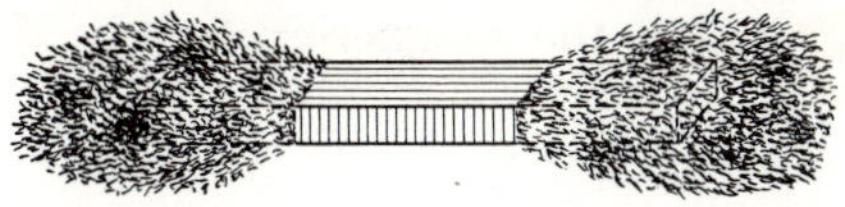

Fig. 5–1 A Bar Magnet Dipped into Iron Filings Emerges with Most of the Filings Clinging to Its Poles

25°W of the North Pole.) This particular pole of the magnet is designated the north-seeking, or simply *north pole* (abbreviated N), and the opposite pole is designated the south-seeking, or *south pole* (abbreviated S). The ability of a magnet to line itself up in this manner is the property utilized in the mariner's compass.

A bar magnet may be twisted into a U, angular, or horseshoe shape, but its tips still retain their N and S identity.

When a bar magnet is broken into numerous pieces, each piece becomes an independent magnet with its own north and south pole like the parent magnet.

5.3 MAGNETIC FIELD

Magnetism and electricity are similar in many respects. Around an electric charge, for example, there is a region in which the charge has influence over other charges and neutral bodies. There is a similar region, the *magnetic field,* surrounding a magnet. Just as the electric field is filled with electrostatic lines of force constituting electrostatic flux (see Section 1.4, Chapter 1), the magnetic field is filled with *magnetic lines of force* constituting *magnetic flux.*

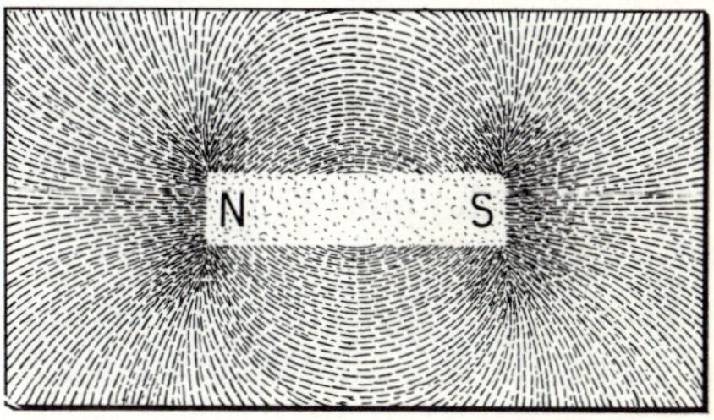

Fig. 5–2 Magnetic Lines of Force Traced by Iron Filings

The lines of force in one plane of the magnetic field may be mapped in a simple way: Cover the magnet with a sheet of paper and sprinkle the top of the paper with iron filings. When the paper is tapped, the filings will arrange themselves along the lines of force, as shown in Figure 5–2.

The lines of force are considered to come out of the north pole and to re-enter the magnet at its south pole, as shown in Figure 5–3. They are said to extend around the *magnetic circuit.* But the lines do not

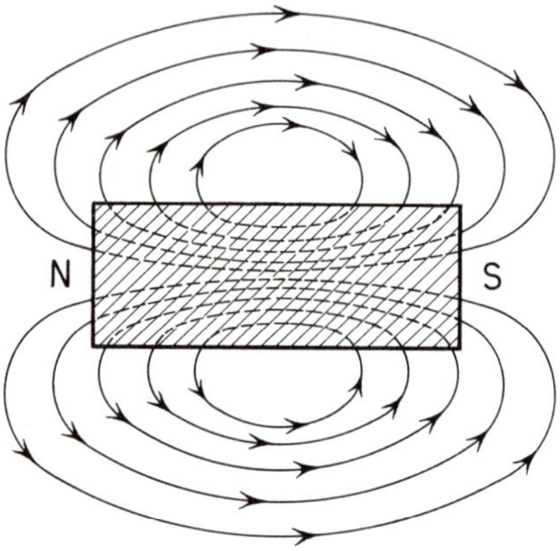

Fig. 5–3 A Portion of the Field of a Bar Magnet

constitute a current flow. They merely indicate the direction and strength of the magnetic field. Note from Figure 5–3 that the lines pass *through* the magnet, as well as extending into the space around it. In this respect they resemble the flow of electric current, which passes through the generator as well as around the external circuit.

5.4 MAGNETIC ATTRACTION AND REPULSION

The forces acting between two magnets are similar to those between two electric charges. Thus, *like poles repel each other,* while *unlike poles attract.* This is illustrated in Figure 5–4. A south pole will repel another south pole, and a north pole will repel another north pole. But a north pole will attract a south pole, and vice versa.

These facts make it possible to identify the polarity of a magnet if a known pole is available as a test tool. The mariner's compass can function as such a tool. When the north pole of the compass needle turns toward the pole of a magnet, that pole is south. When the north pole of the needle turns away from the pole, the latter is north.

Figure 5–5 illustrates the reason for magnetic attraction and repulsion. For clarity, only four magnetic lines are shown in the sketches, but it is understood that there are a great many such lines. In Figure 5–5(A), opposite poles are placed close together. Here, lines of force leave the north pole of magnet *X*, as shown by the arrowheads, enter the south pole of magnet *Y*, extend through *Y*, leave the north pole of *Y*, enter the south pole of *X*, and extend through *X*. The two magnets are pulled together by the action of these lines of force which link them. We see in Figure 5–5(B) that no lines of force are common to both

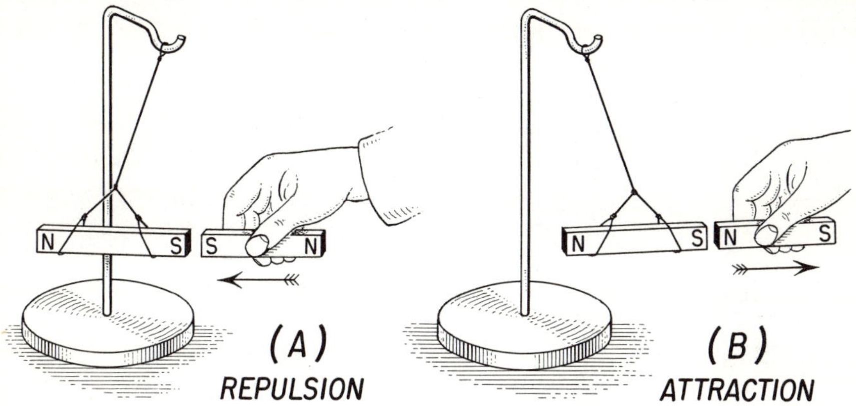

Fig. 5–4 Forces between Two Magnets

magnets, but that the field around the south pole of *X* is pushed back and its lines are crowded by the corresponding field around the south pole of *Y*. This effect results in the repulsion of one magnet by the other.

Note the similarity of the field patterns in Figure 5–5 to the electric fields depicted for attraction and repulsion in Figure 1–2, Chapter 1.

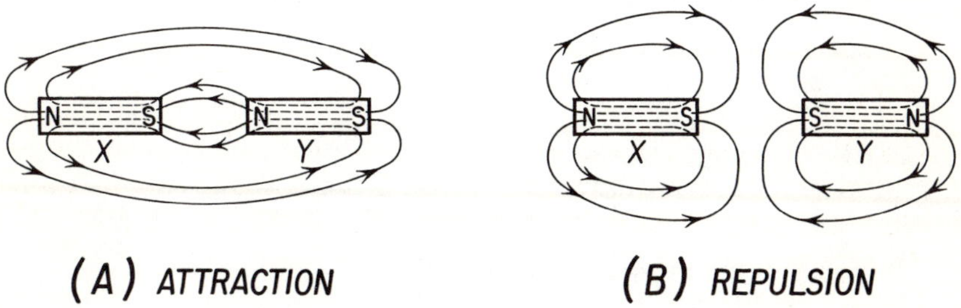

Fig. 5–5 Line Arrangement in Attraction and Repulsion

The previous case deals with attraction between two magnets. Figure 5–6 illustrates the attraction of unmagnetized iron or steel by a magnet. In Figure 5–6(A), the lines of force extend from the north pole of the magnet, enter one end of the iron, extend through the iron, emerge from the other end, and return to the south pole of the magnet. Since the lines enter one end of the iron, extend through it, and leave the other end, the front end becomes the north pole and the other end the south pole of a temporary magnet. This process whereby the

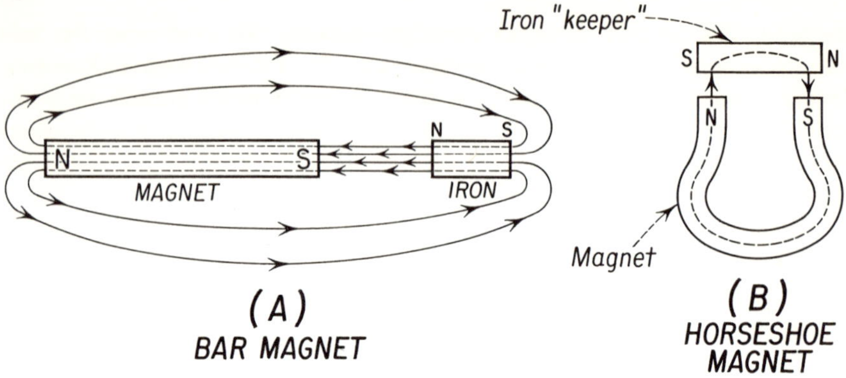

Fig. 5–6 Magnet Attracting Iron

nearby iron becomes a magnet is called *induction.* Since there now are two separate magnets and their *unlike* poles face each other, the iron is attracted to the magnet, or vice versa. Figure 5–6(B) gives the same illustration for a horseshoe magnet and its iron "keeper." For simplicity, only one line of force is shown in the latter illustration.

5.5 MAGNETIC SHIELDING

Magnetic lines pass more easily through some metals than they do through air. This is especially true of ferrous metals when we are dealing with a steady magnetic field around a magnet.

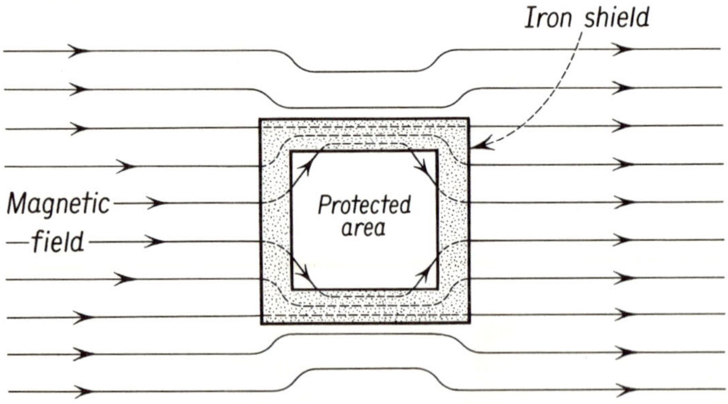

Fig. 5–7 Magnetic Shield

When an area must be protected from a magnetic field, it may be surrounded by a suitable metallic box or case. The box is called a *shield* from the fact that it serves to shield the protected area from the field. The shield does not somehow insulate the area from the field but serves to detour the lines through its walls and thus around the area. This is shown in Figure 5–7. While a large number of the lines are shunted around the area by the walls of the shield, a few of them succeed in penetrating into the area, as seen in Figure 5–7, since the shield is not completely perfect in its performance.

Shields of the type shown here are used to protect sensitive electrical instruments from disturbance by nearby strong magnets.

5.6 MAGNETIC FIELD SURROUNDING A CONDUCTOR

In 1820, Hans Christian Oersted (1777–1851), a young Danish professor, observed that the needle of a compass deflected whenever he closed or opened a nearby current-carrying circuit.

This observation led to the discovery that a magnetic field surrounds every current-carrying conductor as long as the current flows. This field is circular in shape, as shown in Figure 5–8(A). It is at right angles to the length of the conductor and is whirling. The lines of force extend around a straight wire in concentric circles, not spirals.

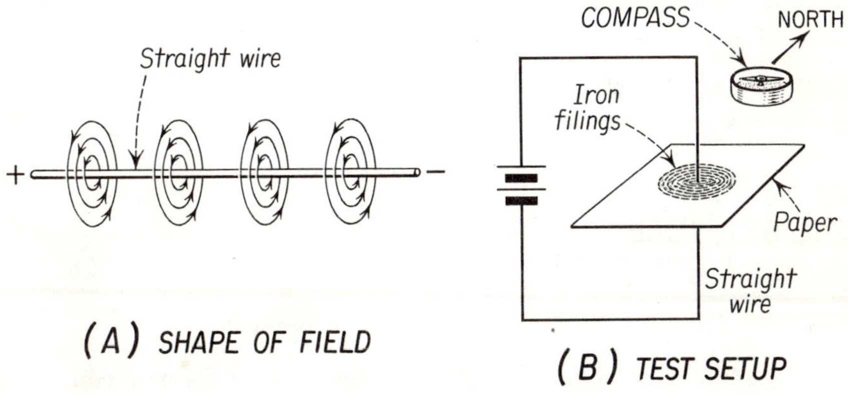

Fig. 5–8 Demonstration of Field around Wire

The field whirls around the wire in the direction indicated by the arrowheads in Figure 5–8(A). If we look down the wire in the direction of the negative terminal, the field is whirling in a clockwise direction.

The field about a wire may be demonstrated with the arrangement shown in Figure 5–8(B), where a straight, current-carrying conductor is passed through a sheet of paper sprinkled with iron filings. The filings will arrange themselves in a circular pattern indicating the shape of the field. The needle of a nearby compass will be deflected from its normal north-south position. A north pole suspended close to the end of the wire will rotate around the latter under the influence of the whirling magnetic field. If the battery is reversed, the pole will swing in the opposite direction.

5.7 THE ELECTROMAGNET

If the single, current-carrying conductor is twisted into a loop or ring, as in Figure 5–9(A), the field whirls are seen to enter one flat face of the ring and to leave via the other face. (If we look directly down on the edge of the ring, the whirls are seen to enter under the loop and leave above it.) This resembles the way in which lines leave one pole of a permanent magnet and enter its other pole.

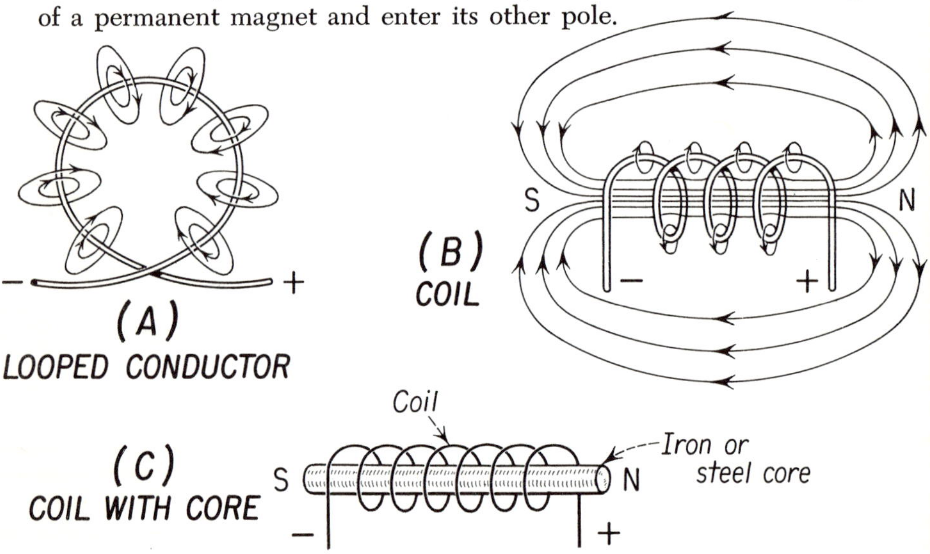

Fig. 5–9 Evolution of the Electromagnet

If additional loops are made in the wire to form a coil, as in Figure 5–9(B), and the turns or loops of the coil are placed close together, less magnetic lines will surround individual turns and more will pass lengthwise through the coil. The end at which these lines leave the coil becomes a north magnetic pole, and the opposite end where they re-enter the coil becomes a south magnetic pole. The more turns in the coil and the higher the current flowing through it, the stronger the magnetic force that is generated. When the current is interrupted, the coil ceases to be a magnet. This is the principle of the *electromagnet.*

If a core of iron or steel is placed inside the coil, as shown in Figure 5–9(C), the magnetic lines will pass more readily through this core than through air. Consequently, an enormous number of lines now pass lengthwise through the coil, as compared to the small number in the *air-core* coil of Figure 5–9(B). This increases the strength of the magnet tremendously.

Electromagnets are true *temporary magnets.* Many electrical machines, such as motors and generators, and numerous devices and accessories make use of the principle of electromagnetism. Electromagnets are used in industry to lift tons of ferrous metal.

5.8 MAGNETIC CONSTANTS AND UNITS

In order to deal quantitatively with magnetic materials and devices, a number of constants and units for use in descriptions and calculations have been worked out. These are defined in the following paragraphs.

Ampere-Turns. Ampere-turns are the product, NI, of the number of turns (N) in a coil and the current (I), in amperes, flowing through the coil.

Flux. The magnetic lines of force are known collectively as *flux.* Magnetic flux is comparable to electric current, but it is not a flow in the sense that current is. Flux is measured as the number of lines passing through a magnetic circuit. The unit of flux is the *maxwell,* named in honor of James Clerk Maxwell (1831–1879). Its symbol is ϕ, the Greek letter phi.

Flux Density. This is the degree of magnetization, expressed as the number of lines per square centimeter. The unit of flux density is the *gauss,* named in honor of Karl Friedrich Gauss (1777–1855). 1 gauss = 1 line per square centimeter. The symbol for flux density is B. Total flux $\phi = BA$ maxwells, where A is the area (in square centimeters) of the magnetic material.

Reluctance. Magnetic lines of force pass through some materials more readily than through air, while they pass through other materials less readily. The property whereby these materials either aid or obstruct the magnetic flux is termed *reluctance.* Reluctance is comparable to electrical resistance and often has been called magnetic resistance. Its magnitude depends upon the kind of material and the size of the magnetic circuit it forms. The symbol for reluctance is $\mathcal{R}$, the script R. The unit of reluctance is the *oersted,* named in honor of Hans Christian Oersted. (See Section 5.6.)

For nonmagnetic materials, $\mathcal{R} = l/A$, where l is the length of the magnetic circuit (in centimeters) and A is the cross-sectional area of the circuit (in square centimeters).

Magnetomotive Force. This is "magnetic pressure" and is similar to emf in an electric circuit. It is essential to the establishment of flux in a circuit. The unit of magnetomotive force (mmf) is the *gilbert,* named in honor of William Gilbert (1540–1603). Its symbol is F. In an electromagnet, $F = 1.26NI$, where NI is the ampere-turns of the coil. (See *Ampere-Turns* defined earlier in this section.)

Magnetomotive force is the work required to carry a unit magnetic pole around the magnetic circuit. (Compare with electromotive force, the work required to carry a unit electric charge around the electric circuit.) A unit magnetic pole has such strength that it attracts or repels an equal pole, 1 cm distant in vacuum, with a force of 1 dyne.

Magnetic Field. A unit magnetic field has such strength that a unit pole, when placed in this field, will be acted upon by the latter with a force of 1 dyne. Field strength is expressed in gausses. Its symbol is $\mathcal{H}$, the script H.

Magnetizing Force. This characteristic is based upon the number of gilberts required to pass a given number of lines of force through a magnetic circuit of known length; thus, the mmf per unit length. The unit of magnetizing force is the *oersted.* Its symbol is H. 1 oersted = 1 gilbert per centimeter.

Permeability. The ratio, B/H, of the flux density to field strength is called magnetic *permeability.* The permeability of any material is equal to the number of magnetic lines of force set up in a 1-cm cube of the material when a mmf of 1 gilbert is applied between opposite faces. The unit of permeability has no name. Its symbol is μ, the Greek letter mu.

The permeability of magnetic materials (iron, steel, etc.) is higher than that of nonmagnetic substances. For magnetic materials, μ has a range of several thousand to one. The permeability of air or a vacuum is 1. Materials with μ less than 1 are *diamagnetic;* those with μ higher than 1 are *paramagnetic.* A paramagnetic material affords an easier path for magnetic lines than does air, with the result that lines concentrate through the material. A diamagnetic material has the opposite effect, spreading the lines apart and conducting only a few.

The permeability of magnetic materials is not a constant figure for any one material but varies with magnetization. At low degrees of magnetization, the *initial permeability* is fairly constant, but, as the magnetization is increased, the permeability rises to a peak value and then decreases as the magnetization is increased further.

Saturation. Every magnetic material has a certain magnetic capacity; that is, there is a limit to the amount of magnetization it will accept. In a given material, the flux density increases as the applied magnetizing force is increased. This increase is very rapid up to a certain point, as shown in Figure 5–10. Beyond this point, further increases in magnetizing force, although they may be large, add very few more lines to the total flux. The level at which this tapering-off effect begins to be evident is called the *saturation point.*

For a given magnetizing force, the saturation point varies with the kind of material and its dimensions.

Hysteresis. A small amount of magnetism remains in steel after the external magnetizing force has been removed. This is called *residual magnetism.* Iron possesses some of this *retentive* power (*retentivity*) but not so much as steel.

Residual magnetism combines with the phenomenon of saturation, illustrated in Figure 5–10, to give an effect known as *hysteresis.* The curve in Figure 5–11 illustrates hysteresis. From this curve, we see that the magnetization (flux density, B) lags behind the magnetizing force, H.

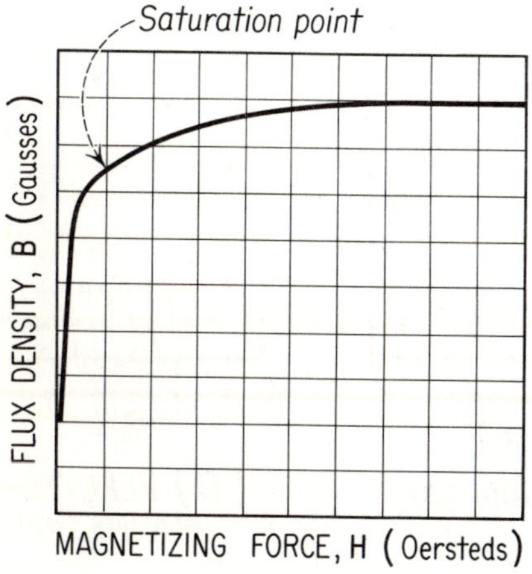

Fig. 5–10 Magnetization Curve

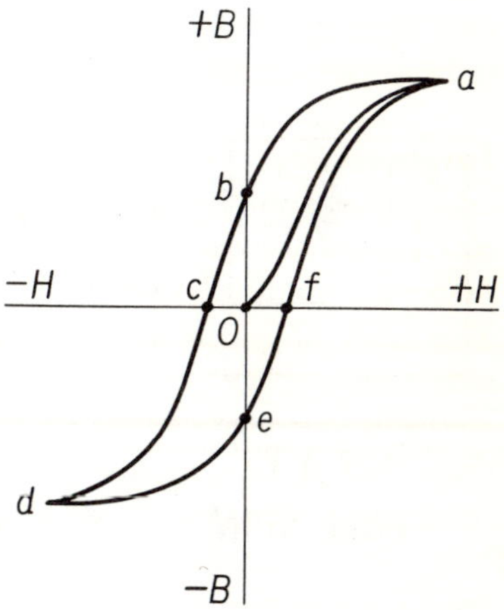

Fig. 5–11 Typical Hysteresis Loop

When H is increased in the positive direction, B increases positively from zero to a. If H then is decreased to zero, B decreases to b. Here, residual magnetism prevents B from returning to zero along with H. The magnetizing force must be increased *negatively* to c, in order to reduce B to zero. This value, represented by the length from 0 to c, is called the *coercive force,* H_c. As H is increased further negatively, B

increases negatively to *d*. If, then, *H* is decreased to zero, *B* decreases to *e*, this lag behind *H* again being due to residual magnetism. A positive coercive force (0 to *f*) now is needed to reduce *B* to zero. Finally, as *H* is increased further positively, *B* increases to *a*, and the full cycle *abcdefa* may be repeated.

In electrical apparatus containing iron or steel, a definite amount of electrical energy is required to take the material through a complete cycle of magnetization from positive to negative and return when alternating currents are used. This energy is expended in heating the material and is called *hysteresis loss.* This loss is encountered in all iron-containing apparatus in which the magnetization is varied, and it constitutes a significant part of the total losses in the apparatus.

5.9 THE MAGNETIC CIRCUIT

In a magnetic circuit, the lines of force generated by the magnetomotive force pass from the north to the south pole through the system. This is similar to the electric circuit, in which electron current generated by the electromotive force flows through the system. In the magnetic circuit, however, there is no current flow in the sense of the electron drift.

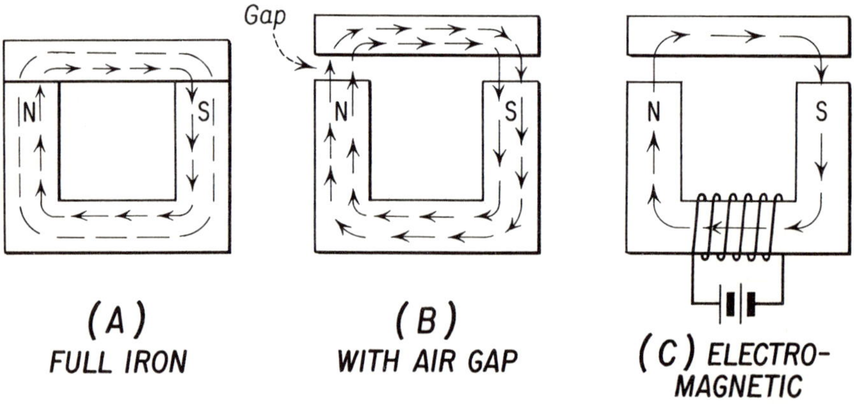

Fig. 5–12 Magnetic Circuits

The basic laws of the magnetic circuit are analogous to those of the electric circuit. The equations for making calculations are in the same form in both cases and differ only in the interchange of electrical units and magnetic units in them.

Figure 5–12 shows several types of magnetic circuit. In Figure 5–12(A), the lines pass through a circuit comprised of a permanent magnet and an iron keeper held to the poles by attraction. In Figure

5–12(B), the keeper has been separated from each pole by a small air gap. In Figure 5–12(C), an electromagnet is the source of the lines of force.

5.10 LAW OF THE MAGNETIC CIRCUIT

Ohm's law for the magnetic circuit states that *total magnetic flux is directly proportional to magnetomotive force and inversely proportional to reluctance.* This is expressed by the equation

(5–1) $$\phi = F/\mathcal{R}$$

where ϕ = total flux (in maxwells)
F = magnetomotive force (in gilberts)
$\mathcal{R}$ = reluctance (in oersteds)

Also

(5–2) $$F = \phi\mathcal{R}$$

and

(5–3) $$\mathcal{R} = F/\phi$$

Notice that there is great similarity between these expressions and Ohm's law for direct current [Equations (2–8), (2–9), and (2–10), Chapter 2].

Example: A certain magnetic circuit has a reluctance of 3.2 oersteds. When the mmf = 756 gilberts, what is the total flux?

From Equation (5–1),

$$\phi = F/\mathcal{R} = 756/3.2 = 236 \text{ maxwells (236 lines)}$$

Computations may be made on electromagnetic circuits of the types similar to Figure 5–12(C) provided the mmf in gilberts is calculated beforehand.

(5–4) $$\text{mmf} = 1.26NI$$

where N = number of turns in the coil
I = current (in amperes) flowing through the coil

Example: A current of 5 amp through a 1,000-turn coil sets up a flux of 2,500 lines. What is the reluctance of the circuit?

From Equation (5–4),

$$\text{mmf} = 1.26NI = 1.26(1{,}000)5 = 6{,}300 \text{ gilberts}$$

From Equation (5–3),

$$\mathcal{R} = F/\phi = 6{,}300/2{,}500 = 2.52 \text{ oersteds}$$

5.11 CALCULATING RELUCTANCE

When the dimensions of a magnetic circuit and the materials used in it are known, the reluctance of the circuit may be calculated as readily as the resistance of an electric circuit.

$$\mathcal{R} = l/(\mu A) \qquad (5\text{–}5)$$

where $\mathcal{R}$ = reluctance (in oersteds)
μ = permeability of the material (obtained from a table of metals)
A = area (in square centimeters)
l = length (in centimeters)

Example: The permeability of an iron sample is found from iron tables to be 900. The length of a magnetic circuit employing this material is 25 cm and its cross-sectional area is 2 sq cm. What is the reluctance of this circuit?

$$\mathcal{R} = l/(\mu A) = 25/(900 \times 2) = 25/1{,}800 = 0.014 \text{ oersteds}$$

When air gaps are present in the magnetic circuit, the reluctance of each gap must be added to the reluctance calculated for the metallic path in the circuit. For air

$$\mathcal{R} = l/A \qquad (5\text{–}6)$$

where l = gap length (in centimeters)
A = gap area (in square centimeters)

Example: Two air gaps, each 0.1 cm long and 1¼ sq cm in area, are located in a magnetic circuit. What is the total reluctance due to air which must be added to the value of reluctance calculated for the metal in the circuit?

$$\mathcal{R} = l/A = 0.1/1.25 = 0.08 \text{ oersted per gap}$$
$$= 2(0.08) = 0.16 \text{ total}$$

5.12 SOLENOID

A solenoid is a special form of electromagnet. Its basic construction is shown in Figure 5–13.

The coil consists of many turns of wire wound around a tube or bobbin, made of insulating material, with a central hole slightly larger in diameter than the iron core. When Switch S is closed, current flows from Battery *B* through the coil, setting up magnetic lines of force in and around it. The magnetic force attracts the core which then is sucked up into the coil. The disc at the bottom of the core acts as a stop as it strikes the coil. If it were not present, the core would pass through the coil and be ejected from the other end ("electric cannon" effect). When the switch is opened, the magnetism is interrupted, and the core drops out of the coil.

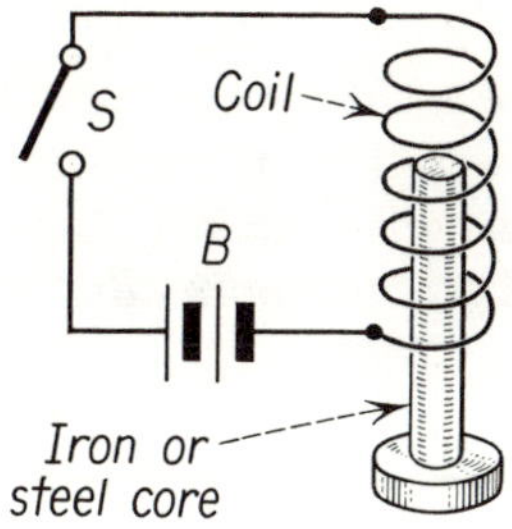

Fig. 5–13 Solenoid

Solenoids are used widely in industry. If a contact is mounted on the core, it will engage a stationary contact mounted close by when the core is sucked into the coil and will close a circuit. *Solenoid switches* of this type permit the use of a small current to control large currents. Special switching devices, such as *circuit breakers,* often use the solenoid principle.

5.13 RELAY

Another widely used type of electromagnetic switch is shown in Figure 5–14. This is the *relay.*

Unlike the solenoid, the electromagnet in the relay contains a core that is set permanently in place. Mounted close to one end of this core is a flat, often springy, blade (the *armature*) made of magnetic

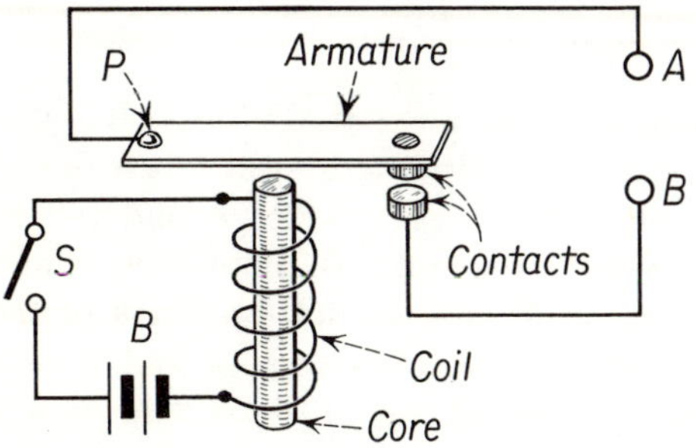

Fig. 5–14 Relay

material, such as iron. It is pivoted at one end (P) and has a contact, made of good-conducting metal (such as silver or tungsten), attached to its free, movable end. A second, stationary contact is mounted close to this movable contact.

When Switch S is closed, current flows from Battery *B* through the coil and magnetizes the core. The magnetic force attracts the armature, and this movement closes the two contacts. This completes the circuit between Terminals *A* and *B*. When the switch is opened, this circuit is broken.

The armature and stationary contact of the relay thus are seen to correspond to the pole and contact of a simple spst switch. Relay armatures and contacts may be ganged together in the manner of switches, to provide multipole and multiposition switching operations. As in switches, the contacts of some relays are *normally open* and others *normally closed.*

5.14 BUZZER

Figure 5–15 shows the construction of a buzzer, a device which employs interrupted magnetization to produce a signal tone.

The buzzer is seen to resemble the relay, except that the armature is connected to one end of the coil, and an external contact (*A*) normally rests against the contact on the armature.

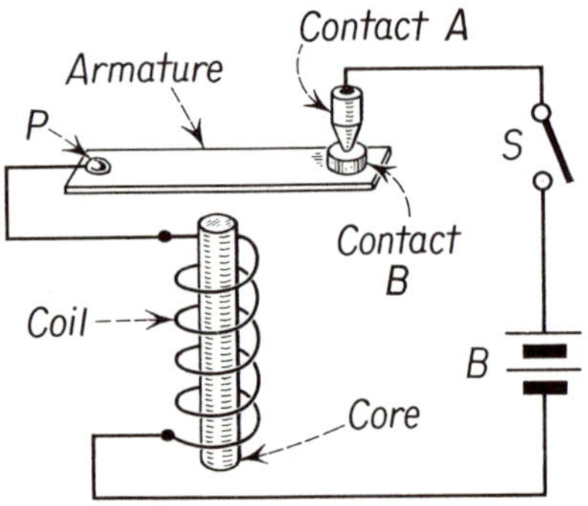

Fig. 5–15 Buzzer

When Switch S is closed, current from the battery magnetizes the core. This causes the armature to be attracted. But when the armature moves toward the core, Contact *B* leaves Contact *A* and the circuit is opened, interrupting current flow in the coil. Without current in the coil, the core now no longer is a magnet, and the spring action of the armature returns the latter to its original position with *A* and *B* in contact. With these contacts together, current again flows, the armature is attracted, and the cycle is repeated.

The armature vibrates back and forth, as a result of this action, at a rate (*frequency*) determined by its dimensions and its distance from the end of the core. This vibration sets up an audible buzz, from which the device takes its name. Contact *A* often is a setscrew which enables the armature to be positioned nearer or farther from the core, and this adjustment allows the buzzer pitch, or tone, to be adjusted at will. Buzzer action continues as long as Switch S remains closed.

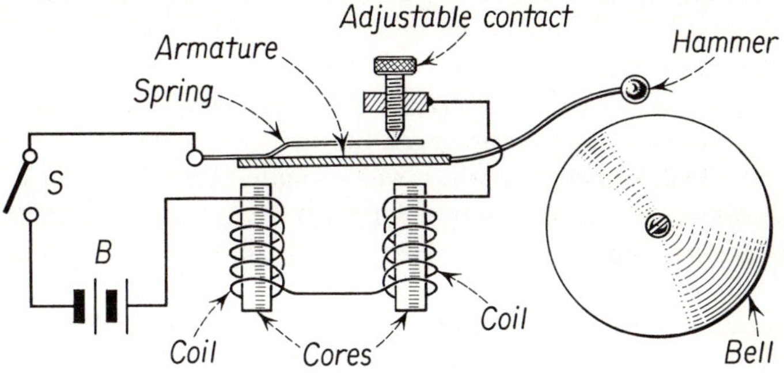

Fig. 5–16 Electric Bell

5.15 ELECTRIC BELL

In Figure 5–16, the buzzer has been taken a step further by attaching an arm and hammer to the free end of the armature and positioning a bell or gong so that it will be struck by the hammer each time the armature moves toward the core.

The electric bell is one of the oldest of household electrical appliances.

SUMMARY OF CHAPTER 5

Permanent Magnets retain their magnetism for some time after the original magnetizing force has been removed. **Temporary Magnets** lose their magnetism when the external magnetizing force is removed.

Every magnet, whether permanent or temporary, has a **North Pole** and **South Pole.** If the magnet is suspended horizontally, its north pole will point in the general direction of the earth's North Pole. The mariner's compass is based on this property.

The **Magnetic Field** is made up of magnetic lines of force filling the space in the immediate vicinity of the magnet. Collectively, these lines are called **Magnetic Flux.** The magnetic field is analogous to the electrostatic field surrounding an electric charge. Lines of force in the magnetic field are considered to extend through the magnet lengthwise, to leave it via the north pole and re-enter via the south pole.

Magnetic Attraction and **Repulsion** resemble electrostatic attraction and repulsion. *Like* magnetic poles repel each other; *unlike* poles attract.

A **Current-Carrying Conductor** is surrounded by a magnetic field which is whirling and circular. The field about a conductor may be intensified by looping the conductor or by winding it into the form of a coil. This is the basis of the **Electromagnet.** An electromagnet may be intensified still further by inserting an iron core into the coil. The electromagnet is a true *temporary* magnet.

Ampere-Turns are the product, NI, of the turns in an electromagnetic coil and the current flowing through the coil.

The unit of **Flux** is the **Maxwell,** also the **Line.** Its symbol is ϕ. Flux is expressed as the total number of lines.

The unit of **Flux Density** is the **Gauss.** Its symbol is B. 1 gauss = 1 line per square centimeter.

Reluctance is magnetic resistance. The unit of reluctance is the oersted. Its symbol is $\mathcal{R}$. Like electrical resistance, the magnetic reluctance varies for different materials and with their physical dimensions.

Magnetomotive Force is magnetic pressure, similar to emf in electrical circuits. The unit of mmf is the **Gilbert.** Its symbol is F.

Magnetizing Force is the magnetomotive force per unit length of a magnetic circuit. The unit of magnetizing force is the **Oersted.** Its symbol is H. 1 oersted = 1 gilbert per centimeter.

The **Permeability** of any material is a figure showing the number of lines set up in a 1-cm cube of the material by an mmf of 1 gilbert. Permeability = B/H. The permeability of air is 1, that of *diamagnetic* materials less than 1, and of *paramagnetic* materials greater than 1. The unit of permeability has no name. Its symbol is μ.

Each magnetic material has a **Saturation Point.** This may be thought of as "magnetic capacity." Beyond this point, any further increase in the applied magnetizing force will add few, if any, additional lines to the total flux.

Hysteresis is the combined effect of saturation and **Residual Magnetism** (the small amount of magnetism remaining in a material after the magnetizing force has been removed). Hysteresis causes the changes in magnetization due to variations in the applied magnetizing force to lag behind the latter.

Ohm's Law for the Magnetic Circuit

$$\text{Total flux } \phi = \frac{\text{magnetomotive force}}{\text{reluctance}} = \frac{\text{gilberts}}{\text{oersteds}} = \frac{F}{\mathcal{R}}$$

The **Reluctance of a Magnetic Circuit** is equal to $l/(\mu A)$, where l is the circuit length (in centimeters), A is the cross-sectional area (in square centimeters), and μ is its permeability. The reluctance of an **Air Gap** is l/A.

QUESTIONS AND EXERCISES

1. Explain three simple ways to identify the north pole of a magnet.
2. Describe the difference between a permanent magnet and a temporary magnet.
3. What is the rule for magnetic attraction and repulsion?
4. Why doe a magnet attract iron?
5. Why does coiling a current-carrying conductor increase its magnetic strength?
6. What is residual magnetism?
7. Does magnetism *flow* through a magnetic circuit as current flows through an electric circuit?
8. Explain the operation of a buzzer.
9. A certain magnetic circuit has a reluctance of 5.4 oersteds. When the applied mmf = 1,000 gilberts, what will be the total flux?
10. The permeability of an iron sample is found from tables to be 1,000. It is used to construct a magnetic circuit 50 cm long, 1½ cm square, and 1½ cm high. What is the reluctance of this circuit?

Chapter **6**

Introduction to Alternating Currents

6.1 PRODUCTION OF EMF BY MAGNETIC FIELD

In Section 5.6, Chapter 5, it was explained that a magnetic field surrounds any conductor carrying current. Now we are to learn that the reverse also is true; that is, a magnetic field can produce a current in a conductor which is part of a circuit. However, this is a dynamic effect,

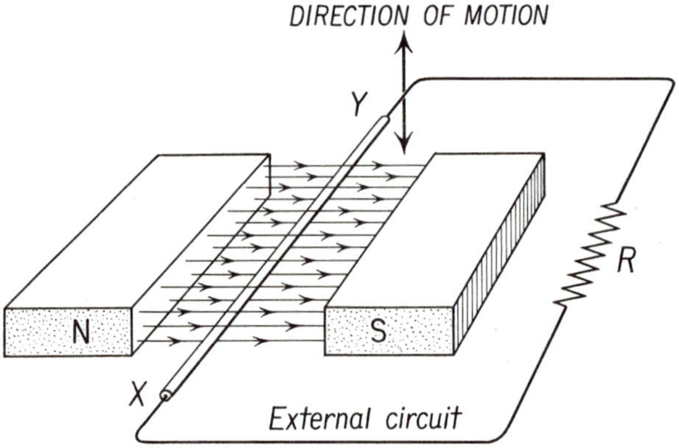

Fig. 6–1 Moving Conductor in Magnetic Field

not a static one. By this is meant that either the conductor *or* the field must be in *motion* in such a way that the conductor cuts lines of force or is cut by them. When a conductor simply rests in a field, there is no generation of current.

Figure 6–1 illustrates this effect. Here, a straight wire, *XY*, is placed perpendicular to the lines of force in a uniform magnetic field between the north and south poles. The wire is movable up and down in the direction indicated by the arrows labeled *direction of motion.* When the wire is moved upward, an emf is set up making End *Y* negative and End *X* positive. We call this an induced *emf.* As a result of this voltage, current will flow in the external circuit containing the load resistance, *R*, since electrons flow from *Y* back around to *X* through this circuit. This current will continue as long as the wire is moving but will fall to zero when the movement halts.

When the wire is moved downward, the reverse action occurs; that is, the polarity of the induced emf shifts, making End *X* of the wire negative and End *Y* positive. Current in the external circuit then flows in the opposite direction, from *X* to *Y*.

If wire *XY* is moved rapidly up and down in the field, the current in the external circuit will flow in first one direction and then the other. Thus, an *alternating current* will be generated. This simple effect is the basis of all *generators,* machines for the production of electricity from magnetism. All that is needed to complete the mechanical system is a means of moving the conductor in the field, or moving the magnet past the conductor.

6.2 VALUE OF INDUCED EMF

The strength of the electromotive force induced in the conductor is governed by the number of lines cut *per second.* Thus, a wire cutting 2,000 lines per second would show twice the voltage of one cutting only 1,000 lines per second. When a wire cuts 100 million (that is, 10^8) lines per second, 1 v is generated.

Example: A single wire passes at the rate of 25 times per second between the poles of a magnet having a flux density of 22,000 lines per square inch. The face of each pole is 1×2 in. What is the value of voltage generated?

(*a*) Total flux = area of magnet × flux density = $1 \times 2 \times 22{,}000$ = 44,000 lines

(*b*) Lines cut per second = $44{,}000 \times 25 = 1{,}100{,}000$

(*c*) Emf = 1,100,000/100,000,000 = 11/1,000 = 0.011 v = 11 mv

From the preceding explanation and example, we can see that the induced voltage may be raised by increasing the speed of movement of the conductor or by increasing the flux density, or both.

6.3 SINGLE-COIL A-C GENERATOR

The simple arrangement shown in Figure 6–1 serves to demonstrate the experimental production of emf, but it is not the best mechanical arrangement. It is far more feasible, mechanically and electrically, to rotate a looped conductor in the magnetic field. This latter arrangement is illustrated by Figure 6–2(A).

Here, the conductor has been bent into a single-turn coil, termed the *armature,* and is considered to be mounted on a horizontal shaft in such a way that it may be spun in the field. The poles of the magnet are cut in a curved manner, so that the separation between the pole and the armature is constant.

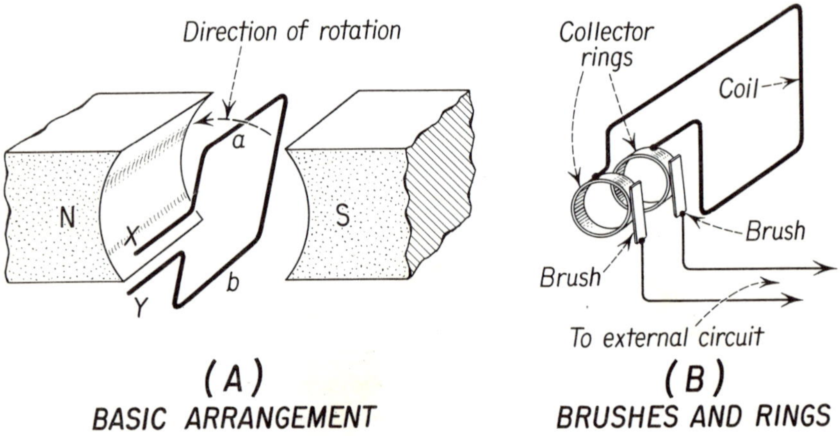

Fig. 6–2 Single-Coil Alternating-Current Generator

The upper half-turn, *a,* and the lower half-turn, *b,* thus cut lines of force except when the armature passes through its vertical position. When the armature moves through its vertical position between the poles, it is cutting lines at the lowest rate since it is traveling almost *along* lines rather than *across* them. The armature is cutting lines at the maximum rate when it moves through its horizontal position between poles, since in this position it is cutting directly across lines.

It is easy to see that the Leads, *X* and *Y,* will become hopelessly twisted after the armature has turned several times. To prevent this and still make contact with the ends of the coil, the arrangement shown in Figure 6–2(B) is used. Here, each end of the coil is connected to a metal ring (*collector ring* or *slip ring*) mounted axially on the shaft but insulated from it. Pressing against each ring is a conducting wiper, or *brush* (in its simplest form, a flat, metallic spring) connected to the external circuit. The outer brush thus always is connected to the same half-turn, and the inner brush to the same half-turn, regardless of the position of the coil.

Consider first that the coil is standing vertically with half-turn *a* on top. This half-turn is at rest and cutting no lines, so no emf will be present at *X* and *Y*. The lower half-turn, *b*, also is at rest. As the coil rotates counterclockwise, as indicated by the curved arrow, *a* begins to cut lines and so does *b*, so an emf appears at *X* and *Y*. But since *a* and *b* now are moving almost *along* the lines, very few lines are cut per second, and the emf is low. As the movement progresses, *a* and *b* cut across progressively more lines per second, and the emf increases proportionately. Maximum emf is induced when the coil passes through its horizontal position (at right angles to its initial starting position).

From this point, *a* and *b* are moving toward the second vertical position of the coil. With *a* at the bottom and *b* at the top, the armature coil has moved two right angles, or 180 deg, from its starting position. As the half-turns move toward this position, progressively less lines per second are cut, and the emf falls off proportionately until it again reaches zero when *a* is at the bottom and *b* on top.

As the rotation continues, *a* now moves into the field near the south pole, and *b* near the north pole. The polarity of the emf at *X* and *Y* reverses, and the voltage builds up, as before, to a maximum when the coil is horizontal with *a* adjacent to the south pole, and *b* adjacent to the north pole. Then the emf decreases progressively to zero, as *a* continues to its original top position and *b* to its original bottom position.

From this action, we see that the emf at *XY* is an *alternating voltage*. It (1) starts at zero, (2) increases to a maximum positive value, (3) decreases to zero, (4) increases to a maximum negative value, and (5) decreases to zero, thereby returning to its original starting point. This rhythmic rise, fall, and reversal will be repeated as long as the coil is rotated.

6.4 VOLTAGE VARIATION IN GENERATED ALTERNATING CURRENT

The voltage generated by the rotating coil [Figure 6–2(A)] passes through an infinite number of positive and negative values, depending upon the position of the coil, and has the value of zero three times during each complete revolution. This is shown in Figure 6–3.

When half-turn *a* [Figure 6–2(A)] is on top, no voltage is generated, and this corresponds to the left-hand zero in Figure 6–3. As *a* rotates into the field toward the north pole, *X* becomes positive; and when *a* has passed through 30 deg, voltage e_1 is generated. At 60 deg, e_2 is generated. At 90 deg (*a* horizontally opposite the north pole), the voltage has reached its maximum positive value, *E*. At 120 deg, the voltage has fallen to e_3 which has the same value as e_2 at 60 deg; and at 150 deg, e_4 has the same value as e_1 at 30 deg. Finally at 180 deg, the emf again is zero, since *a* is vertical once more but now is at the bottom position. As the rotation progresses in the same direction, the polarity of X changes to negative, and, at 210 deg, we have $-e_1$ which has the same numerical value as e_1. The voltage now rises through similar steps to a negative maximum, $-E$, equal numerically to the positive maximum, E. It then falls through $-e_3$ and $-e_4$, reaching zero at 360 deg, with a in its original vertical, top position. If rotation of the coil is continued, the voltage rises again through its continuous chain of positive and negative humps.

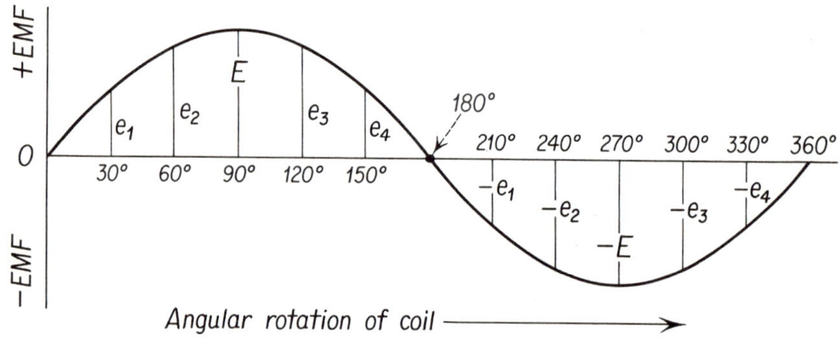

Fig. 6–3 Voltage Variation in Generated EMF

In the light of these variations, how is it possible to speak of a certain value, such as 100 v, when the voltage constantly is changing? Does the average of all of the values the voltage assumes during the 360-deg excursion of the rotating coil give the actual voltage that does work in a circuit? These questions and others like them are reasonable and can be answered best by examining the next section in detail and then studying carefully Sections 6.7 to 6.11.

6.5 A-C DEFINITIONS

Figure 6–3 reveals that an alternating voltage has a large number of possible positive and negative values. Also apparent is the fact that the interval from 0 to 360 deg represents a definite time during which the coil turns through one complete revolution. Thus, if the speed of rotation is known very accurately, the horizontal axis of the graph in Figure 6–3 might be marked off in seconds instead of degrees. Certain descriptive definitions are needed to define the information which may be determined from Figure 6–3. The most important of these definitions follow.

Maximum, or **Peak, Voltage** (E_m). The highest value, E or $-E$, reached by the voltage.

Instantaneous Voltage (e_i). Any intermediate value, such as e_1, e_2, $-e_1$, $-e_2$, etc., attained by the voltage as the coil rotates in the field.

Cycle (~). One complete excursion of the voltage. A single cycle consists of a positive half-cycle followed by a negative half-cycle. Figure 6–3 shows one cycle. The positive half-cycle extends from 0 to 180 deg, and the negative half-cycle from 180 to 360 deg.

Phase Angle (ϕ). In a simple generator of the type shown in Figure 6–2, having two magnetic poles, the *phase angle* is the same as the angle of rotation of the coil. It will be shown later that this is not true in more complicated machines, the term then having a somewhat different meaning. Phase angle is expressed in degrees.

Frequency (f). The number of complete cycles generated in 1 sec is called the *frequency*. This is similar to pitch in sound. Frequency is expressed in *cycles per second* (abbreviated ~ or cps). Most commercial power is supplied at a frequency of 60 cps.

Period (t). The time interval, in seconds, required to complete one full cycle is called the *period*. Period is equal to the reciprocal of frequency. That is $t = 1/f$. The period of the 60-cycle power line is $1/60 = 0.0166$ sec $= 16.6$ millisec.

Wave. The characteristic figure shown in Figure 6–3 is called an *a-c wave,* mainly because of its resemblance to a water wave. There are both voltage waves and current waves.

Wave Form. The shape of a wave figure is termed *wave form* or *wave shape*. Not all a-c waves have the smooth, symmetrical shape shown in Figure 6–3. Many are more complex in appearance.

6.6 SINE WAVE

When the generator coil travels around a circular path between magnetic poles cut to the circular shape shown in Figure 6–2, the generated voltage has the smooth, symmetrical wave form shown in Figure 6–3.

This figure has the same shape obtained when a graph is drawn with angles from 0 to 180 deg along the horizontal axis, and the corresponding positive and negative values of the *sine* of these angles along the vertical axis. Figure 6–4 shows what is meant by the sine of an angle, an important function contributed by trigonometry: In a right triangle

[Figure 6–4(A)], the sine of Angle A (written *sin* A) is determined by dividing the length of side a by the length of side c. It does not matter

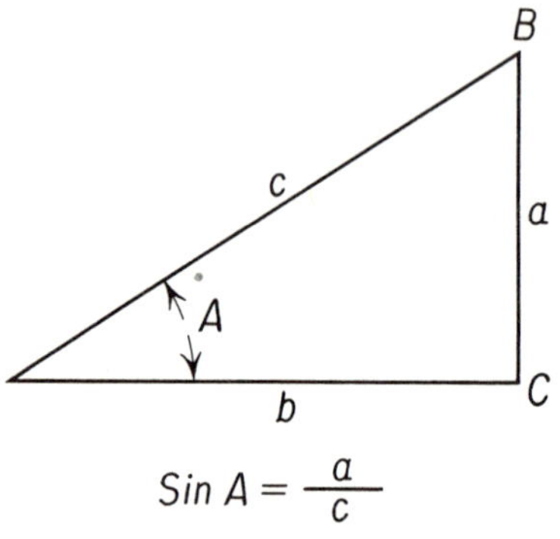

(A) RIGHT TRIANGLE

ANGLE	SINE
0°	0.0000
30°	0.5000
60°	0.8660
90°	1.0000
120°	0.8660
150°	0.5000
180°	0.0000
210°	−0.5000
240°	−0.8660
270°	−1.0000
300°	−0.8660
330°	−0.5000
360°	0.0000

(B) TABLE OF SINES

Fig. 6–4 Sine Relationships

what units are used to express these lengths, as long as both sides are measured in the same units (inches, centimeters, etc.). Figure 6–4(B) lists the sines of the angles shown for the a-c wave in Figure 6–3.

An a-c wave having this characteristic shape is termed a *sine wave*, or it is said to be sinusoidal.

6.7 VOLTAGE RELATIONS IN ALTERNATING CURRENT

When working with electrical circuits and machinery, it is necessary to be able to calculate the various values of voltage associated with the a-c cycle. The following sections concern these voltages and their relationships. The student should familiarize himself thoroughly with the arithmetic of these calculations.

6.8 INSTANTANEOUS VOLTAGE

The value of any instantaneous voltage (e_i) may be computed if the maximum, or peak, value, E_m, and the phase angle, ϕ, are known.

(6–1) $$e_i = E_m \sin \phi$$

where e_i and E_m are in volts, and ϕ is in degrees.

Example: When a generator delivers a peak voltage of 150 v, what is its instantaneous voltage output at phase angle of 120 deg?

From Figure 6–4(B),

$$\sin 120 \text{ deg} = 0.8660$$

$$e_i = E_m \sin \phi = 150(0.8660) = 129.9 \text{ v}$$

6.9 AVERAGE VOLTAGE

The *average voltage* is the calculated average (or *mean*) of all the instantaneous values in one complete cycle, regardless of the sign of these values. To find the average value, a large number of instantaneous values between 0 and 360 deg are added, ignoring the positive and negative signs and using only the numbers, and then this sum is divided by the number of instantaneous values taken. The average value obtained in this manner, or by means of advanced mathematics, is found to be equal to 63.6 per cent of the peak value. Thus

$$E_{avg} = 0.636E_m \tag{6–2}$$

Example: The peak value of an a-c voltage wave is 387 v. What is the average value of this voltage?

$$E_{avg} = 0.636E_m = 0.636(387) = 246.1 \text{ v}$$

6.10 EFFECTIVE, OR RMS, VOLTAGE

From an inspection of Figure 6–3, it is clear that a circuit is exposed to the peak voltage, or to zero, or to any instantaneous value for only a short instant of time. Naturally, neither of these values can be credited with the production of a net electrical action in a circuit, since all are present for such short times.

In this connection, it is convenient to have a reliable means for comparing an a-c voltage, which varies over a range of values, with a d-c voltage, which has a single, steady value. The means is the *heating effect* produced in the same resistance load by alternating current and direct current. When an a-c voltage will produce the same heating effect as a certain value of direct current, the alternating current is said to be equivalent to the direct current. This value of a-c voltage is called the *effective* value because it is the value that does work.

Heating is a power effect, and electrical power (P) varies directly as the square (E^2) of the voltage. (See Section 2.8, Chapter 2.) Hence, the heating effect of an a-c voltage may be determined by computing the average of the *squares* of all the instantaneous voltage values in one complete cycle.

The square root of the sum of the squares of all the instantaneous voltage values is called the *root mean square* voltage, since it is numerically the square root of the mean (average) squares of the e_i values. This is the value of the effective voltage mentioned two paragraphs ago.

The root mean square value (abbreviated rms) obtained in this manner, or by advanced mathematics, is found to be equal to 70.7 per cent of the peak value. Thus

$$E_{rms} = 0.707E_m \tag{6–3}$$

Example: A certain a-c generator delivers a peak voltage of 162.7 v across a resistance load. What is the rms (effective) voltage in this instance?

$$E_{rms} = 0.707(162.7) = 115 \text{ v}$$

The rms, or effective, value is the commonest voltage specified in a-c practice. Unless specifically designated otherwise, this is the voltage value meant. Thus, when speaking of a 110-v power line, we mean that the rms value of the voltage is 110. The peak value here is 155.5 v. The rms value is used so widely because this is the "work" value. Ordinary a-c meters read the rms value.

6.11 A-C VOLTAGE CONVERSIONS

The following equations show how one value of a-c voltage may be determined from another with the aid of simple arithmetic.

$$E_m = 1.414E_{rms} = 1.57E_{avg} \tag{6–4}$$

$$E_{rms} = 0.707E_m = 1.11E_{avg} \tag{6–5}$$

$$E_{avg} = 0.636E_m = 0.90E_{rms} \tag{6–6}$$

6.12 ALTERNATING CURRENT IN RESISTIVE CIRCUIT

Alternating current flows in a circuit as the result of application of an alternating voltage. If the circuit contains *resistance only,* the current will pass through zero and maximum at the same instants as the voltage, as shown in Figure 6–5(B). Current and voltage thus are in step here. This step is called *phase.* E and I are in phase normally only in a resistive circuit

Alternating *current* has maximum (peak), instantaneous, average, and rms (effective) values identical with the same-named values already described for voltage. In fact, current I may be substituted for voltage E in Equations (6–1), (6–2), (6–3), (6–4), (6–5), and (6–6). Unless specified otherwise, the rms value is assumed when a certain current is given. Ordinary a-c ammeters read the rms value.

For the resistive circuit, Ohm's law is the same as for direct current; that is, $R = E/I$, $E = IR$, and $I = E/R$.

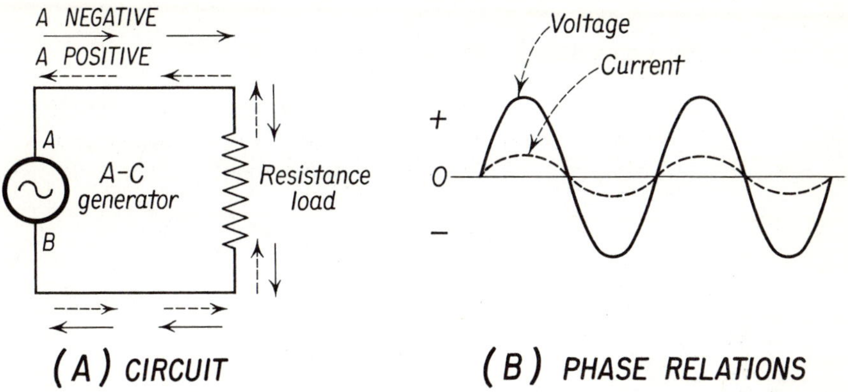

Fig. 6–5 Alternating Current Flow in Resistive Circuit

6.13 A-C POWER IN RESISTIVE CIRCUIT

In a circuit containing resistance only, the power at any instant is equal to the product of the resistance and the square of the instantaneous current. Thus, $P = R \times i_i^2$. The average power for the a-c cycle is the average of all the instantaneous power levels. But, since the average of all i_i^2 terms is i_{rms}, the average power $P_{avg} = R \times i_{rms}^2$. Other useful equations are:

(6–7) $$P = E^2/R$$

(6–8) $$P = EI$$

where E = rms voltage (in volts)
R = resistance (in ohms)
I = rms current (in amperes)

6.14 COMPLEX GENERATORS

Practical electromechanical generators do not have the simple form of the single-coil generator described in Section 6.3 and Figure 6–2. Modern generators, as we shall see in a later chapter, have many more loops of wire in their armatures. Some have a number of magnetic poles, as well. In others, the coil (*stator*) is stationary, while the magnet poles (*rotor*) whirl around it.

In addition to rotating machines, certain nonmechanical apparatus also is used to produce alternating currents. This includes vacuum-tube oscillators and transistor oscillators. A mechanical generator of alternating current, not a member of the family of *rotating* machines, is the vibrator-transformer combination.

6.15 NONSINUSOIDAL WAVE FORMS

Not all a-c wave forms are intended to be sinusoidal. Figure 6–6 shows several of the principal nonsinusoidal shapes. Many hundreds of special wave shapes are employed in specialized applications, espe-

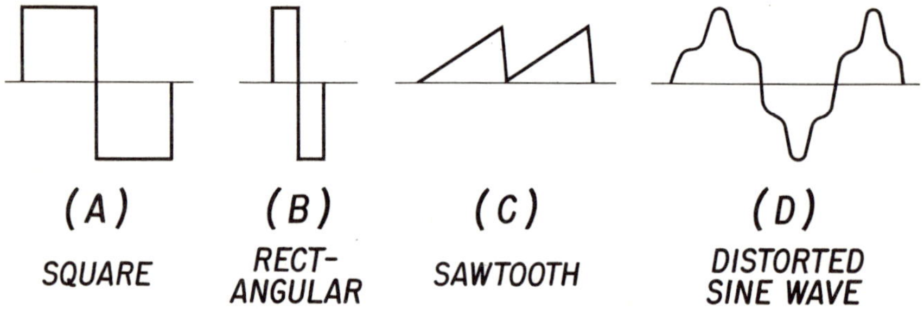

Fig. 6–6 Nonsinusoidal Wave Forms

cially in the electronics field. Shown here are a square wave, rectangular wave (pulse), saw-tooth wave, and distorted sine wave. The latter often results from faulty operation of equipment, although it may be generated intentionally.

6.16 HARMONICS

An important characteristic of any nonsinusoidal wave is its content of *harmonics.* Harmonics are voltages or currents having frequencies that are exact multiples (even or odd) of the main frequency (called the *fundamental* frequency of *first harmonic*). A perfect sine wave contains no harmonics. Mathematical and experimental analysis shows all nonsinusoidal waves to be composed of a number of separate sine waves which include a fundamental and several harmonics.

Harmonics in a-c electricity are comparable to overtones in sound. The second harmonic has a frequency equal to twice the fundamental frequency. Thus, it is one octave removed from the fundamental. The third harmonic frequency is three times the fundamental, fourth harmonic four times, etc. The twelfth harmonic of the 60-cycle power-line frequency, for example, is 720 cycles per second (cps).

Harmonics usually are small in voltage and current in comparison with the fundamental frequency.

6.17 RANGE OF FREQUENCIES

A-C frequencies employed in practical electrical and electronic applications extend from less than 1 cps to many thousand million cycles per second.

As the number of cycles becomes larger, the figures become more cumbersome to handle. For this reason, other units have been created: 1 kilocycle (kc) = 1,000 cycles, 1 megacycle = 1,000,000 cycles, 1 kilomegacycle = 1,000 megacycles.

Frequencies in the region between 20 and 15,000 cps are called *audio* frequencies from the fact that their currents will produce audible tones in earphones and loud-speakers. Below this lie the *subaudible* or *subsonic* frequencies. Just above the audio spectrum lie the *supersonic* frequencies. Extending from 10 kc upward are *radio* frequencies, so called because of their importance in wireless communication. Radio engineers have further divided this latter region into *long waves, short waves, ultrashort waves,* and *microwaves.*

SUMMARY OF CHAPTER 6

An **Induced EMF** is set up whenever a conductor is moved in a magnetic field, cutting lines of force, or when the field is moved past the conductor.

An **Alternating EMF** is produced when the conductor or the field moves *back and forth.* This emf causes an **Alternating Current** to flow in an external circuit connected to the conductor.

The **Value of Induced EMF** is governed by the number of lines cut per second.

A **Single-Coil A-C Generator** consists of a loop of wire spinning in a magnetic field. The voltage value at any instant depends upon the angular position of the loop in the field. The voltage passes from zero to a positive maximum, through zero to a negative maximum, and back to zero, in one complete revolution of the loop. The voltage describes a sine wave if the loop rotates in a circular path between circular-cut poles. Multicoil and multipole generators are employed in industry.

The **Instantaneous EMF** $e_i = E_m \sin \phi$.

A **Cycle** is a complete excursion of voltage or current from zero, through positive maximum, zero, negative maximum, and return to zero.

Frequency is the number of complete cycles per second.

Period is the time required to complete one cycle.

Instantaneous Voltage or **EMF** is the value of the voltage at any instant in the cycle.

Maximum (Peak) Voltage is the highest value that is attained during the positive half-cycle or negative half-cycle.

Wave Form refers to the shape of the cyclic pattern when drawn as a graph. If it duplicates the graph for the sine of an angle, it is called a **Sine Wave.** Sine waves are noted for their freedom from harmonics. **Harmonics** are extraneous frequencies which are exact multiples (even or odd) of the principal (*fundamental*) frequency.

Peak (E_m), **Instantaneous** (e_i), **Average** (E_{avg}), and **RMS (Effective)** (E_{rms}) **Voltages** may be determined from each other:

$$e_i = E_m \sin \phi = 1.414E_{rms} \sin \phi = 1.57E_{avg} \sin \phi$$
$$E_m = 1.414E_{rms} = 1.57E_{avg}$$
$$E_{rms} = 0.707E_m = 1.11E_{avg}$$
$$E_{avg} = 0.636E_m = 0.90E_{rms}$$

Peak (I_m), **Instantaneous** (i_i), **Average** (I_{avg}), and **RMS (Effective)** (I_{rms}) **Current** values may be determined in a similar manner:

$$i_i = I_m \sin \phi = 1.414I_{rms} \sin \phi = 1.57I_{avg} \sin \phi$$
$$I_m = 1.414I_{rms} = 1.57I_{avg}$$
$$I_{rms} = 0.707I_m = 1.11I_{avg}$$
$$I_{avg} = 0.636I_m = 0.90I_{rms}$$

When specifying an a-c voltage or current, the **RMS** or **Effective** value is understood unless the peak, instantaneous, or average value is stated.

Current and **Voltage** are **In Phase** in a circuit containing *resistance only.* Ohm's law for direct current may be applied directly to an a-c circuit containing resistance only.

Average Power in an a-c circuit containing resistance only is equal to E^2/R, I^2R, or EI, where E and I are rms voltage (volts) and current (amperes), respectively, and R is the resistance in ohms.

A-C Frequencies in practical use extend from less than 1 cps to several thousand million cycles per second. All those above about 10,000 cycles are allocated as *radio frequencies.* 1,000 cps = 1 kc. 1,000 kc = 1 megacycle = 1,000,000 cps. 1 kilomegacycle = 1,000 megacycles.

QUESTIONS AND EXERCISES

1. Why must a conductor be in *motion* in a magnetic field in order to produce an emf?

2. What emf is produced in a conductor passing 50 times per second through a flux density of 30,000 lines per square inch, when the pole faces measure 1½ × 1½ in.?

3. The maximum (peak) voltage delivered by a certain single-coil generator is supposed to be 330 v. What is the instantaneous voltage when the coil has moved through 150 deg?

4. A certain generator delivers 1,580 peak volts. What is the average value of this voltage?

5. In Question 4, what is the rms voltage?

6. An average value of 40 v corresponds to what rms value?

7. Calculate the average power (in watts) in an a-c circuit of 550 ohms when the applied voltage is 120 v peak.

8. What is the fifth harmonic of 400 cps?

9. How many kilocycles equal 37 megacycles?

10. Convert 3 kilomegacycles to kilocycles.

Chapter **7**

Capacitors and Capacitance

7.1 BASIC CAPACITOR

A capacitor essentially is a device for storing an electrostatic charge. Although this is its basic property, the capacitor performs many circuit functions which, unless analyzed, seem to bear no relation to charge storage. The simplest capacitor consists of two flat, parallel, metallic plates of the same shape separated by air or a dielectric material, either solid, liquid, or gaseous. Figure 7–1(A) shows this type of construction.

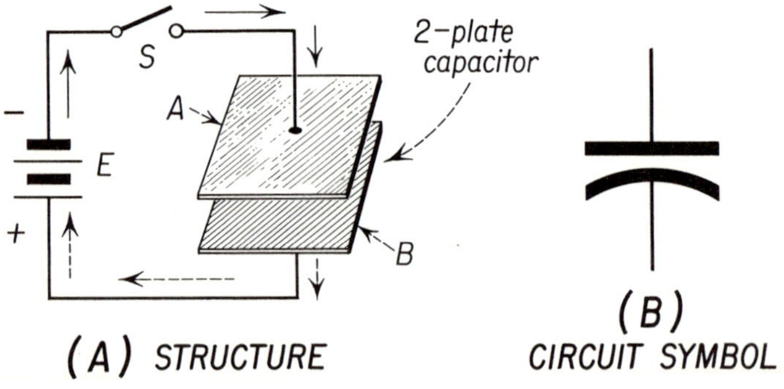

Fig. 7–1 Basic Capacitor

The capacitor is shown in a simple circuit with a battery. When Switch S is closed, the positive pole of Battery *E* attracts free electrons from the atoms of the lower plate, *B*. At the same time, an electrostatic field is set up between the plates, since one is connected directly to the positive terminal of a generator (Battery *E*), while the other is connected to the negative pole. These electrons flow through the lower wire in the direction of the dotted arrows, leaving Plate *B* with a net positive charge. Electrons also flow from the negative pole of the battery, through the upper wire, to the top plate, *A*, in the direction of the solid arrows. This gives Plate *A* a net negative charge. This electron current, called the *charging current,* continues to flow until the potential between Plates *A* and *B* is identical with that of the battery, whereupon current ceases. No current can pass across the gap between the plates. Note that this charging current flows *into* the capacitor and not *through* it, as it might appear. As a result of this charging action, the two plates which formerly were neutral electrically acquire positive and negative polarity.

If the switch then is opened and the capacitor is tested with a suitable d-c voltmeter, the capacitor will be found to possess a plate-to-plate voltage equal to the battery voltage, *E*. This voltage will drop quickly as the charge leaks off through the meter, discharging the capacitor.

Figure 7–2 illustrates the charge and discharge of a capacitor. When the capacitor is connected across the battery, electron current flows into the capacitor in the direction shown by the arrows. [See Figure 7–2(A).] This current is high at first, but it decreases gradually as the

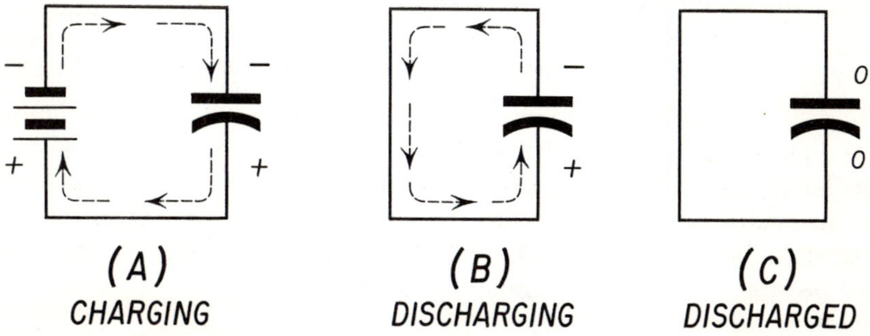

Fig. 7–2 Charge and Discharge of Capacitor

top plate fills with electrons, and finally drops to zero when the capacitor has become fully charged. The more resistance present in the circuit, as in the connecting leads, the slower will be this decay of current. When the battery is removed and replaced by a short circuit, as in Figure 7–2(B), electron current flows *in the opposite direction*, as electrons move, through the short circuit, from the top plate to the electron-deficient bottom plate. This *discharge current* continues to flow until there is an equal number of electrons on both plates, whereupon it ceases and the potential difference between the plates falls to zero. [See Figure 7–2(C).] The rate at which this discharge current decays also depends upon the amount of resistance in the circuit.

The capacitor is seen to employ the principles of electrostatics discussed in Chapter 1. A high-quality capacitor will hold a charge for some time after being disconnected from the source of emf. All capacitors eventually lose their charge through slow leakage through the dielectric between the plates and leakage between the terminals.

The structure of a two-plate capacitor need not be restricted to the square, flat-plate configuration illustrated by Figure 7–1(A). The same effect may be obtained when the flat plates in effect are rolled up into

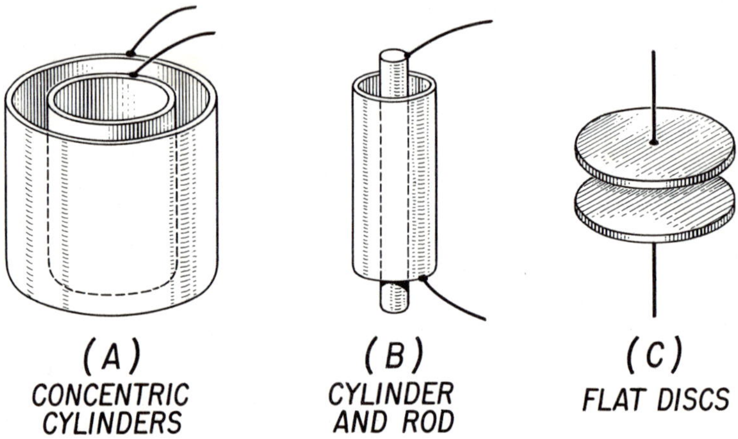

Fig. 7–3 Miscellaneous 2-Plate Capacitor Structures

cylinders mounted concentrically with respect to each other [Figure 7–3(A)], or with a concentric cylinder and rod [Figure 7–3(B)]. Flat, circular plates also may be used [Figure 7–3(C)].

In each case, it is necessary to keep the plate area large and the separation between plates small if the maximum charge, in coulombs, is to be stored by a given applied voltage.

7.2 CAPACITANCE

The property whereby we may express the ability of a capacitor to store a charge is called *capacitance.* The unit of capacitance is the *farad* (fd), named in honor of Michael Faraday (1791–1867). Its symbol is C. Actually, the farad is such a large unit that its millionth part, the *microfarad* (μfd), and its trillionth part, the *micromicrofarad* ($\mu\mu$fd), are used in practice.

Capacitance may be determined from quantity (Q) and potential (E).

(7–1) $$C = Q/E$$

where C = capacitance (in farads)
Q = quantity of charge (in coulombs)
E = charging potential (in volts)

From this

(7–2) $$Q = CE$$

and

(7–3) $$E = Q/C$$

A capacitor has a capacitance of 1 fd when it is charged to a potential of 1 v by 1 coulomb of electricity. The higher the capacitance, the larger will be the amount of charge stored in a capacitor by a given applied voltage.

Example: What is the capacitance of a capacitor which is charged to a potential of 1 v by 1 microcoulomb (1 millionth of a coulomb) of electricity?

$$C = Q/E = 0.000001/1 = 0.000001 \text{ fd} = 1\ \mu\text{fd}$$

7.3 FIELD AND ENERGY OF CAPACITOR

Electrostatic lines of force fill the space between the two charged plates of a capacitor, as between any other two oppositely charged

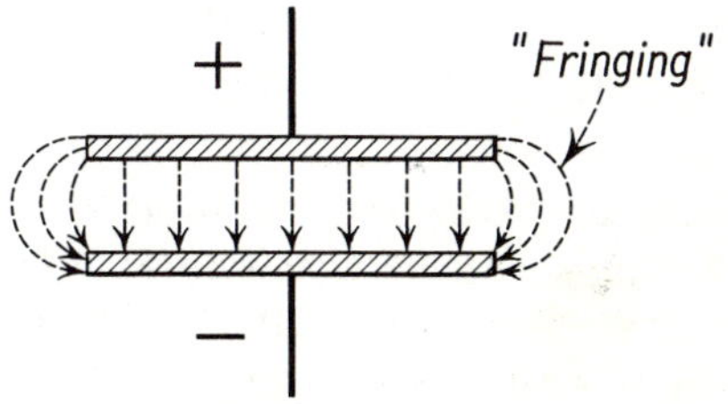

Fig. 7–4 Electrostatic Field between Capacitor Plates

bodies. This is illustrated by Figure 7–4. A few of these lines bend out at the edges of the plates. Except for this "fringing" effect, the electrostatic field is seen to be confined to the space between the plates.

Energy is stored in this field in the form of potential energy. Thus

(7–4) $$W = \tfrac{1}{2}CE^2$$

where W = energy (in joules)
C = capacitance (in farads)
E = potential (in volts)

7.4 CAPACITANCE OF TWO-PLATE AIR CAPACITOR

When air is the only dielectric between the identical plates of a two-plate capacitor, the capacitance is directly proportional to the area of one plate, and inversely proportional to their separation.

(7–5) $$C = A/(4.45d)$$

where C = capacitance (in micromicrofarads)
A = area of one plate (in square inches)
d = plate separation (in inches)

Example: What is the capacitance of a two-plate air capacitor having 2 in. × 2 in. plates separated by 10 thousandths of an inch?

$$C = A/(4.45d) = (2 \times 2)/(4.45 \times 0.01)$$
$$= 4/0.0445 = 89.9\ \mu\mu\text{fd}$$

When either capacitance, area, or separation is known, the other parameter may be calculated.

(7–6) $$A = 4.45Cd$$

(7–7) $$d = A/(4.45C)$$

7.5 EFFECT OF DIELECTRIC

Other insulating materials than air usually fill the space between capacitor plates. Various advantages are obtained from the use of these materials.

When a dielectric material is inserted between the plates of a capacitor, the capacitance of the latter *increases.* The dielectric thus appears to possess a property which is somewhat of an electrostatic analogy of permeability in magnetics. This property is called *dielectric constant.* Its symbol is k.

In magnetics, the permeability of air is taken as 1. (See Section 5.8, Chapter 5.) Similarly, the dielectric constant of air is taken as 1. All other dielectrics have a k value higher than 1. If the simple two-plate air capacitor is immersed in castor oil ($k = 4$), for example, its capacitance is quadrupled. The dielectric constant of a material thus is equal to the ratio of the capacitance of a capacitor having the material as a dielectric to the capacitance of the same capacitor with air as a dielectric.

When a voltage is applied to a capacitor, no current flows through the dielectric itself (except an exceedingly tiny *leakage current,* due to imperfection in the dielectric, which may be neglected in most practical work). However, the electric field between the plates sets up a strain in the dielectric because electrons in the dielectric structure are repelled by the negative plate and attracted by the positive plate. This causes an internal distortion of the electron bonds in the direction of the positive plate. This strain persists as long as the capacitor is charged, and it is relieved upon discharge.

The small movement of the electrons into the strained position upon application of an emf to the capacitor plates constitutes a current flow, the *displacement current,* which ceases when the electrons settle in their new position. Conversely, the relaxation of the electrons to their position of rest when the capacitor is discharged constitutes a current flow in the opposite direction. (See Figure 7–2.) These displacement currents cease when the electrons settle at their new positions.

After the initial application of an emf to a capacitor containing certain dielectric materials, displacement current continues to flow for a short interval beyond the time normally expected for the charging of the capacitor. This additional time is required for the electric strain to be completed in the dielectric. The dielectric is thought to absorb some of the charge during this period.

Likewise, when the capacitor subsequently is short-circuited externally, it does not discharge completely. Displacement current will flow in lesser amounts each time the capacitor is short-circuited again, until finally the charge is dissipated completely. This is due to the additional time taken for the electric strain to become relaxed in the dielectric. This effect can be troublesome, and in some applications dangerous, since one cannot be sure that such capacitors are fully discharged unless they are tested carefully.

This phenomenon is known as *dielectric absorption.* It is more pronounced in some dielectrics, such as bakelite and glass, than in others such as mica, paper, castor oil, and mineral oil. In practical circuits, dielectric absorption is accompanied by the generation of heat in the capacitor, and, since this constitutes a source of loss, it leads to inefficient performance of the capacitor.

7.6 DIELECTRIC STRENGTH

The insulating property of a dielectric is indicated by its *dielectric strength* rating. This is a figure, usually given as the minimum *volts-per-centimeter* or *volts-per-mil* required to rupture the dielectric.

This means simply that if the applied emf is made high enough, the dielectric will be ionized and will become a conductor, since a high potential will tear electrons from the dielectric atoms. This conduction results in permanent damage, since the dielectric becomes punctured or burned.

Figure 7–5 is a table listing a few common dielectric materials together with dielectric constant and dielectric strength ratings.

DIELECTRIC MATERIAL	DIELECTRIC CONSTANT (k)	DIELECTRIC STRENGTH (volts/mil)
Air	1.0	19–23
Castor Oil	4.3–4.7	380
Cellulose Acetate	6–8	300–1000
Glass, Pyrex	4.2–4.9	335
Lucite	2.5–3.0	480–500
Mica, India	6.4–7.5	600–1500
Paper	2.0–2.6	1250
Polyethylene	2.2	600
Polystyrene	2.6	500–2500

Fig. 7–5 Typical Dielectric Characteristics

7.7 CAPACITANCE OF TWO-PLATE CAPACITOR WITH DIELECTRIC

The method given in Equation (7–5) for the calculation of the capacitance of a two-plate air capacitor is changed only to the extent of multiplying by the dielectric constant, *k*, when a dielectric other than air is placed between the plates.

(7–8) $$C = (kA)/(4.45d)$$

where k = dielectric constant of the material

All other constants (C, A, and d), are the same as in Equation (7–5).

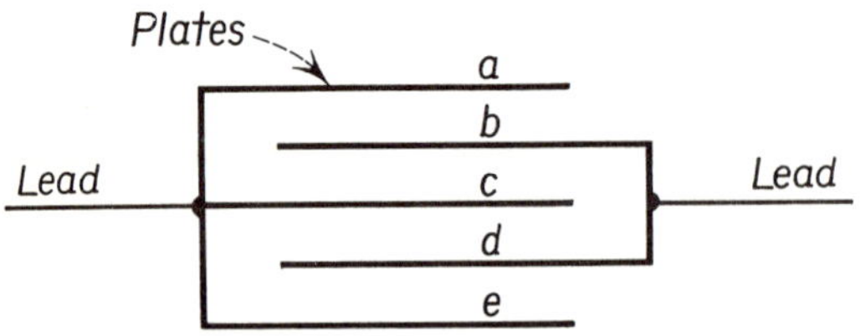

Fig. 7–6 Structure of a Multiplate Capacitor

7.8 MULTIPLATE CAPACITOR

Higher capacitance may be obtained by employing more than a single pair of plates of a given area. In gaining capacitance the capacitor thus grows along its thickness dimension, rather than in length and width.

The plates of a multiplate capacitor are interleaved in the manner shown in Figure 7–6. Alternate plates *a*, *c*, and *e* on one side are connected together in parallel; alternate plates *b* and *d* on the other side are paralleled. Any desired number of plates, within practical limitations, may be stacked in this manner.

7.9 CAPACITANCE OF ANY CAPACITOR

The capacitance of *any* capacitor with identical plates may be calculated from the number of plates, plate area, constant plate separation (dielectric thickness), and dielectric constant. Thus

(7–9) $$C = \frac{kA(n-1)}{4.45d}$$

where C = capacitance (in micromicrofarads)
k = dielectric constant
A = area of one plate (in square inches)
d = separation of any pair of plates (dielectric thickness) (in inches)
n = number of plates

Example: What is the capacitance of a 23-plate capacitor employing 5-mil-thick mica dielectric, with ½ in. × ½ in. plates?

From Figure 7–5, the average value of k for mica = 6.95.

$$C = \frac{6.95(0.5 \times 0.5)(23-1)}{4.45(0.005)} = \frac{6.95(0.25)22}{0.0222}$$

$$= \frac{1.74(22)}{0.0222} = 38.28\ \mu\mu fd$$

7.10 CAPACITORS IN SERIES

[See Figure 7–7(A).] The formula for the total capacitance (C_t) of capacitors in series has the same form as the formula for resistors in parallel. Thus

(7–10) $$C_t = \frac{1}{\dfrac{1}{C_1} + \dfrac{1}{C_2} + \dfrac{1}{C_3}}$$

Example: What is the total capacitance of a circuit containing one each 0.001-, 0.03-, and 0.1-μfd capacitors in series?

$$C_t = \frac{1}{\dfrac{1}{0.001} + \dfrac{1}{0.03} + \dfrac{1}{0.1}} = \frac{1}{1{,}000 + 33.3 + 10}$$

$$= \frac{1}{1{,}043.3} = 0.00096\ \mu fd$$

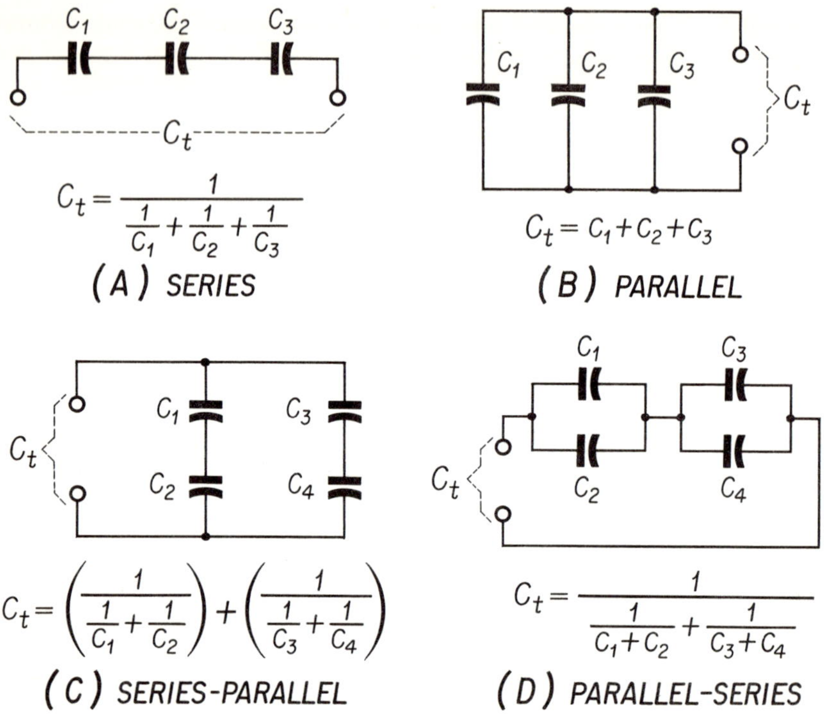

Fig. 7–7 Common Capacitor Networks

7.11 CAPACITORS IN PARALLEL

[See Figure 7–7(B).] The formula for the total capacitance (C_t) of capacitors in parallel has the same form as the formula for resistors in series. Thus

(7–11) $$C_t = C_1 + C_2 + C_3$$

Example: What is the total capacitance of a circuit containing one each 150-, 500-, and 750-$\mu\mu$fd capacitors in parallel?

$$C_t = 150 + 500 + 750 = 650 + 750 = 1{,}400\ \mu\mu\text{fd}$$

7.12 CAPACITORS IN SERIES-PARALLEL

[See Figure 7–7(C).] The formula for the total capacitance of capacitors in series-parallel has the same form as the formula for resistors in parallel-series. Thus

(7–12) $$C_t = \left(\frac{1}{\frac{1}{C_1} + \frac{1}{C_2}}\right) + \left(\frac{1}{\frac{1}{C_3} + \frac{1}{C_4}}\right)$$

Example: What is the total capacitance of a series-parallel circuit with one each 0.0025- and 0.0005-μfd capacitors in series in one leg, and one each 0.002- and 0.005-μfd capacitors in series in the other leg?

$$C_t = \left(\frac{1}{\frac{1}{0.0025} + \frac{1}{0.0005}}\right) + \left(\frac{1}{\frac{1}{0.002} + \frac{1}{0.005}}\right)$$

$$= \left(\frac{1}{400 + 2{,}000}\right) + \left(\frac{1}{500 + 200}\right) = \frac{1}{2{,}400} + \frac{1}{700}$$

$$= 0.000416 + 0.00143 = 0.00184\ \mu\text{fd}$$

7.13 CAPACITORS IN PARALLEL-SERIES

[See Figure 7–7(D).] The formula for the total capacitance (C_t) of capacitors in parallel-series has the same form as the formula for resistors in series-parallel. Thus

(7–13) $$C_t = \frac{1}{\left(\frac{1}{C_1 + C_2}\right) + \left(\frac{1}{C_3 + C_4}\right)}$$

Example: What is the total capacitance of a parallel-series circuit containing one each 10- and 40-μfd capacitors in parallel in the first leg, and one each 50- and 100-μfd capacitors in parallel in the second leg?

$$C_t = \frac{1}{\left(\frac{1}{10 + 40}\right) + \left(\frac{1}{50 + 100}\right)} = \frac{1}{\frac{1}{50} + \frac{1}{150}}$$

$$= \frac{1}{0.02 + 0.0066} = \frac{1}{0.0266} = 37.59\ \mu\text{fd}$$

7.14 VARIABLE CAPACITORS

The capacitors discussed up to this point are *fixed* capacitors, so called because their capacitance has a constant, nonvariable value. In some applications, it is convenient to be able to adjust capacitance to a desired value. This is accomplished with *variable* capacitors.

Like variable resistors, variable capacitors are of the *step type* and *smooth type*. In the step type [Figure 7–8(A)], a selector switch (S) connects either of several capacitors (C_1, C_2, C_3, C_4) to a pair of access terminals, *X, Y*. If each capacitor has a higher capacitance than the preceding one by some predetermined amount, the capacitance at Terminals *XY* will increase in steps as the switch is rotated through its four points from 1 to 4.

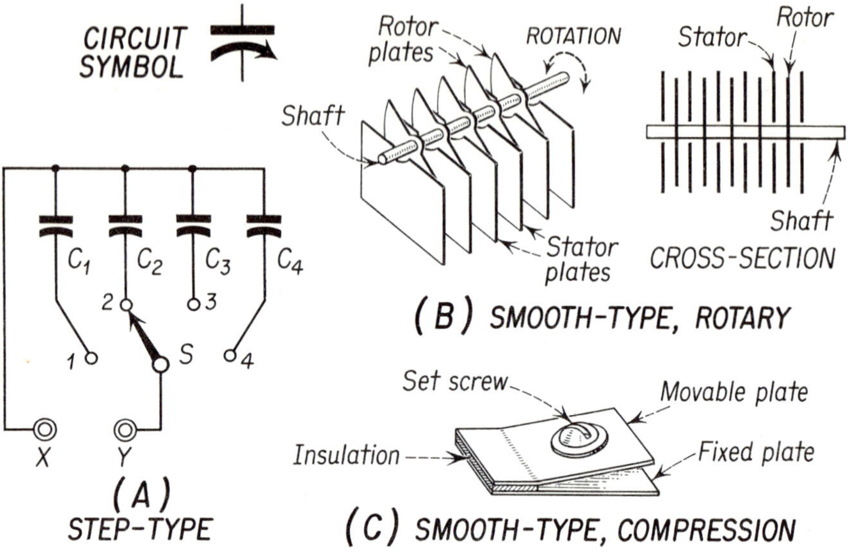

Fig. 7–8 Variable Capacitors

The rotary variety of smooth-type variable capacitor with multiple plates has the general construction shown in Figure 7–8(B). Several semicircular metal *rotor* plates are attached to a metal shaft, and these plates are spaced between a set of stationary *stator* plates which are attached together. When the shaft is turned, the rotors are moved in or out between the stators, thus exposing more or less of their area to the stators. Since capacitance varies directly as the area [Equation (7–9)], the capacitance is least when the rotors are turned completely out of the stators; it is maximum when they are turned in completely opposite the stators.

Another style of smooth-control variable capacitor, having limited application, is shown in Figure 7–8(C). Here, the plate area is constant, but the separation between the two plates may be adjusted to vary the capacitance. In this simple, *compression-type* unit, the plates are insulated from each other by a thin sheet of mica. A setscrew passes through the movable top plate and is insulated from the stationary bottom plate. As the screw is tightened, the top plate is pressed down closer to the bottom plate, decreasing the separation and increasing the capacitance. The protecting sheet of mica prevents the plates from touching and short-circuiting when they are pressed together.

There are many varieties of smooth-type variable capacitors other than the two shown in Figure 7–8. Most of the smooth-type units are employed in electronics and instrumentation. The step type [Figure 7–8(A)] is the more prevalent in the electrical industry.

7.15 A-C BEHAVIOR OF CAPACITOR

When a capacitor is connected to a source of d-c voltage, it becomes charged to a voltage equal to that of the source and with the same polarity. When the charging voltage is removed, the capacitor is able to act somewhat as a generator and will deliver a current until it has been completely discharged. If the capacitor is of high quality (that is, its losses are low), the quantity of electricity delivered during discharge will be equal to the quantity stored during charge.

When a capacitor is connected to an a-c source, it is alternately charged and discharged as the voltage changes polarity. While the capacitor is charging during one half-cycle, electrons flow through the circuit in one direction; while it is discharging during the next half-cycle, electrons flow in the reverse direction. The direction of electron flow thus changes with the a-c cycle, and an alternating current is found to flow in the circuit between the generator and capacitor. This current seems to flow *through* the capacitor, while actually it is only flowing *in* and *out* of it. This phenomenon has given rise to the erroneous belief among laymen that alternating current flows through a capacitor, while direct current does not. In many applications this certainly is the effect which is utilized, but there is no actual flow of current through the capacitor.

Figure 7–9 illustrates a-c capacitor action. Assume here that at the start the capacitor is empty and that the top terminal of the generator is positive at that instant. [See Figure 7–9(A).] The capacitor will

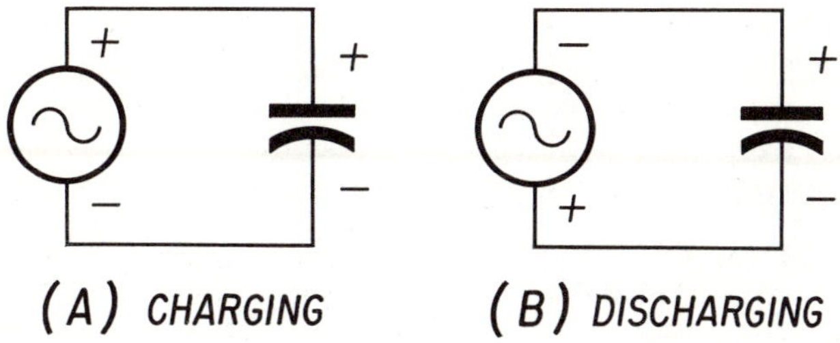

Fig. 7–9 Capacitor Action on Alternating Current

charge, its upper terminal becoming positive, and its lower terminal negative. When the a-c cycle reverses [Figure 7–9(B)], the top terminal of the generator swings negative, while the top terminal of the capacitor, to which it is connected, remains positive for the time being since the charge is still in the capacitor. The capacitor and generator now act like two batteries connected in series. The capacitor begins to discharge through the generator.

Next, the generator again reverses polarity, returning to the state shown in Figure 7–9(A). If the capacitor has not had time fully to discharge, its voltage has the same polarity as the generator and will oppose the flow of charging current from the generator. *Only when the capacitor has been able to discharge fully during the previous half-cycle will the charging current be maximum.* The bucking voltage which is responsible for this opposition to the build-up of charging current is the charged voltage of the capacitor. Conversely, we note from Figure 7–9 that the generator voltage also alternately aids and opposes the flow of current.

From this action, it is clear that a capacitor can oppose the flow of alternating current. This opposition is called *capacitive reactance,* the symbol for which is X_c. The unit of reactance is the ohm.

The strength of current (I) flowing in a circuit containing a capacitor is proportional to the capacitance (C), the applied voltage (E), and the *rate of change* of the voltage. The rate of change is indicated by the expression $2\pi f$, where π (Greek letter *pi*) = 3.1416, and f is the frequency in cycles. Thus

$$I = 2\pi fCE \tag{7–14}$$

where I is in amperes, f in cycles per second, C in farads, and E in volts.

Ohm's law may be applied to determine the extent of the opposing action (reactance) of the capacitor, since this opposition must be equal to E/I.

$$E/I = E/(2\pi fCE) = 1/(2\pi fC) = X_c \tag{7–15}$$

where X_c = capacitive reactance (in ohms)
C = capacitance (in farads)
f = frequency (in cycles per second)
I = current (in amperes)

The equations for Ohm's law for alternating current, with respect to capacitive reactance, are $I = E/X_c$, $E = IX_c$, and $X_c = E/I$.

Equation (7–15) shows that: (1) For a given capacitance, the higher the frequency the lower will be the reactance. (2) For a given frequency, the higher the capacitance the lower will be the reactance.

7.16 RATE OF CHANGE, ANGULAR VELOCITY

The expression $2\pi f$, introduced in the preceding section as rate of change, often appears in a-c formulas, not alone the one for capacitive reactance. It is important to understand the meaning of this expression and to know how it is obtained.

For the answer, we turn first to the circle and its measurement, since the conductor in a generator travels around a circular path through the magnetic field. The diameter (D), radius (r), and circumference (C) of a circle have the following relationships:

(7–16) $$D = 2r$$

(7–17) $$C = 2\pi r$$

(7–18) $$C = 360°$$

(7–19) $$C/D = 3.1416 = \pi$$

In Figure 7–10(A), the lines OA and OB both are radii of the circle which has its center at point O. When OA and OB are separated, as shown, so that the arc AB also is equal to OA and to OB (the radius of the circle), the angle between OA and OB is called 1 *radian*. AB is

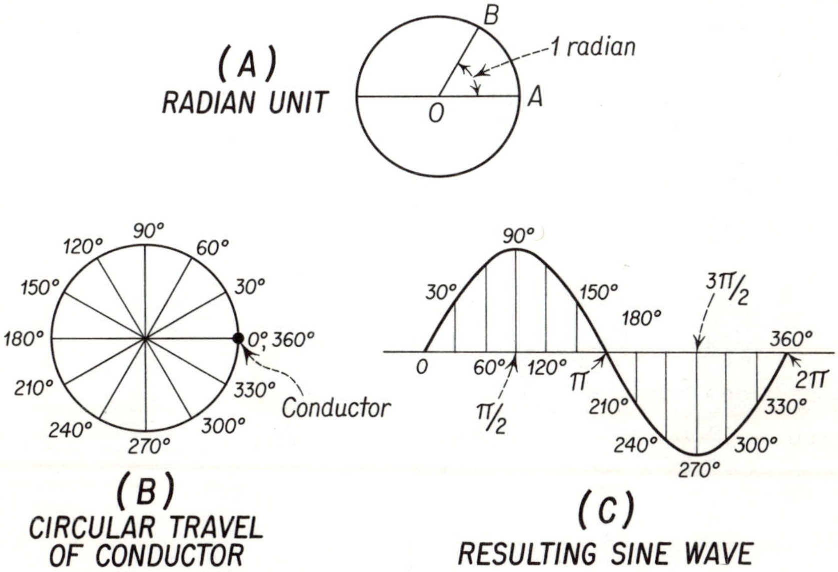

Fig. 7–10 Circular Relations in Alternating-Current Generation

a part of the circumference, and, since $AB = r$, and $C = 2\pi r$, then there are 2π radians in the 360 deg contained in a circle. In 180 deg, therefore, there are π radians. Thus, 1 radian = 57.3 deg, or 1 deg = 0.0174 radian. The radian is the unit of angular rotation. In one complete revolution, a wheel describes 2π radians.

In a generator, the moving conductor in passing through 360 deg [Figure 7–10(B)] thus completes 2π radians. A single cycle is shown in Figure 7–10(C) with corresponding principal degrees and radians identified. In the case of a wheel, the *angular velocity* is 2π radians multiplied by the revolutions per second. With the a-c generator, the revolutions per unit time are replaced with cycles per second. Thus in the generator, angular velocity $\omega = 2\pi f$. Angular velocity is expressed in *radians per second* and has the symbol, ω (lower-case Greek letter *omega*).

7.17 COMMERCIAL CAPACITORS

The commonest commercial fixed capacitor is the film type. In this type, two usually long sheets of tin foil or aluminum foil (used as the capacitor plates) are separated by a thin film of paper impregnated with wax or oil, or a film of plastic. Rolling such a capacitor doubles its capacitance. The rolled-up type usually is enclosed in a cardboard,

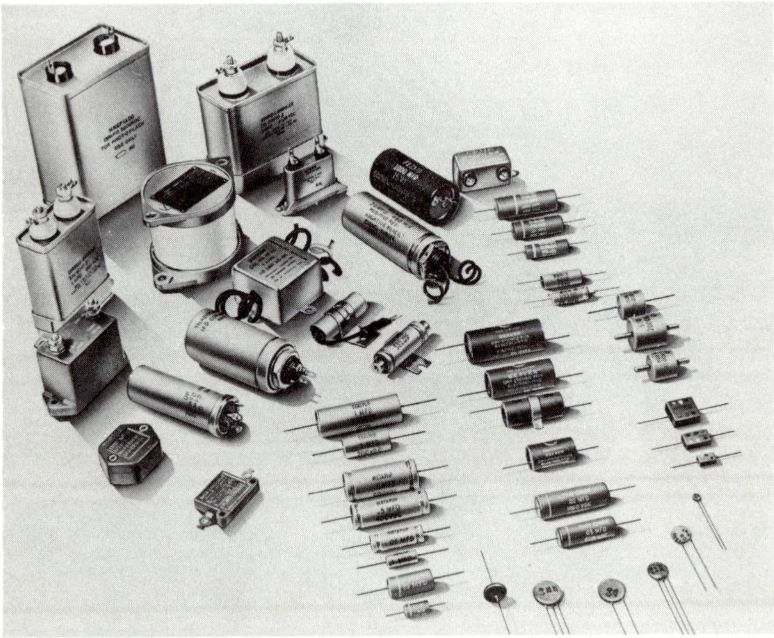

Plate 7–1 Commercial Capacitors. Shown here are paper, oil, mica, electrolytic, and ceramic types. (Courtesy, Cornell-Dubilier Electric Corp.)

plastic, or metal tube and is called a *tubular* capacitor. Sometimes it is housed in an oil-filled can. The unrolled type is enclosed in a flat case. In the *metallized paper* type, a metallic film is deposited on each side of the dielectric film.

Another widely used type is the stacked capacitor. In this, alternate sheets of metal foil and dielectric material are stacked and connected in the manner illustrated in Figure 7–6. Mica, glass, and composition films are employed as dielectrics. In a special version of the stacked capacitor, metal films are deposited (by firing, evaporation, or sputtering) on each side of mica films for close bonding of the metal plate to the dielectric. This is the *silvered mica* type widely used in electronic equipment.

A special kind is the *electrolytic capacitor.* In this, very high capacitances (up to several thousand microfarads) are obtained in a small space. Figure 7–11 shows the basic arrangement of the electrolytic

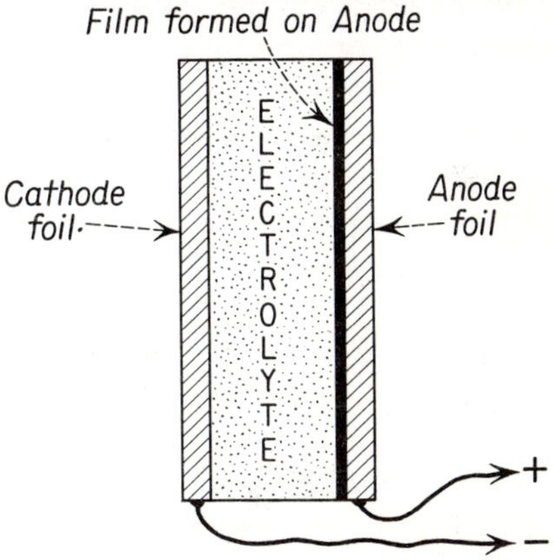

Fig. 7–11 Basic Arrangement of Electrolytic Capacitor

capacitor. The essential parts are two high-purity aluminum foils between which is an electrolyte, such as boric acid. A very thin film is *electro*formed on one foil (the anode) by passing a direct current through the capacitor after it has been manufactured. This film serves as the dielectric. The anode foil is one plate and the electrolyte acts as the other plate of the capacitor. The cathode foil makes contact with the electrolyte. The thinness of the film on the anode foil is responsible for the large capacitance obtainable with this type.

The conventional electrolytic capacitor has polarity and cannot be used in a-c applications; that is, the anode always must be connected to the positive d-c terminal in a circuit, and the cathode to the negative, otherwise the capacitor will become "de-formed," the dielectric film being destroyed. Early capacitors of this type contained liquid electrolytes, but modern electrolytic capacitors contain a pasty ("dry") electrolyte.

Ceramic capacitors are widely used in electronic circuits, especially at radio frequencies. These units afford reasonably high capacitances (up to 0.02 μfd) in very small spaces. In this type, silver plates are fired or deposited on the surfaces of tubes, discs, or plates of such ceramic materials as titanium dioxide and barium titanate.

7.18 INHERENT CAPACITANCE IN CIRCUITS AND DEVICES

We have seen that capacitance is an effect evidenced by two separated conductors connected to a source of emf. From this, it is evident that *any* two separated conductors connected to a voltage source will form a capacitor of a sort.

Unplanned capacitance of this sort is known as *stray capacitance, inherent capacitance,* and by other descriptive terms. It is encountered often in electric circuits and must be reckoned with. At very high frequencies, stray capacitance often causes mischief by providing unsuspected paths through which currents may leak. Stray capacitance is found, for example, between leads. The longer these leads and the closer they are together, the higher is the capacitance.

SUMMARY OF CHAPTER 7

The **Capacitor** is a device for storing an electric charge. Charging and discharge current flow *in* and *out* of a capacitor, not *through* it. In a perfect capacitor the quantity of electricity delivered during discharge would be equal to the full quantity stored during charge.

Capacitance (C) expresses the degree of capacitor action. The unit of capacitance is the farad (fd). 1 microfarad (μfd) = 0.000001 fd. 1 micromicrofarad ($\mu\mu$fd) = 1×10^{-12} fd. In terms of applied voltage and quantity of electricity stored, $C = Q/E$. A 1-fd capacitor is charged to a potential of 1 v by 1 coulomb of electricity.

The **Electrostatic Energy** stored in a capacitor = $W = \frac{1}{2}CE^2$. The capacitance of a two-plate capacitor = $C = A/(4.45d)$.

A **Dielectric Material,** other than air, between the plates increases the capacitance by a factor k which is called the **Dielectric Constant** of the material. The capacitance of a two-plate capacitor containing the material = $C = (kA)/(4.45d)$.

The capacitance of a **Multiplate Capacitor** is

$$C = \frac{kA(n - 1)}{4.45d}$$

Total Capacitance of Capacitors in Series

$$C_t = 1/(1/C_1 + 1/C_2 + 1/C_3)$$

Total Capacitance of Capacitors in Parallel

$$C_t = C_1 + C_2 + C_3$$

Total Capacitance of Capacitors in Series-Parallel

$$C_t = \left(\frac{1}{\dfrac{1}{C_1} + \dfrac{1}{C_2}}\right) + \left(\frac{1}{\dfrac{1}{C_3} + \dfrac{1}{C_4}}\right)$$

Total Capacitance of Capacitors in Parallel-Series

$$C_t = \frac{1}{\left(\dfrac{1}{C_1 + C_2}\right) + \left(\dfrac{1}{C_3 + C_4}\right)}$$

Variable Capacitors are of the *step type,* in which the capacitance is adjustable in steps, and the *smooth type,* in which the capacitance is continuously adjustable.

An **Alternating EMF** alternately charges and discharges a capacitor. Unless the capacitor can discharge completely, its charged voltage will buck the applied voltage on successive charging half-cycles and oppose the flow of charging current. This opposition is **Capacitive Reactance,** X_c. Capacitive reactance varies inversely with frequency and capacitance.

The **Rate of Change** or **Angular Velocity** of an alternating current or voltage is expressed in *radians per second* and is equal to $2\pi f$, or ω. 1 radian = 57.3 deg; 1 deg = 0.0174 radian. The current "through" a capacitor is proportional to the applied voltage and the rate of change (angular velocity).

Stray Capacitance results from capacitor action between any two closely separated conductors or metallic objects which are connected to a source of emf.

QUESTIONS AND EXERCISES

1. Explain why alternating current flows in and out of a capacitor and not *through* it.

2. Convert 0.1 fd to microfarads.

3. What quantity of electricity is required to charge a 0.5-μfd capacitor to 10 v?

4. A two-plate air capacitor has 2 in. $\times$ 2 in. plates separated by 10/1,000 of an inch. What is its capacitance?

5. What is the capacitance of the same capacitor if mica is employed as the dielectric?

6. A 13-plate air capacitor has 1 in. $\times$ 2¼ in. plates with 0.0625-in. separation. What is its capacitance?

7. In Figure 7–7(A) $C_1 = 10$ μfd, $C_2 = 40$ μfd, and $C_3 = 50$ μfd. What is the total capacitance of this series circuit?

8. In Figure 7–7(B) $C_1 = 0.0015$ μfd, $C_2 = 0.00025$ μfd, and $C_3 = 0.0005$ μfd. What is the total capacitance of this parallel circuit?

9. In Figure 7–7(C) $C_1 = 0.0001$ μfd, $C_2 = 0.0025$ μfd, $C_3 = 0.0008$ μfd, and $C_4 = 0.001$ μfd. What is the total capacitance of this series-parallel circuit?

10. In Figure 7–7(D) $C_1 = 50$ μfd, $C_2 = 30$ μfd, $C_3 = 25$ μfd, and $C_4 = 10$ μfd. What is the total capacitance of this parallel-series circuit?

11. What is the reactance of a 500-$\mu\mu$f capacitor at 1 megacycle?

12. What angular velocity corresponds to 60 cps?

Chapter **8**

Inductors and Inductance

8.1 PHENOMENON OF INDUCTION

Two important basic facts concerning electromagnetism were learned from the study of earlier chapters: (1) A magnetic field surrounds every conductor or coil carrying current (Sections 5.6 and 5.7, Chapter 5), and (2) the reverse also is true—a changing magnetic field will generate an emf in a conductor cut by its lines of force (Section 6.1, Chapter 6). These facts now may be put together to explain the unique action of coils carrying alternating or changing current.

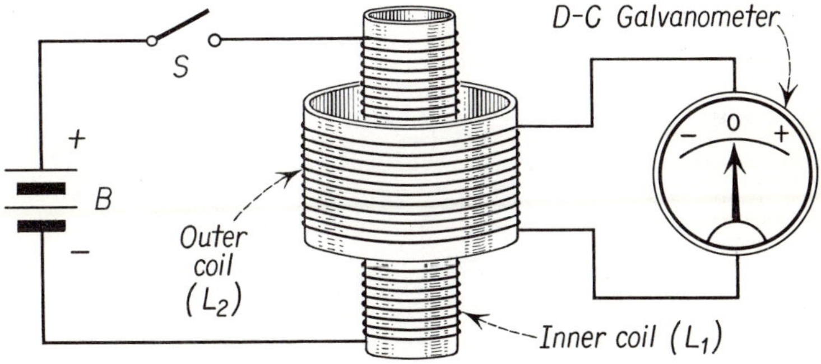

Fig. 8–1 Demonstration of Induction

We have encountered the phenomenon of induction before. In Chapter 6, for example, it was explained that an emf is *induced* in a conductor if the latter is moving in such a way as to cut lines of force, or if the field is moving so that its lines of force cut the conductor. Either the field or the conductor must be moving. This is how a generator works.

There are many ways to illustrate induction. Figure 6–2, Chapter 6, is one way. Figure 8–1 is a demonstration of another sort. Here, two coils are mounted coaxially. This arrangement guarantees that the lines of force surrounding one *inductor* or coil, owing to the flow of current through it, must cut the turns of the other coil. The term *coaxial* arises from the fact that each coil has the same axis.

The outer coil (L_2) is connected to a d-c galvanometer. The latter is an indicating instrument which shows both the direction and relative strength of current flow. When no current is flowing through the galvanometer, its pointer rests at the center of the scale, this point being marked *zero*. The instrument is assumed to be connected into the circuit of Figure 8–1 so that the pointer will be deflected to the right when the upper terminal of the galvanometer receives a positive emf, and to the left when the upper terminal is negative.

When Switch *S* is closed, current flows from Battery *B* through the inner coil, L_1. This current causes a magnetic field to spread around the inner coil and also to include in its path the turns of the outer coil, L_2. When the switch is opened, the current ceases and the magnetic field disappears. As long as the current is held to a steady value and does not reverse, this field is nonvarying.

There are two times, however, when the current is *changing* and the lines of force of the magnetic field are in motion. One of these times is immediately after closing the switch. The full value of current does not flow instantaneously, but the current rises from zero to maximum over a brief interval. During this *transient* interval, the lines of force expand outward from coil L_1 and pass through the turns of L_2. The other time is immediately after opening the switch. The current does not fall to zero immediately at this time, but rather "coasts to a stop" rapidly over an interval. While the current thus is *decaying*, the lines of force are collapsing into Coil L_1, again cutting through the turns of L_2.

From this explanation, we see that there are two brief intervals during which the turns of the coils are being cut by magnetic lines in motion—when the switch is opened and when it is closed. During these intervals, an emf is induced in Coil L_2, but it drops to zero when the current reaches its final steady value or when it falls to zero. When the switch is closed, the emf across L_2, owing to expanding lines, will have one polarity (say, upper terminal positive and lower terminal negative), and the galvanometer, accordingly, will be deflected momentarily to the right. When the switch is opened, the collapsing lines reverse the polarity of the emf, and the galvanometer is deflected momentarily to the left.

Thus, with no physical connection between the two coils, the making or breaking of current flow in one coil causes current to flow in a circuit connected to the other. If an alternating current is passed through Coil L_1, a continuous alternating emf will be set up in L_2 by the alternating magnetic field.

The inductive effect illustrated by Figure 8–1 will be obtained whether the current is passed through the inner or outer coil, as long as the two have the mutual relationship shown. It is customary to speak of the coil through which the initiating current is passed as the *primary coil* or *primary winding,* and the coil in which the emf is induced as the *secondary coil* or *secondary winding.* When the two coils are wound on a core of iron or other magnetic material, as shown in Figure 8–2(A), the inductive effect is intensified because the concentration of flux through the core causes more lines of force to link the coils. The two coils may be wound on top of each other on the core, as well as side-by-side in the manner shown in Figure 8–2(A).

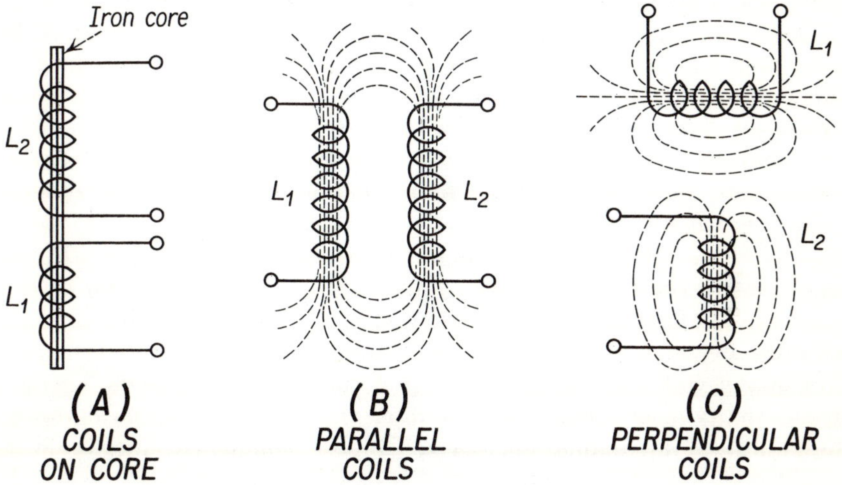

Fig. 8–2 Primary-Secondary Relationships

Induction also is obtained between parallel coils. [See Figure 8–2(B).] The only requirement is that the primary and secondary be close enough together for good flux linkage. When the coils are widely separated, the number of lines common to both coils decreases, and eventually all inductive effects disappear. When closely spaced coils are oriented at right angles to each other [Figure 8–2(C)], flux linkage becomes very low, as shown by the dotted lines, and induction is reduced to a minimum.

Coils are said to be *coupled* when they are so arranged physically that induction can take place between them.

8.2 SELF-INDUCTION

Induction is not a property possessed exclusively by *pairs* of coils. A single coil, carrying current, also can exhibit inductive effects.

Inductive action in a single coil is due to the effect of the coil's own magnetic field. The lines of force which expand and collapse about the coil, as the current is made, broken, or changed, cut the turns of the coil and develop an emf across the same coil.

8.3 COUNTER EMF

The electromotive force generated as the result of self-induction has a polarity opposite to that of the emf originally applied to the coil. This second emf accordingly opposes or bucks the original emf and therefore is called a *counter emf.* It is referred to also as a *back voltage* or *bucking voltage.*

Lenz's law (after H. F. E. Lenz, 1804–1865, who expressed it in 1833) states that *the direction of a current set up in a circuit by a change of magnetic flux must be such as to oppose the phenomenon causing it.* Thus, a counter emf will oppose an applied emf, or a conductor moved through lines of force has generated in it a current which, in turn, sets up an opposite magnetic field which opposes the motion of the conductor.

The efficacy of Lenz's law is demonstrated when a copper disc is spun between the poles of a strong magnet. The rotary motion of the disc is rapidly braked by the opposing magnetic field set up by currents generated in the disc by the latter's motion in the original field.

Counter emf *opposes* the build-up of current through a coil. It also opposes decay of current. Thus, counter emf imparts to a coil carrying alternating or fluctuating current a property similar to inertia in a moving body. Just as a body at rest is hard to move and, once started, resists attempts to stop it, so counter emf impedes the starting and stopping of current.

8.4 INDUCTANCE

Counter emf is proportional to the rate of change of current. It is *equal* to this rate of change *times* a constant, *L,* called the *inductance* of the coil. Thus

$$E = \left(\frac{I_2 - I_1}{t_2 - t_1}\right) L \qquad (8\text{–}1)$$

where L = inductance
I_1 = current at time instant t_1 (in seconds)
I_2 = current at later time instant t_2 (in seconds)

Inductance has been called *electrical inertia.* The property of inductance may be considered to describe the ability of an inductor or coil to store energy in an electromagnetic field. This is somewhat similar to capacitance, which expresses the ability of a capacitor to store energy in an electrostatic field.

The inductance of a coil varies directly as the square of the number of turns it contains and inversely as its length from end to end. Thus, doubling the number of turns, while keeping all other dimensions constant, quadruples the inductance; while doubling its length, with all other dimensions constant, halves the inductance of a given coil.

Self-inductance refers to the inductance of a single inductor or coil, while *mutual inductance* refers to the inductive interaction of two coupled coils.

8.5 UNITS OF INDUCTANCE

The unit of inductance is the *henry,* named in honor of Joseph Henry (1797–1878). The symbol for self-inductance is L. A coil or circuit has a self-inductance of 1 henry (h) when a change of 1 amp per second [see Equation (8–1)] induces an emf of 1 v.

The symbol of mutual inductance is M. Mutual inductance also is expressed in henrys. The mutual inductance between two coils is 1 h when a change of current in the primary coil at the rate of 1 amp per second induces in the secondary coil an emf of 1 v.

While the henry is the practical unit of inductance, this unit sometimes is too large for practical use and its subdivisions are used. For example: 1 millihenry (mh) = 1/1,000 h, while 1 microhenry (μh) = 1/1,000,000 h.

In an earlier chapter, it was explained that energy is stored in the electrostatic field in a capacitor. Similarly, energy is stored in the electromagnetic field in an inductor. The energy stored in the magnetic field is

$$W = \tfrac{1}{2}LI^2 \tag{8–2}$$

where W = energy (in watt-seconds)
L = inductance (in henrys)
I = current (in amperes)

8.6 INDUCTIVE EFFECTS IN CONDUCTORS CARRYING CURRENT

Since every current-carrying conductor is surrounded by a magnetic field (Section 5.6, Chapter 5), self-induction and self-inductance are exhibited by all such conductors.

Inductive effects are lowest in straight conductors and are lower in *short* straight conductors than in long ones. In a conductor of given length, inductive effects are lower in thicker conductors. Inductive effects increase when conductors are bent or coiled, since coiling tends to concentrate the magnetic field. (See Section 5.7, Chapter 5.) For these reasons, short, straight, heavy conductors often are mandatory in many types of circuits.

8.7 SINGLE-LAYER INDUCTOR

The simplest practical inductor is a single-layer coil without core. In this type the turns of wire are wound in a single layer and usually in the shape of a cylinder. Insulated wire (enameled or fabric-covered) is employed to prevent adjacent turns from short-circuiting.

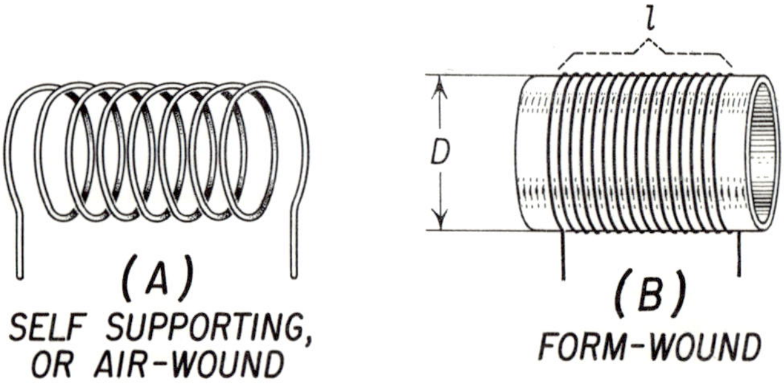

Fig. 8–3 Coreless Single-Layer Inductors

If only a very few turns are required and the coil is wound with thick wire, no supporting material will be needed. A coil of this type [Figure 8–3(A)] is self-supporting and is called "air-wound" (from its lack of any core other than air). When many turns are required, especially if fine wire is used, the coil cannot support itself, and the turns must be wound around a form of insulating material, usually cylindrical, as shown in Figure 8–3(B).

In practical sizes for efficient operation, single-layer coils cannot provide large amounts of inductance. They seldom are encountered outside of the electronic field where their inductance ratings are in the range between 1 μh and 1 mh.

The important dimensions of the single-layer coil are number of turns (N), length of winding (l), and diameter of coil (D). From these factors, the inductance may be calculated with the aid of the following equation.

$$L = \frac{0.2D^2N^2}{3D + 9l} \tag{8-3}$$

where L = inductance (in microhenrys)
D = diameter (in inches)
l = length (in inches)
N = number of turns

Example: A single-layer coil consists of 230 turns of No. 32 enameled wire on a 1½-in.-diameter bakelite form. The length of the winding is 2 in. What is the inductance?

(*a*) $N = 230, \quad D = 1.5, \quad l = 2$

(*b*) $$L = \frac{0.2(1.5)^2 230^2}{3(1.5) + 9(2)} = \frac{0.2(2.25)52{,}900}{4.5 + 18}$$

$$= \frac{23{,}805}{22.5} = 1{,}058\ \mu\text{h} = 1.06\ \text{mh}$$

8.8 MULTILAYER INDUCTOR

Larger inductance values may be obtained by winding a coil in several layers, one on top of the other, thereby procuring a greater number of turns and increased magnetic flux. A multilayer air-core coil is illustrated by Figure 8–4. Here, many turns of wire are wound on a spool or bobbin of insulating material.

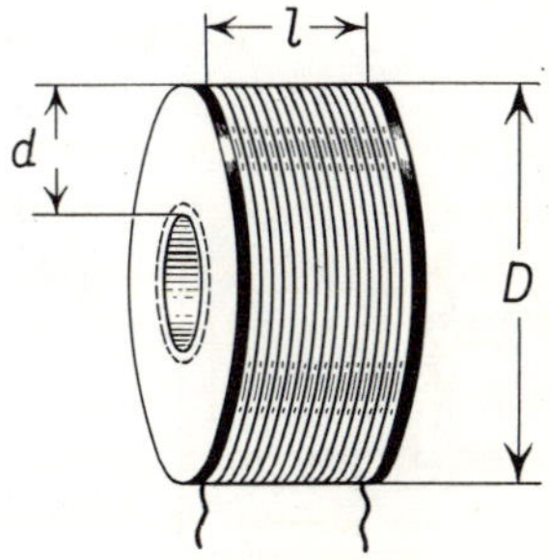

Fig. 8–4 Coreless Multilayer Inductor

The important dimensions of the air-core, multilayer coil are number of turns (N), length of winding (l), diameter (D), and radial depth (d). These dimensions are shown in Figure 8–4. From these factors, the inductance of the coil may be calculated with the aid of the following equation.

$$L = \frac{0.2D^2N^2}{3D + 9l + 10d} \quad (8\text{–}4)$$

where L = inductance (in microhenrys)
D = diameter (in inches)
d = radial depth (in inches)
l = length of winding (in inches)

Example: A ½-in.-diameter bobbin is wound full of No. 34 enameled wire (3,942 turns). The radial depth of the completed coil is ⅜ in. and the length of the winding ½ in. What is the inductance?

$$D = 0.5, \quad d = 0.375, \quad l = 0.5, \quad \text{and} \quad N = 3{,}942$$

$$L = \frac{0.2(0.5)^2(3{,}942)^2}{3(0.5) + 9(0.5) + 10(0.375)} = \frac{0.2(0.25)15{,}539{,}364}{1.5 + 4.5 + 3.75}$$

$$= \frac{0.05(15{,}539{,}364)}{9.75} = \frac{776{,}968}{9.75} = 79{,}689\ \mu\text{h} = 79.69\ \text{mh}$$

Effect of Core. The coils described in the two preceding examples are of the air-core type; that is, they have no magnetic material, only air, within their magnetic fields.

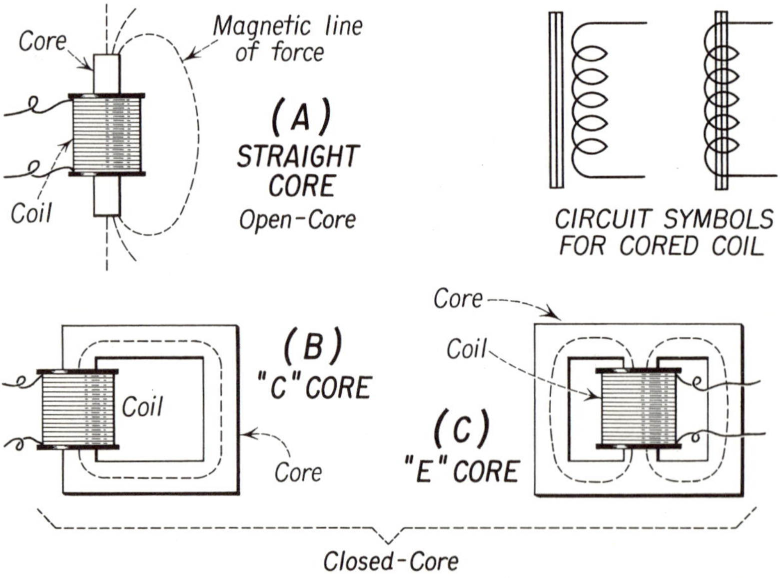

Fig. 8–5 Cored Inductors

We found in our study of the electromagnet (Section 5.7, Chapter 5) that the addition of an iron core to a current-carrying coil intensifies its magnetic effect by concentrating the lines of force. This is true also in an inductor. The addition of a core of magnetic material increases the inductance.

Figure 8–5 shows several common constructions for cored inductors. Iron or silicon-steel cores are used in most conventional inductors, although powdered iron, alloy dust, or ferrites are employed in the cores of many inductors designed for high-frequency electronic applications. In Figure 8–5(A), the core consists of a straight bar or rod, also known as an "I" core. Since many of the lines of force pass through the air and do not link the turns of the coil (as shown by the dotted lines), this simple type of core does not offer the maximum advantage. Figure 8–5(B), called a "C" core, is a better arrangement. Here, a great many lines pass around the magnetic circuit through the core. Figure 8–5(C), the "E" core, is a better arrangement still, through which a maximum number of lines pass on each side of the coil.

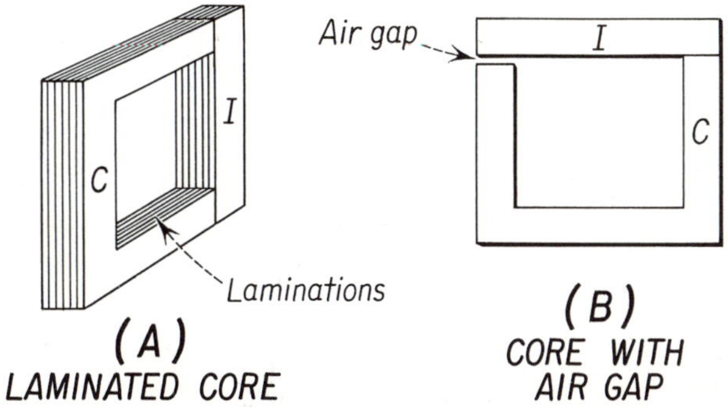

Fig. 8–6 Core Construction

Figure 8–6(A) shows typical construction of an inductor core. The latter consists of a number of thin sheets (*laminations*) of stamped sheet iron or steel stacked together tightly. In this instance, one set of laminations is cut into a square *C* shape, the other set into a straight *I* shape. Stacked together, as shown, they form the rectangular core shown before in Figure 8–5(B).

The laminated type of core is preferred to a solid one for the following reason. When magnetic lines of force cut the core, the latter (like any other conductor in a moving magnetic field) has current generated in it. These currents cannot be used externally but flow around inside the core, heating it. Power is lost in maintaining these *eddy currents.* Maximum efficiency and minimum heating demand that eddy currents be minimized. This is accomplished by separating the core into many laminations insulated from each other by a coat of lacquer. Thus, the paths are shortened through which eddy currents might flow, although this technique does not degrade the magnetic circuit. The thinnest laminations are the most effective.

When the coil of an inductor is to carry heavy currents, the core is split at one point to provide a small air gap in the magnetic circuit. This introduction of air reluctance in series with the iron reluctance prevents magnetic saturation at the high current level. One such method of obtaining an air gap is shown in Figure 8–6(B).

The important dimensions of an iron-cored coil are number of turns (N), core length (l), cross-sectional area of core (A), and permeability of the core material (μ). From these factors, the self-inductance of the cored coil may be calculated with the aid of the following equation.

(8–5) $$L = \frac{4.06N^2\mu A}{1.27(10^8)l}$$

where L = inductance (in henrys)
N = number of turns
μ = permeability of core material
A = cross-sectional area of core (in square inches)
l = total length of core (in inches)
10^8 = 10,000,000

Example: A 1,000-turn coil is wound on a 10-in.-long core (2½ in. per leg) having a cross-sectional area of 2¼ sq in. and a permeability of 2,500. What is the inductance?

$$N = 1{,}000, \quad \mu = 2{,}500, \quad A = 2.25, \quad \text{and} \quad l = 10$$

$$L = \frac{4.06(1{,}000)^2 2{,}500(2.25)}{1.27(10^8)10} = \frac{4.06(10^6)5{,}625}{12.7(10^8)}$$

$$= \frac{22{,}837(10^6)}{12.7(10^8)} = \frac{22{,}837}{12.7(100)} = \frac{22{,}837}{1{,}270} = 17.98 \text{ h}$$

Calculation of Mutual Inductance. Mutual inductance is exhibited whenever two coils are situated closely enough that their fields interact. The value of mutual inductance for two coils with identical cores (or wound on the same core) may be calculated in the following manner.

(8–6) $$M = \frac{4.06N_1N_2\mu A}{1.27(10^8)l}$$

where M = mutual inductance (in henrys)
N_1 = number of turns in first coil
N_2 = number of turns in second coil

μ = permeability of core material
A = cross-sectional area of core (in square inches)
l = length of core (in inches)

8.9 DISTRIBUTED CAPACITANCE IN INDUCTORS

Any two closely spaced metallic bodies can act as the plates of a capacitor. Section 7.18, Chapter 7, explained how this action causes "sneak" capacitances to appear in various parts of electrical circuits.

The adjacent turns of a coil are no exception. They lie close to each other and act as the plates of small capacitors. This is illustrated in Figure 8–7. The effect is the same as if a small capacitor (C_1, C_2, C_3) were connected between each pair of adjacent turns. In addition to

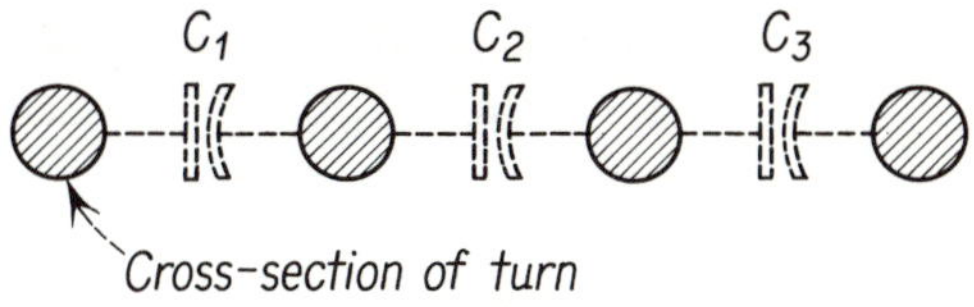

Fig. 8–7 Capacitance between Turns of Coil

the capacitances between turns, a multilayer coil will also exhibit capacitance between layers. In a large, multilayer inductor, the layer-to-layer capacitance is so much higher than the turn-to-turn capacitance that the latter usually is of lesser consequence.

The combined effect of all the inherent capacitances in a coil is termed the *distributed capacitance,* for which the symbol is C_d. Distributed capacitance is expressed in $\mu\mu$f and may be measured. The total distributed capacitance acts like a capacitor of the same capacitance value connected in parallel with the inductance of the coil.

Distributed capacitance must be taken into account in the design of inductors. There are various winding techniques which reduce the distributed capacitance of a coil. All of these attempt to destroy the adjacency of the turns. In one type of winding, often found in radio and television inductors, the turns are wound zigzag on a form so that successive turns do not lie side by side.

8.10 SPECIAL TYPES OF INDUCTORS

The commonest kinds of inductors are the "solenoid" coils shown in Figures 8–3 and 8–4. Many other constructions have been devised, however, to meet particular electrical and mechanical demands. Figure 8–8 shows several special types.

The *flat-spiral* type [Figure 8–8(A)], nicknamed "pancake coil," is suitable for very small values of inductance where physical thinness is desired and high distributed capacitance can be tolerated.

The *flat-rectangular* type [Figure 8–8(B)] is similar to the flat spiral.

The *rectangular solenoid* [Figure 8–8(C)] is obtained whenever a coil is wound tightly around a core or form having a square or rectangular cross section.

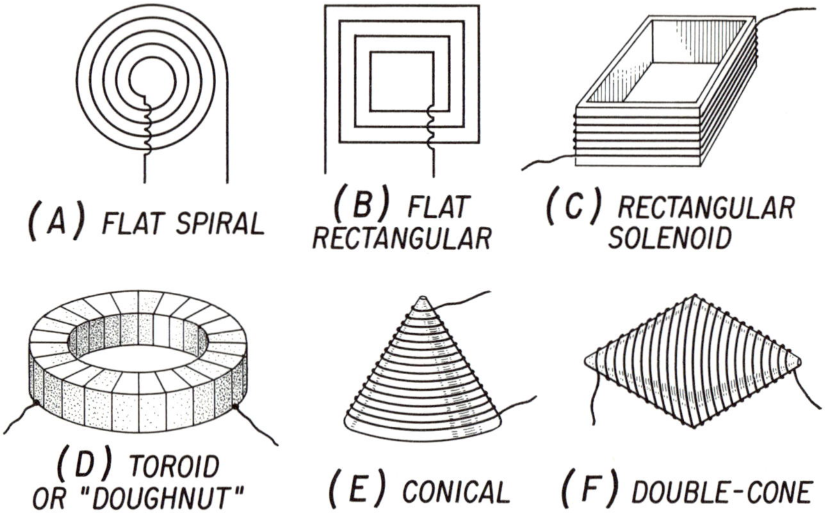

Fig. 8–8 Various Special Types of Inductors

In the *toroid* or "doughnut" coil [Figure 8–8(D)], the ring-shaped core completely encloses the magnetic field, permitting high values of flux linkage. When high-permeability core material is employed, large inductance values are obtained with relatively few turns on small-sized cores.

The *conical* [Figure 8–8(E)] and *double-cone* [Figure 8–8(F)] designs are aimed toward minimizing distributed capacitance by reducing the diameter of each successive turn. These types are suitable only for very low inductance values.

It is easy to see from inspection of these special types that only the rectangular solenoid [Figure 8–8(C)] and toroid [Figure 8–8(D)] are suitable for practical multilayer inductors. All others are single layer.

8.11 VARIABLE INDUCTORS

It is convenient to be able to vary the self-inductance of an inductor. This permits the inductance to be set exactly to a desired value.

There are several practical methods of varying inductance. Like variable resistors and variable capacitors, variable inductors may be of the step type or smooth type. Figure 8–9 shows several examples.

In Figure 8–9(A), the coil has been tapped at various numbers of turns, and these taps have been connected to the contacts of a selector switch. As the switch is rotated, the inductance of the coil is varied by changing the number of turns in use. This is a step-type

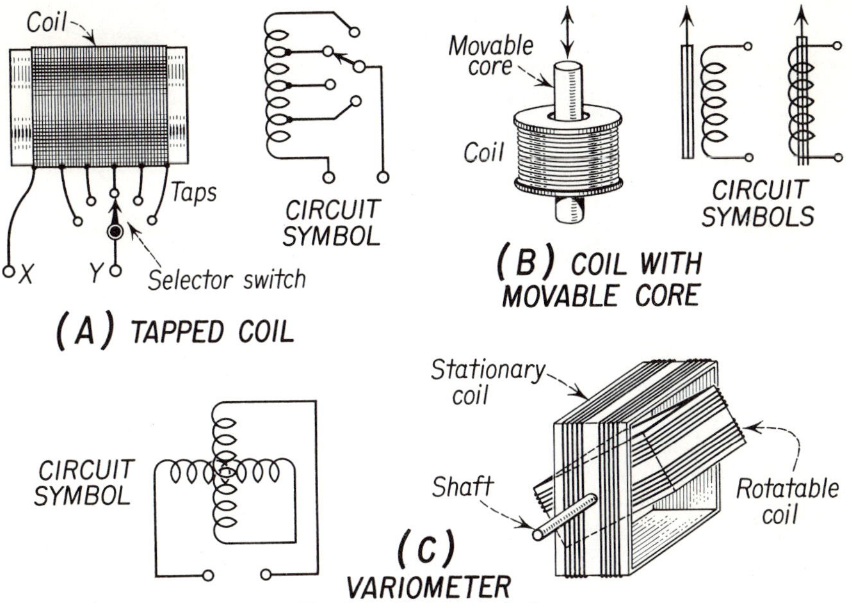

Fig. 8–9 Variable Inductors

variable inductor. The closeness with which it permits inductance adjustment depends upon the closeness of tapped-turn spacing. In an inductor of this type, the steps cannot be too closely spaced, since the number of switch points required soon would become impractically high.

Figure 8–9(B) shows an inductor in which the core of magnetic material may be moved in and out of the coil. The coil has maximum inductance when the core is inserted completely within the coil, and minimum inductance when the core is all of the way out. This provides a smooth, continuously variable change of inductance. Adjustable cores of this type often are threaded so as to be screwed in and out of the coil. This permits very close adjustment of the inductance.

Figure 8–9(C) shows a smooth-type variable inductor known as a *variometer* and sometimes called an *inductometer*. It consists of a rotatable coil mounted inside a stationary coil and connected in series with the latter. When the rotatable coil is turned to one extreme position at which it is completely enclosed by the stationary coil and parallel to the latter, the magnetic fields of the two coils are in opposition and cancel almost completely, reducing the inductance to a very low minimum. The mutual inductance between the two coils is considered negative (thus, $-M$) in this situation. When the rotor is positioned at its other extreme, 180 deg away, so that once again it is completely enclosed within the stator, the fields are aiding rather than opposing, M is positive, and the device exhibits its maximum inductance.

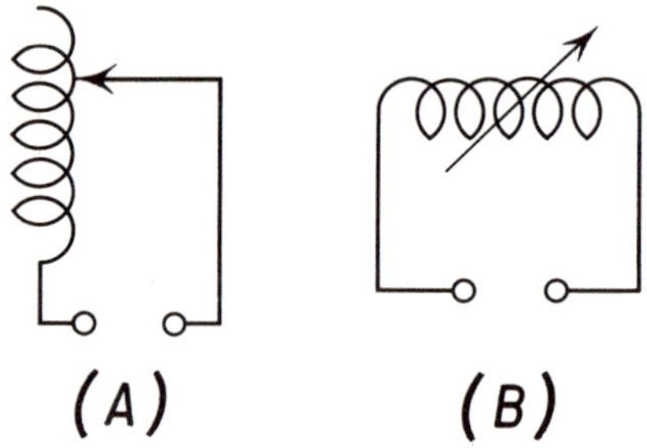

Fig. 8–10 Generalized Circuit Symbols for Variable Inductor

8.12 INDUCTANCES IN SERIES

There are two ways of connecting inductors in series: (1) the end of the first coil may be connected to the start of the second coil, or (2) the end of the first coil may be connected to the end of the second coil. This makes little difference except when the coils are situated so close together that their magnetic fields interact. Then, in one instance, the fields will be aiding and the inductance of the combination will be high. In the other instance, the fields will be bucking and the inductance will almost be nullified.

When coils are spaced closely, the mutual inductance must be considered when calculating the total inductance of the circuit. The total inductance of a coupled series combination thus is

For aiding fields

(8–7)
$$L_t = L_1 + L_2 + 2M$$

For bucking fields

(8–8)
$$L_t = L_1 + L_2 - 2M$$

where L_t = total inductance
L_1 = self-inductance of first coil
L_2 = self-inductance of second coil
M = mutual inductance

L_1, L_2, and M may be in henrys, millihenrys, or microhenrys, provided all three are in the same units.

When inductors are completely shielded or are widely spaced so as to eliminate interaction between their fields, the total inductance (L_t) of a series combination (Figure 8–11) is

(8–9) $$L_t = L_1 + L_2 + L_3$$

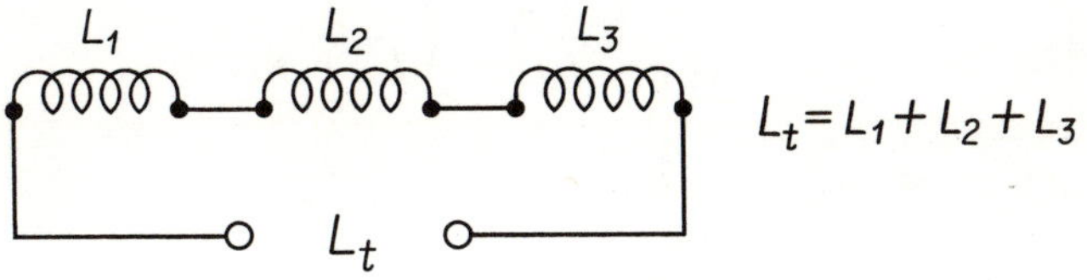

Fig. 8–11 Inductances in Series

Example: 10-, 45½-, 60-, and 90-h inductors are connected in series. What is the total inductance (L_t) of the circuit?

$$L_t = 10 + 45.5 + 60 + 90 = 205.5 \text{ h}$$

8.13 INDUCTANCES IN PARALLEL

When inductors are completely shielded or are oriented so as to eliminate interaction between their fields, the total inductance (L_t) of a parallel combination (Figure 8–12) is

(8–10) $$L_t = \frac{1}{\dfrac{1}{L_1} + \dfrac{1}{L_2} + \dfrac{1}{L_3}}$$

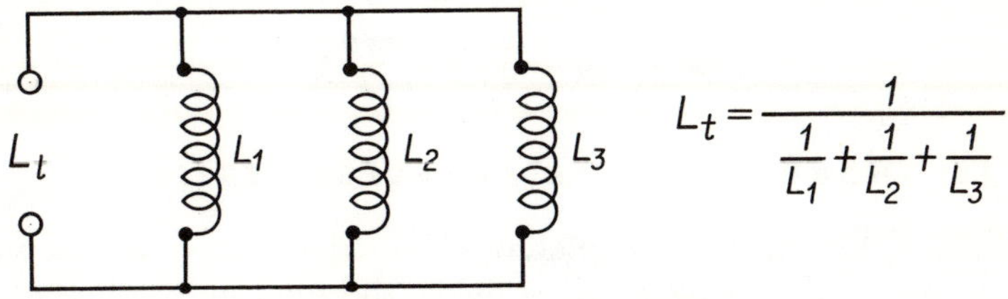

Fig. 8–12 Inductances in Parallel

Example: 1-, 50-, and 200-mh inductors are connected in parallel. What is the total inductance in henrys?

$$L_t = \frac{1}{\dfrac{1}{1} + \dfrac{1}{50} + \dfrac{1}{200}} = \frac{1}{1 + 0.02 + 0.005}$$

$$= 1/1.025 = 0.976 \text{ mh} = 0.000976 \text{ h}$$

8.14 INDUCTANCES IN SERIES-PARALLEL

When inductors are completely shielded or are oriented so as to eliminate interaction between their fields, the total inductance (L_t) of a series-parallel combination (Figure 8–13) is

(8–11)
$$L_t = \frac{1}{\left(\frac{1}{L_1 + L_2}\right) + \left(\frac{1}{L_3 + L_4}\right)}$$

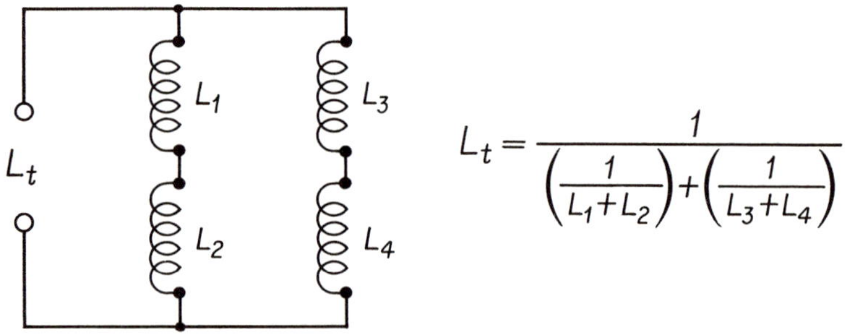

Fig. 8–13 Inductances in Series-Parallel

Example: 20- and 30-h inductors are connected in series in the first leg of a series-parallel circuit. 25- and 75-h inductors are connected in series in the second leg. What is the total inductance of the circuit?

$$L_t = \frac{1}{\left(\frac{1}{20 + 30}\right) + \left(\frac{1}{25 + 75}\right)} = \frac{1}{\frac{1}{50} + \frac{1}{100}}$$

$$= \frac{1}{0.02 + 0.01} = \frac{1}{0.03} = 33.3 \text{ h}$$

8.15 INDUCTANCES IN PARALLEL-SERIES

When inductors are completely shielded or are oriented so as to eliminate interaction between their fields, the total inductance (L_t) of a parallel-series combination (Figure 8–14) is

(8–12)
$$L_t = \left(\frac{1}{\frac{1}{L_1} + \frac{1}{L_2}}\right) + \left(\frac{1}{\frac{1}{L_3} + \frac{1}{L_4}}\right)$$

Example: 2- and 5-mh inductors are connected in parallel in the first leg of a parallel-series circuit. 30- and 80-mh inductors are connected in parallel in the second leg. What is the total inductance?

$$L_t = \left(\frac{1}{\frac{1}{2}+\frac{1}{5}}\right) + \left(\frac{1}{\frac{1}{30}+\frac{1}{80}}\right) = \frac{1}{0.5+0.2} + \frac{1}{0.033+0.0125}$$

$$= 1/0.7 + 1/0.0455 = 1.43 + 21.98 = 23.41 \text{ mh}$$

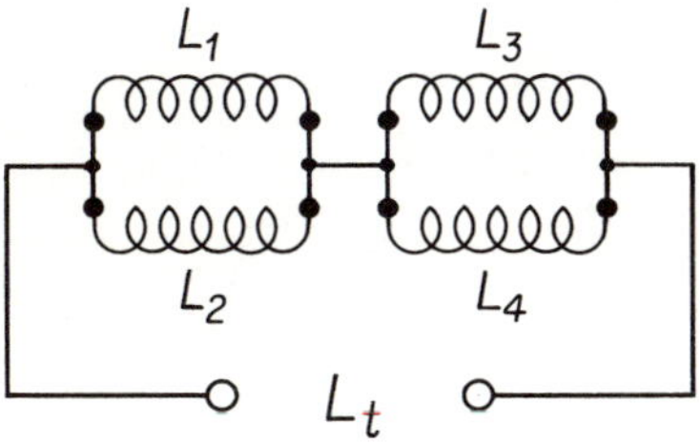

$$L_t = \left(\frac{1}{\frac{1}{L_1}+\frac{1}{L_2}}\right) + \left(\frac{1}{\frac{1}{L_3}+\frac{1}{L_4}}\right)$$

Fig. 8–14 Inductances in Parallel-Series

8.16 INHERENT INDUCTANCE IN CIRCUITS AND DEVICES

Just as stray capacitance evidences itself in electrical circuits and devices (Section 7.18, Chapter 7), so does *stray inductance.*

Every current-carrying lead is surrounded by a magnetic field and accordingly exhibits inductance. (See also Section 8.6.) Capacitors possess stray inductance, just as inductors have distributed capacitance.

Although stray inductance, like stray capacitance, usually is small, it can cause faulty operation of some circuits, especially those operating at high frequencies where inductive reactance values are high, and it cannot be neglected in circuit analysis.

8.17 A-C BEHAVIOR OF INDUCTORS

When an alternating voltage is applied to an inductor, the counter emf generated by induction opposes the cyclic build-up of alternating current through the coil. With a given inductance, the value of the counter emf increases with frequency, since this voltage is proportional to the angular velocity, $2\pi f$.

The net effect of the opposition offered by the counter emf is to reduce the flow of alternating current through the coil. The value of this current is directly proportional to the applied voltage and inversely proportional to the product of the inductance and the angular velocity. Thus

$$I = \frac{E}{2\pi f L} \tag{8–13}$$

where I = current (in amperes)
E = applied voltage (in volts)
$2\pi = 6.28$
f = frequency (in cycles per second)
L = inductance (in henrys)

The term $2\pi fL$ is called the *inductive reactance*, X_L, of the inductor. Inductive reactance is expressed in ohms and is similar to resistance and to capacitive reactance in that it limits current flow.

$$X_L = \omega L = 2\pi fL \tag{8–14}$$

where X_L = inductive reactance (in ohms)
$\omega = 2\pi f = 6.28f$
f = frequency (in cycles per second)
L = inductance (in henrys)

Example: A 150-mh inductor is operated at 100 kc. What is its reactance at this frequency?

$$L = 150 \text{ mh} = 0.15 \text{ h}; f = 100 \text{ kc} = 100{,}000 \text{ cps.}$$

$$X_L = 6.28(100{,}000)0.15 = 628{,}000(0.15) = 94{,}200 \text{ ohms}$$

Inductive reactance is seen to act in the same *way* but to be opposite in *nature* to capacitive reactance, X_c. (See Section 7.15, Chapter 7.) The larger the capacitance or higher the frequency, the lower the value of X_c; while the larger the inductance or higher the frequency, the *higher* the value of X_L.

At a given frequency, an inductance value may be chosen such that X_L is high enough, to all practical intents, to block the passage of alternating current through the inductor. However, steady direct currents still can pass through the coil. If a mixture of alternating current and direct current is applied to the coil, direct current will pass through more or less readily, but alternating current will be obstructed. This principle is utilized frequently when alternating current and direct current must be separated in a circuit. An inductor used for this purpose is called a *choke coil,* from the choking action it exerts on the alternating current.

Note that the circuit behavior of the inductor is the opposite of that of the capacitor. When a mixture of alternating current and direct current is applied to a capacitor, it is the alternating current that flows "through" while the direct current is blocked.

The equations for Ohm's law for alternating current, with respect to inductive reactance, are $I = E/X_L$, $E = IX_L$, and $X_L = E/I$.

SUMMARY OF CHAPTER 8

The phenomenon of **Induction** results from the cutting of a conductor by *moving* magnetic lines of force. Induction between coupled coils (inductors) causes an emf to be induced across one coil as the result of current fluctuating in the other.

In **Self-Induction,** a *counter emf,* or *back voltage,* is generated by the lines of force cutting the turns of the coil which produced the field. This phenomenon gives the inductor a sort of electrical inertia whereby the start and stop of current flow is opposed. **Lenz's Law** states that *the direction of a current set up in a circuit by a change of magnetic flux must be such as to oppose the phenomenon causing it.*

Energy is stored in the magnetic field of an inductor. Energy $W = \frac{1}{2}LI^2$.

Self-Inductance describes the ability of an inductor to store energy in its electromagnetic field. Self-inductance (L) is expressed in *henrys, millihenrys,* and *microhenrys.* A coil or circuit has an inductance of 1 h when a change of current at the rate of 1 amp per second induces an emf of 1 v across the coil.

Mutual Inductance exists between coils spaced close enough together for their magnetic fields to interact. Mutual inductance (M) is expressed in henrys; it is positive when the fields are aiding and negative ($-M$) when they are opposing, or bucking. The mutual inductance between two coils is 1 h when a current change of 1 amp per second in one coil induces an emf of 1 v across the other.

Inductance Calculations

(*a*) *Single-Layer Air-Core Inductor*

$$L = \frac{0.2D^2N^2}{3D + 9l} \quad \mu\text{h}$$

(*b*) *Multilayer Air-Core Inductor*

$$L = \frac{0.2D^2N^2}{3D + 9l + 10d} \quad \mu\text{h}$$

(*c*) *Inductor with Core of Magnetic Material*

$$L = \frac{4.06N^2\mu A}{1.27(10^8)l} \quad \text{h}$$

(*d*) *Mutual Inductance*

$$M = \frac{4.06N_1N_2\mu A}{1.27(10^8)l} \quad \text{h}$$

Distributed Capacitance is the combined capacitance effect between turns and between layers of an inductor. This capacitance (C_d) effectively shunts the inductance of the coil.

Basic Inductance Circuits. When inductors are completely shielded or are oriented to eliminate interaction between their fields, the total inductance (L_t) of circuits made up with these inductors may be calculated in the same manner as for resistances.

Inductances in Series

$$L_t = L_1 + L_2 + L_3$$

Inductances in Parallel

$$L_t = \frac{1}{\dfrac{1}{L_1} + \dfrac{1}{L_2} + \dfrac{1}{L_3}}$$

Inductances in Series-Parallel

$$L_t = \frac{1}{\left(\dfrac{1}{L_1 + L_2}\right) + \left(\dfrac{1}{L_3 + L_4}\right)}$$

Inductances in Parallel-Series

$$L_t = \left(\frac{1}{\dfrac{1}{L_1} + \dfrac{1}{L_2}}\right) + \left(\frac{1}{\dfrac{1}{L_3} + \dfrac{1}{L_4}}\right)$$

Stray Inductance in circuits and devices is similar to stray capacitance. All current-carrying leads are surrounded by magnetic fields and, accordingly, exhibit inductive effects.

Inductive Reactance (X_L) is the impeding effect offered by an inductor to the flow of alternating current. $X_L = \omega L = 2\pi f L$.

Ohm's Law for Alternating Current (with respect to inductive reactance) is expressed $I = E/X_L$, $E = IX_L$, and $X_L = E/I$.

QUESTIONS AND EXERCISES

1. Why does not a *steady* current in a primary coil induce a voltage across a secondary coil?
2. State Lenz's law.
3. Calculate the inductance of a single-layer air-core coil having 110 turns, a diameter of 2 in., and a length of 2¼ in.
4. Calculate the inductance of a multilayer, air-core coil having 3,000 turns, diameter = 2½ in., length = 1 in., and radial depth = ¾ in.

5. Calculate the inductance of a 980-turn coil wound on an iron core having a total length of 7¼ in. and a cross-sectional area of 1¼ sq in. The permeability of the core material is 3,000.

6. Calculate the total inductance in henrys of 10-, 25-, and 80-mh inductors connected in series.

7. Calculate the total inductance of four 50-h inductors connected in parallel.

8. In a series-parallel circuit (Figure 8–13) $L_1 = 30$ h, $L_2 = 10$ h, $L_3 = 2$ h, and $L_4 = 100$ h. Calculate the total inductance.

9. In a parallel-series circuit (Figure 8–14) $L_1 = 2\frac{1}{2}$ mh, $L_2 = 1$ mh, $L_3 = 10$ mh, and $L_4 = 80$ mh. Calculate the total inductance.

10. Calculate the inductive reactance of a 5-mh choke coil at 27 megacycles.

Chapter **9**

Phase Relations and Impedance

9.1 PHASE RELATIONS IN INDUCTIVE CIRCUIT

When an a-c voltage is applied to a circuit containing pure resistance only, the resultant alternating current passes through its instantaneous, peak, and zero values in step with the voltage. We say that, in a pure resistance circuit, the current and voltage are in phase. (See Section 6.12, Chapter 6.) Figure 9–1(A) illustrates this situation. Figure 9–1(B) is a *vector diagram* showing the same conditions. The length

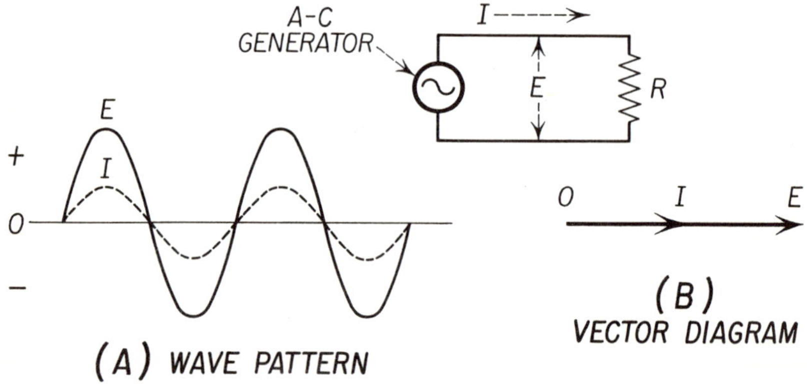

Fig. 9–1 Phase Relations in Resistive Circuit

of Line *E* represents the value of the voltage; for example, 1 in. could equal 1 v. The length of Line *I* represents the current value. Since *E* and *I* are in phase, the two vectors lie upon each other. Since *E* and *I* are alternating, the current and voltage vectors are considered to rotate counterclockwise around point 0, passing through various lengths corresponding to the instantaneous values, and making a complete rotation (360 deg) for each cycle.

These conditions are not the same in a circuit containing inductance or capacitance.

In a circuit containing pure inductance only, current cannot flow immediately upon application of a voltage because its build-up is opposed by the counter emf generated. Current begins to flow *after* the voltage has been applied. In a pure inductive circuit, the alternating current resulting from the application of an a-c voltage goes through its instantaneous, peak, and zero values, always behind the corresponding voltage values. In electrical language, the current is said to *lag* the voltage.

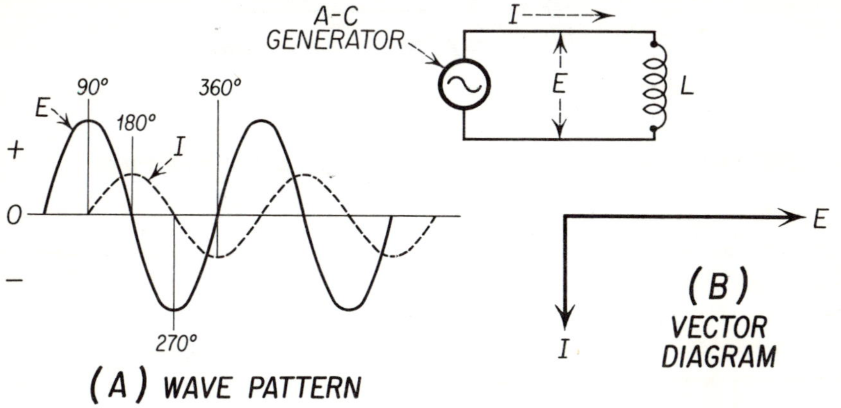

Fig. 9–2 Phase Relations in Inductive Circuit

Figure 9–2(A) illustrates the phase relations in an inductive circuit. Note that the voltage (E) has reached its maximum value (90 deg) before current (I) leaves zero. When the current reaches its first maximum, at 180 deg, the voltage has returned to zero. Thus, current lags behind voltage by 90 deg in an inductive circuit.

This situation is shown also by the vector diagram, Figure 9–2(B). Note that the current vector (I) is displaced a full right angle (90 deg) from the voltage vector (E). It is easy to see that when these vectors rotate counterclockwise, I always trails 90 deg behind E around the circle.

9.2 PHASE RELATIONS IN CAPACITIVE CIRCUIT

The situation in a capacitive circuit is the opposite of that in the inductive circuit. Current flows into a capacitor, continuing until the emf across the capacitor equals the applied voltage. (See Section 7.1, Chapter 7.) The current then ceases, and the voltage is maximum. Thus, the current precedes the voltage in this circuit.

In a circuit containing pure capacitance only, the current *leads* the voltage. The relation between alternating current and voltage in a capacitive circuit is shown in Figure 9–3(A). Note that at 0 deg, current (I) is maximum and voltage (E) is zero, while at 90 deg, current is zero and voltage is maximum. In this capacitive circuit, therefore, current leads voltage by 90 deg.

This situation is illustrated also by the vector diagram, Figure 9–3(B). Here, the current vector (I) is displaced a full right angle (90 deg) from the voltage vector (E). It is easy to see that when these vectors rotate counterclockwise, I always precedes E by 90 deg around the circle.

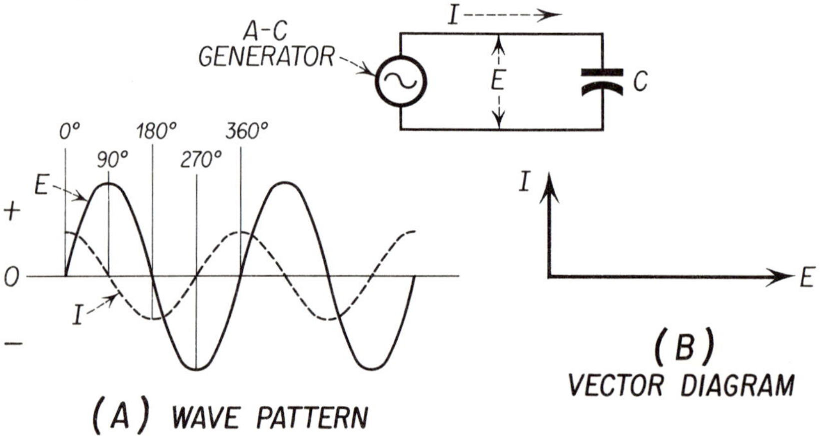

Fig. 9–3 Phase Relations in Capacitive Circuit

9.3 IMPEDANCE OF R-L CIRCUIT

There is no perfect inductor. Pure inductance does not exist in practice, because even the best coil has some resistance in addition to its inductance. At low frequencies, this is due entirely to the resistance of the wire with which the coil is wound, but at high frequencies, as we shall see later, the wire resistance constitutes only a fraction of the total coil resistance. The resistive component appears as a resistor (R) in series with the inductance (L), as shown in Figure 9–4(A). This is a simple *R-L* (resistance-inductance) circuit.

Resistors and inductors often are connected in series in electrical circuits. Figure 9–4(A) represents this arrangement, as well as a coil with internal resistance.

Alternating current flowing through a circuit of this type meets opposition both from the resistance component (R) and the inductive reactance (X_L). The combined effect of this opposition is called *impedance* (Z). Like resistance and reactance, impedance is measured in ohms.

Current (I_r) through a resistance equals E/R, and current (I_x) through an inductive reactance equals E/X_L. But these currents are 90 deg out of phase with respect to each other. [Compare Figures 9–1(B) and 9–2(B).] Therefore, we cannot determine the impedance of an *R-L* circuit simply by adding the R and X_L values, since maximum

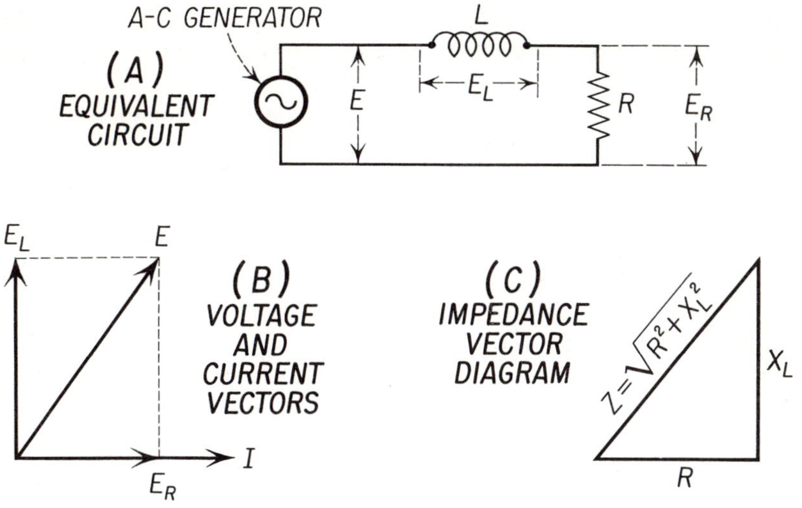

Fig. 9–4 Impedance of R-L Circuit

current is not flowing through R and X_L at the same instants. Phase relations must be considered in the calculation. Figure 9–4(B) shows the important phase relationships. Vector E_L represents the voltage drop across the inductance and leads vector E_R (the voltage drop across the resistance) by 90 deg. The current vector, I, is in phase with E_R but 90 deg behind E_L.

Since E_L and E_R are 90 deg apart, the applied voltage, E, cannot be equal to their simple, arithmetic sum but must be their *vector sum.* From geometry and physics, we recall that a vector sum is the "resultant" of two vectors. Graphically, the resultant may be determined by completing the rectangle between the tips of vectors E_L and E_R, as shown by the dotted lines in Figure 9–4(B), and drawing the vector E as a diagonal in this rectangle. Vector E is the desired resultant, and its length is the vector sum of E_L and E_R.

From this, it is apparent that X_L and R act at 90 deg with respect to each other. If their values are represented as vectors, as in Figure 9–4(C), their vector sum will be the third line drawn between the tips of these vectors and will be equal to the impedance offered by R and X_L in series. The Z vector is seen to be the hypotenuse of a right triangle.

From geometry, *the value of the hypotenuse equals the square root of the sum of the squares of the other two sides of the triangle.* Therefore, impedance equals

$$Z = \sqrt{R^2 + X_L^2} \tag{9–1}$$

where Z = impedance (in ohms)
R = resistance (in ohms)
X_L = inductive reactance (in ohms)

Example: In an *R-L* circuit, a 2-h inductance and 200-ohm resistor are in series. What is the impedance of the circuit at 60 cps?

(*a*) $R = 200$

(*b*) $X_L = 2\pi fL = 6.28(60)2 = 377(2) = 754\Omega$

(*c*) $Z = \sqrt{200^2 + 754^2} = \sqrt{40{,}000 + 568{,}516} = \sqrt{608{,}516} = 780.1\Omega$

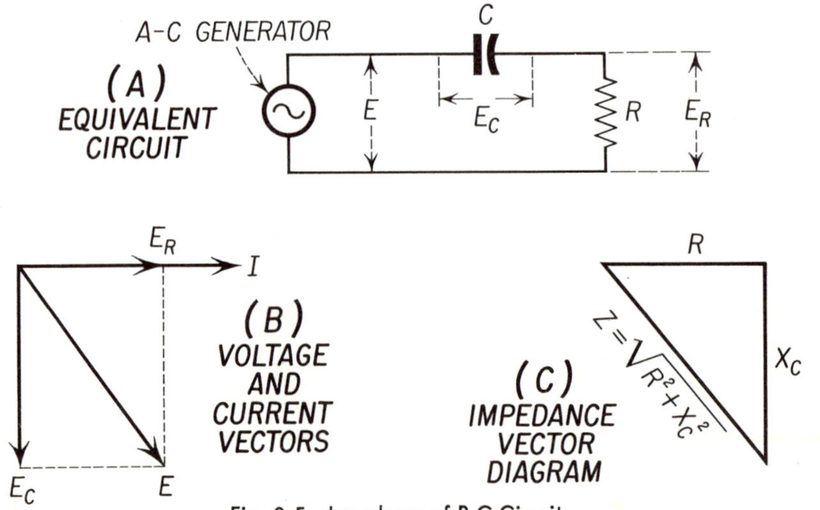

Fig. 9–5 Impedance of R-C Circuit

9.4 IMPEDANCE OF R-C CIRCUIT

There also is no perfect capacitor. In practice, resistance is associated with capacitance in all capacitors. An important component of this resistance appears as a resistor, *R*, in series with the capacitance, *C*, as shown in Figure 9–5(A). This is a simple *R-C* (resistance-capacitance) circuit. Resistors also are connected deliberately in series with capacitors in many electrical circuits.

As in the case of inductance and resistance, vector diagrams may be constructed for the *R-C* circuit, and the impedance of the latter may be determined from measurements on these diagrams or by means of calculations. *R-C* vector diagrams are shown in Figures 9–5(B) and (C). Note that the diagrams are drawn upside down when compared with those for *R-L* circuits. The reason for this is that capacitive reactance is considered negative (that is, $-X_c$). The X_c and E_c vectors accordingly are shown extending downward from the horizontal axis; that is, in the negative direction.

The impedance of the *R-C* circuit, like that of the *R-L* circuit, is equal to the vector sum of the *R* and *X* components. Thus

(9–2) $$Z = \sqrt{R^2 + X_c^2}$$

where Z = impedance (in ohms)
R = resistance (in ohms)
X_c = capacitive reactance (in ohms)

Example: A 1-μf capacitor and 1,000Ω resistor are connected in series. What is the impedance of this combination at 1,000 cps?

(*a*) $R = 1{,}000\Omega$

(*b*) $X_c = 1/(6.28fC) = 1/(6.28 \times 1{,}000 \times 0.000001)$
$= 1/(6{,}280 \times 10^{-6}) = 1/0.00628 = 159\Omega$

(*c*) $Z = \sqrt{R^2 + X_c^2} = \sqrt{1{,}000^2 + 159^2}$
$= \sqrt{1{,}000{,}000 + 25{,}281} = \sqrt{1{,}025{,}281} = 1{,}012.5\Omega$

9.5 APPLICATION OF OHM'S LAW FOR ALTERNATING CURRENT

It was explained in Section 7.15, Chapter 7, and Section 8.17, Chapter 8, that Ohm's law may be expressed in terms of current, voltage, and reactance; that is, $I = E/X_c = E/X_L$.

But the preceding sections of this chapter have indicated that pure inductance and capacitance do not exist in practice. Instead, *L* and *C* occur in combination with resistance, and the net opposition offered to alternating current flow is impedance Z, the vector sum of *R* and *X*.

Ohm's law for alternating current accordingly must be expressed in terms of impedance. In each of the following equations, *E* is in volts, *I* in amperes, and Z in ohms.

(9–3) $$I = E/Z$$

(9–4) $$Z = E/I$$

(9–5) $$E = IZ$$

Example: An emf of 63 v is applied to a circuit having an impedance of 3,100 ohms. What value of current flows?

$$I = E/Z = 63/3{,}100 = 0.0203 \text{ amp} = 20.3 \text{ ma}$$

Impedance is frequency-dependent. Its variation with frequency depends upon whether the reactance in a circuit is inductive or capacitive. In an inductive circuit, Z increases as frequency increases, and vice versa. In a capacitive circuit, Z decreases as frequency increases. In each case, it is assumed that the resistance, *R*, in the circuit remains constant.

9.6 INDUCTANCE AND CAPACITANCE IN COMBINATION, NET REACTANCE

Inductors or capacitors seldom are found alone in circuits. Combinations of these components often are used.

It might be expected that current flowing in a circuit containing both inductance and capacitance is limited by both X_c and X_L. This is indeed true. In a circuit containing inductance (L) and capacitance (C) in series, the *net reactance* of the circuit is the difference between the inductive reactance and capacitive reactance.

$$X = X_L - X_c \tag{9–6}$$

Example: A 50-μf capacitor and 20-h inductor are connected in series. What is the net reactance of this circuit at 500 cps?

(*a*) $X_L = 2\pi fL = 6.28(500)20 = 3{,}140(20) = 62{,}800\Omega$

(*b*) $X_c = 1/(2\pi fC) = 1/(6.28 \times 500 \times 0.000050)$
$= 1/(3{,}140 \times 0.00005) = 1/0.157 = 6.3\Omega$

(*c*) $X = X_L - X_c = 62{,}800 - 6.3 = 62{,}793.7\Omega$

(Here, the capacitive reactance is seen to be negligible.)

9.7 SERIES RESONANCE

Inductive reactance increases with frequency, while capacitive reactance decreases with frequency. For a given inductance and capacitance, there is one frequency at which $X_L = X_c$. Since the sign of X_L is positive and that of X_c negative, the two reactances will cancel each other at this frequency, and the net reactance ($X_L - X_c$) will be zero. An example is a circuit containing 20 mf and 32 mh. The resonant frequency of this combination is 200 cps. At this frequency, $X_L = X_c =$ 40 ohms. Therefore, $X = 40 - 40 = 0$. The frequency at which $X_L = X_c$

is called the *resonant frequency* and is similar to the natural vibration frequency (or pitch) of a spring, reed, or pendulum.

In a *series-resonant* circuit, the capacitor and inductor are connected in series, as shown in Figure 9–6(A). Here, *R* represents the inherent resistance of the circuit. At the resonant frequency, the net reactance is zero and only the resistance, *R*, remains in the circuit to impede current flow. Hence, maximum current flows at this frequency.

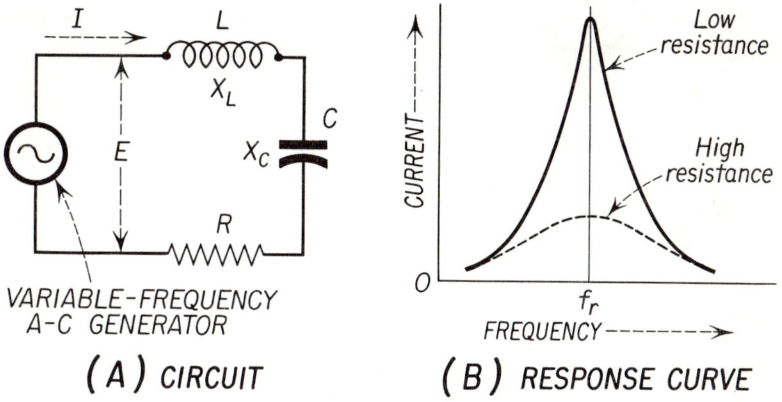

Fig. 9–6 Series-Resonant Circuit

If the applied voltage, *E*, is held constant and only the frequency varies, the current will increase as the resonant frequency is approached, as shown by the curve in Figure 9–6(B), reaching a peak value at the resonant frequency, f_r. As the frequency is increased further, the current will decrease. The net reactance of the circuit is capacitive below the resonant frequency and is inductive above this frequency.

This *resonance curve* will be high and sharp (solid line) if the circuit resistance, *R*, is low, and it will be low and broad (dotted line) if *R* is high. This response shows that the resonant circuit presents high impedance at frequencies above and below resonance, and lowest impedance at resonance.

The same response may be obtained by varying either *C* or *L*, while holding the generator frequency constant. Resonance is the basis of tuning in radio and television. A sharp response curve shows high *selectivity;* that is, the ability to select one desired frequency while rejecting all others. The dotted curve in Figure 9–6(B) shows poor selectivity, or "broad tuning."

In a series-resonant circuit, the generator, inductance, and capacitance are in series. This is true even when *L* and *C* are connected together and the voltage is *induced* into the inductor. The generator, although external physically, is considered to be effectively at the center of the inductor and thus in series with L and C.

If the values of C and L are known, the resonant frequency may be calculated with the aid of the following equation.

$$f_r = \frac{1}{2\pi\sqrt{LC}} \tag{9-7}$$

where f_r = resonant frequency (in cycles per second)
2π = 6.28
L = inductance (in henrys)
C = capacitance (in farads)

Example: What is the resonant frequency of 2 μf and 15 h in series?

(*a*) $2\,\mu\text{fd} = 0.000002$ f

(*b*) $LC = 15(0.000002) = 0.00003$

(*c*) $f_r = 1/(6.28 \times \sqrt{0.00003}) = 1/(6.28 \times 0.00547)$
$= 1/0.0343 = 29.1$ cps

It is interesting to note that the voltage drop across either the capacitor or inductor in a series-resonant circuit is higher than the generator voltage. This is true because the capacitor voltage (E_c) and the inductor voltage (E_L) are out of phase and accordingly oppose each other. The voltage across L and C in series, which is equal to the generator voltage, therefore is less than the larger of these two. For illustration, consider a series-resonant circuit in which $L = 5$ h, $C = 0.5\,\mu$fd, $R = 100$ ohms, $f = 100$ cps, and the applied voltage = 25 v. Here, $X_L = X_c = 3180\Omega$. Since the reactance at resonance is zero, only the 100-ohm resistor is effective in impeding current flow through the circuit. This current, $I = E/R = 25/100 = 0.25$ amp. The capacitor voltage $(E_c) = IX_c = 0.25(3{,}180) = 795$ v. The inductor voltage $(E_L) = IX_L = 0.25(3{,}180) = 795$ v. (Observe that here the inductor or capacitor voltage is 31.8 times the generator voltage.) This phenomenon is referred to as *resonant voltage rise,* or *resonant voltage step up.*

9.8 PARALLEL RESONANCE

Another type of tuned circuit is the *parallel-resonant* circuit. [See Figure 9–7(A).] In this arrangement, the generator, inductor, and capacitor are connected in parallel. The resistance component, R, representing losses in the inductor and capacitor, effectively is in series with L and C. The essential difference between the series- and parallel-resonant circuits is a matter of impedance. The series-resonant circuit has its lowest impedance (highest input current) at resonance, while the parallel-resonant circuit has its highest impedance (lowest input current) at resonance. The tuning, or response curve [Figure 9–7(B)], has the same shape as that of the series-resonant circuit, except that impedance in the parallel-resonant circuit is plotted against frequency. The equation for the resonant frequency of a parallel-resonant circuit is the same as that for the series-resonant circuit. [See Equation (9–7).]

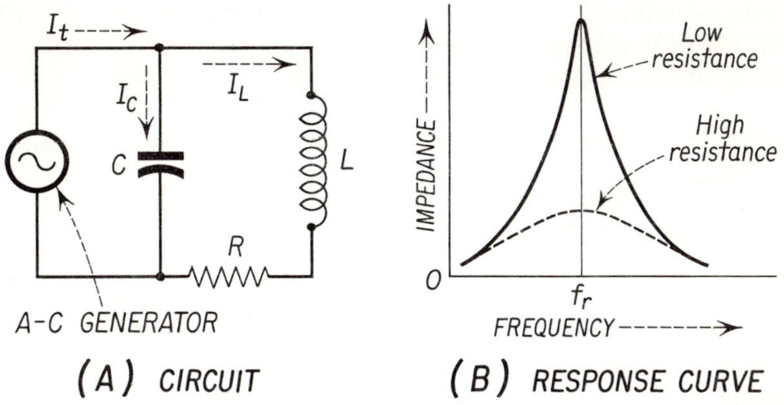

Fig. 9–7 Parallel-Resonant Circuit

Referring to Figure 9–7(A), if we assume the values of C, L, and R to be fixed and the generator voltage to be constant, the resonant frequency of the *LC circuit* may be found by varying the generator frequency. Current, I_t, from the generator divides, one portion (I_c) passing through the capacitor, and the remainder (I_L) through the inductor. Below resonance, I_L is higher than I_c; above resonance, I_c is the higher of the two. At resonance, $I_L = I_c$. At this point, the *line current*, I_t, is at its *minimum* value, because I_t theoretically is the difference between I_L and I_c (which would be zero) but actually has a small value owing to the presence of resistance R in the circuit. A suitable ammeter inserted between the inductor and capacitor will show a large current circulating between L and C, while an ammeter between the generator and the circuit will show a value (I_t) many times smaller.

The voltage across a parallel-resonant circuit rises to a maximum at resonance, just as the current in a series-resonant circuit peaks at resonance. Since $E = IZ$, this action shows that the parallel-resonant impedance is highest at resonance.

In the preceding illustration, it was assumed that L and C are fixed and that the frequency is variable. A parallel-resonant circuit may be tuned also by varying either L or C.

Operation of the parallel-resonant circuit illustrates the behavior of oscillating *tank* circuits. For example, suppose a voltage pulse is applied suddenly to Capacitor C in Figure 9–7(A) and the voltage source is then quickly removed from the circuit. The capacitor will become charged and, when the source is removed, will discharge through Inductor L, storing energy in the magnetic field. When all of the electrons have flowed through the coil, the capacitor will have become recharged with the opposite polarity. It then will discharge, sending electrons through L in the opposite direction and recharging itself with the initial polarity. Thus, during one half-cycle, energy is stored in the magnetic field of the inductor; during the other half-cycle, it is stored in the electrostatic field of the capacitor. This cycle of events is repeated rhythmically, creating an alternating current between L and C.

Owing to the *flywheel* action of the capacitor charging and discharging, this *oscillating* current would continue indefinitely, except that power is lost as current flows through Resistance R, and each half-cycle of current is lower than the preceding one until the current finally dies out. In a conventional system, voltage is applied to the parallel-resonant circuit on each half-cycle. However, the generator is required to supply only enough energy to overcome the losses introduced by Resistance R. This energy, represented by Current I_t in Figure 9–7(A), thus is less than that represented by the high circulating current between the inductor and capacitor. A mechanical analogy is the pendulum. While a stout push is needed to start the pendulum swinging; once it is in motion, a tiny push (in phase with the motion of the pendulum) is all that is required to keep it swinging.

9.9 COMBINED CIRCUITS OF L, C, AND R

Figure 9–8 shows the principal basic circuits comprising combinations of resistance, capacitance, and inductance. With each circuit is given the equation for calculating impedance (Z) and phase angle (ϕ). Also, the equations for calculating time constant (t) are given for Figures 9–8(A), (B), (C), and (D). Impedance has been explained earlier in this chapter. The other two factors are defined below.

Phase Angle. In impedance circuits containing L or C, or both, current is not in phase with the applied voltage. The time difference is measured in degrees and is called the *phase angle* (ϕ). The *tangent* of the phase angle is equal to X/R or to R/X. (See Figure 9–8.) The angle itself is found opposite this value in a trigonometric table of the functions of angles. Figure 9–9(B) is a short table of tangents.

For the significance, refer to Figure 9–9. The tangent (abbreviated

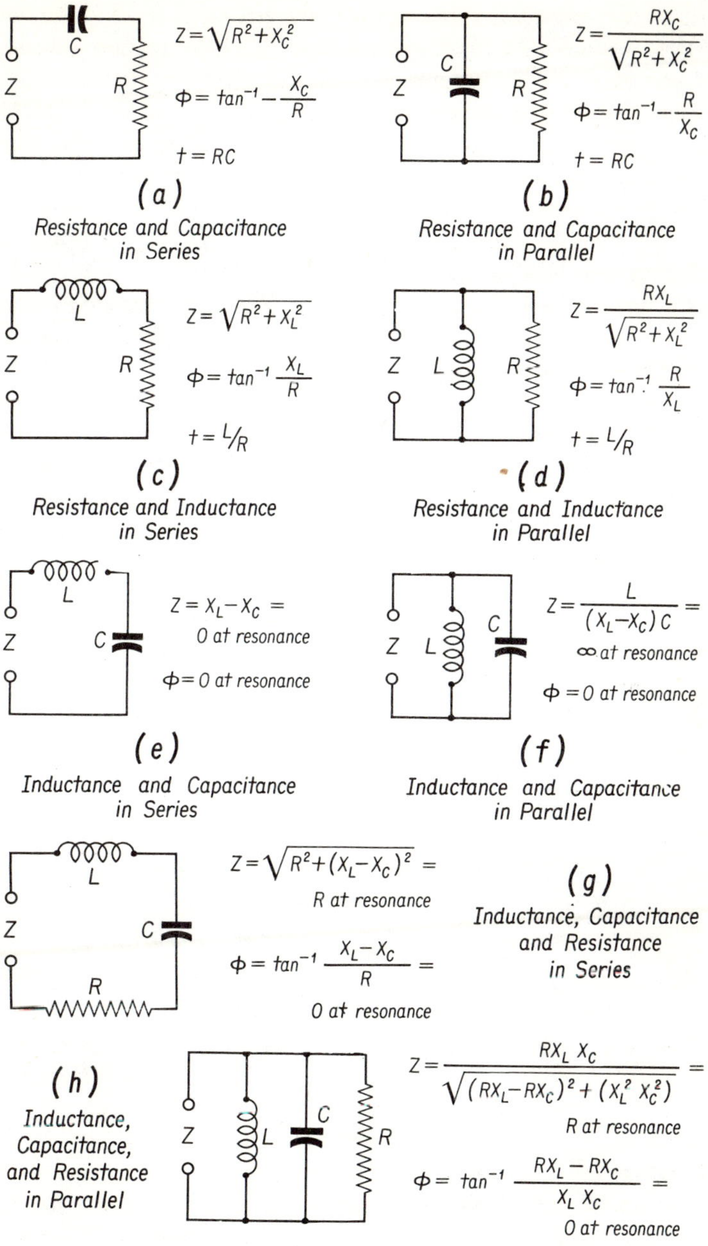

Fig. 9–8 Combined Circuits of L, C, and R

tan) is a trigonometric function representing the ratio a/b [see Figure 9–9(A)] of the opposite side of a right triangle to the adjacent side. Thus, when the conventional acute angle A is designated ϕ to represent the phase angle, $a/b = \tan \phi$.

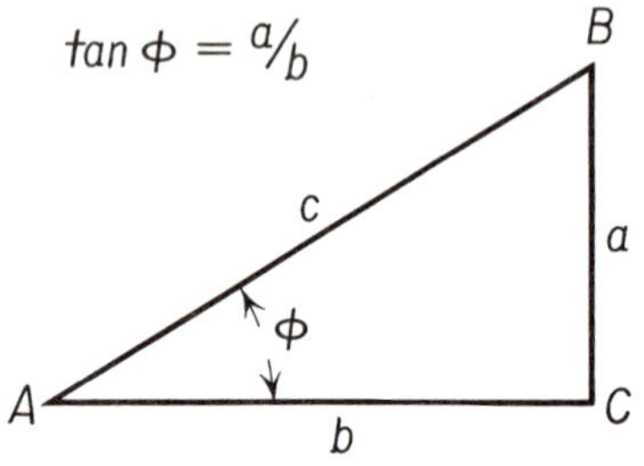

(A) RIGHT-TRIANGLE RELATIONSHIPS

ANGLE	TAN
0°	0.0000
10	0.1763
20	0.3640
30	0.5774
40	0.8391
50	1.1918
60	1.7321
70	2.7475
80	5.6713
90	∞

(B) PARTIAL TABLE OF TANGENTS

Fig. 9–9 Tangent of Angle

In the equations given in Figure 9–8, the term tan^{-1} means *that angle whose tangent has this value.* Thus, the expression $\tan^{-1} X/R$ directs the reader to "look up in the table that angle whose tangent is equal to X/R."

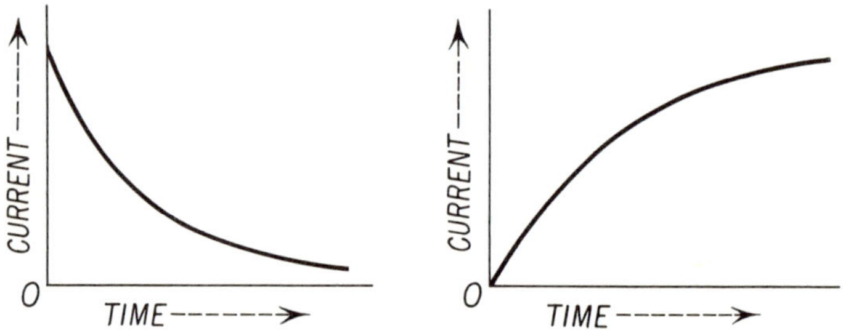

(A) R-C CIRCUIT (B) R-L CIRCUIT

Fig. 9–10 Time-Constant Curves

Time Constant. If a d-c voltage is applied to an R-C circuit [Figure 9–8(A) or (B)], a high value of current will flow at first and then will decrease with time, as the capacitor charges. This effect is illustrated by the curve in Figure 9–10(A). Similarly, if a capacitor is fully charged and then removed from the d-c source and connected to a resistor, it will discharge through the resistor in the same manner as shown by

Figure 9–10(A). The larger the resistance (for a given capacitance), the slower will be the charge or discharge. The *time constant* of an *R-C* circuit is equal to the product of the resistance and capacitance, and is the time required for the current to drop to 37 per cent of its initial value.

(9–8) $$t = RC$$

where t = time constant (in seconds)
R = resistance (in ohms)
C = capacitance (in farads)

Example: A 1-megohm resistor and 0.1-μfd capacitor are connected in series. What is the time constant of the circuit?

(*a*) $R = 1\omega = 1{,}000{,}000\Omega$

(*b*) $C = 0.1\ \mu\text{f} = 0.0000001\ \text{f}$

(*c*) $t = RC = 1{,}000{,}000(0.0000001) = 0.1\ \text{sec}$

If a d-c voltage is applied to an *R-L* circuit [Figure 9–8(C) or (D)], a low value of current will flow at first and then will increase with time, as the counter emf of the inductor is overcome. This effect is illustrated by the curve in Figure 9–10(B). The larger the resistance (for a given inductance), the faster will be the rise of the current to its final steady value. The time constant of an *R-L* circuit is the quotient of the inductance and resistance, and is the time required for the current to rise to 63.2 per cent of its final value.

(9–9) $$t = L/R$$

where t = time constant (in seconds)
L = inductance (in henrys)
R = resistance (in ohms)

Example: A 55-h inductor and 25-ohm resistor are connected in series. What is the time constant of this circuit?

$$t = L/R = 55/25 = 2.2\ \text{sec}$$

9.10 POWER IN A-C CIRCUITS

It will be recalled that d-c power is computed simply by multiplying volts by amperes, $P = EI$. (See Section 2.8, Chapter 2.) Also, $P = I^2R = E^2/R$. The same formulas apply to the calculation of a-c power in a circuit containing pure resistance only (Section 6.13, Chapter 6), since current and voltage are in phase in a resistive circuit.

Current and voltage are out of phase in a circuit containing inductance or capacitance, or both, in combination with resistance, and the simple formulas will not give an accurate solution. The product EI in this instance gives the *volt-amperes* (va) or *kilovolt-amperes* (kva) but not the true a-c watts. The phase angle (ϕ) must be taken into consideration in the calculation.

To find true a-c power, the volt-ampere product, EI, must be multiplied by the *cosine* of the phase angle. (The cosine, abbreviated *cos,* is a trigonometric function equal to the ratio b/c of the adjacent side of a

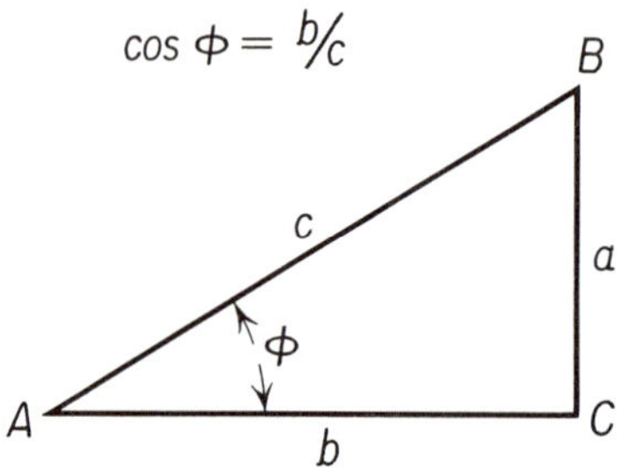

(A) RIGHT-TRIANGLE RELATIONSHIPS

ANGLE	COS
0°	1.0000
10	0.9848
20	0.9397
30	0.8660
40	0.7660
50	0.6428
60	0.5000
70	0.3420
80	0.1736
90	0.0000

(B) PARTIAL TABLE OF COSINES

Fig. 9–11 Cosine of Angle

right triangle to the hypotenuse.) The cosine of any angle between 0 and 90 deg may be found in a table of natural trigonometric functions. Figure 9–11(B) gives an abbreviated table of cosines. The power calculation is

$$P = EI \cos \phi \qquad (9\text{–}10)$$

where P = power (in watts)
E = voltage (in volts)
I = current (in amperes)
ϕ = phase angle (in degrees)

Example: What is the true a-c power when the voltage is 120 v, current is 27 amp, and phase angle is 10 deg?

$\cos 10° = 0.9848$
$P = EI \cos \phi = 120(27)0.9848 = 3{,}240(0.9848) = 3{,}190.7$ w

The term *cos* ϕ is called the *power factor* (pf). Thus: pf = cos ϕ. Power factor often is expressed in per cent. A glance at the table in Figure 9–11(B) will show how this is possible: cos 0° = 100%, cos 40° =

76.6%, cos 80° = 17.36%, etc. Power factor also is expressed as a figure between zero and 1. Here, it is evident that a power factor of 1.0 (or 100 per cent) is the most desirable, since then $P = EI$. For all other power-factor values, the product EI is multiplied by some value less than 1, and the power must be lower than the volt-amperes.

Power in an a-c circuit is highest when the phase angle is zero or low in value. The phase angle is zero ($\cos \phi = 1.0$) only in a resistive circuit. Here, the power factor is unity, or 100 per cent. Power is zero when the phase angle is 90 deg ($\cos \phi = 0.0000$). This could be possible only in a pure reactance circuit (inductive or capacitive), as will be explained in the next paragraph.

Since in a resistive circuit, power $P = I^2R = E^2/R$, one would expect to find a counterpart of these formulas in the study of reactive circuits: $P = I^2X = E^2/X$. Actually, no power is consumed in forcing a current through a pure reactance (X_c or X_L). In the case of a capacitive reactance, the power used to charge the capacitor on one half-cycle is returned to the generator on the next half-cycle as the capacitor discharges. Thus, no power is gained or lost. Similarly, the power used to generate a counter emf in an inductor is returned to the generator during the succeeding half-cycle.

Because no power is consumed in forcing a current (I_x) through a reactance, I_x often is called the *wattless component* of current. Power is consumed in an a-c circuit only when a current, I_r, is forced through the resistive component. Similarly, I^2X or E^2/X may be considered "wattless power," as reactive power often is called.

9.11 FIGURE OF MERIT (Q) OF CAPACITORS AND INDUCTORS

Practical capacitors and inductors are not pure *reactors* because a certain amount of unavoidable ("inherent") resistance occurs in their structures. This was explained in the earlier discussion of impedance. Resistance represents losses in capacitors and inductors and reduces their efficiency as reactors.

Since reactance is the desired property in a reactor and resistance the undesired property, the ratio of reactance to resistance expresses the quality of a capacitor or inductor as a reactor. This ratio expresses a quality factor, Q, which sometimes is called the *figure of merit*.

(9–11) $$Q = X/R$$

where X = reactance (in ohms)
R = resistance (in ohms)

For a Capacitor

(9–12)
$$Q = \frac{1}{2\pi fCR}$$

where f = frequency (in cycles per second)
C = capacitance (in farads)
R = resistance (in ohms)

Example: What is the Q at 100 kc of a 0.005-μfd capacitor having a loss (resistance) component of 1/10 ohm?

(*a*) $f = 100{,}000$ cps

(*b*) $C = 0.000000005$ fd

(*c*) $R = 0.1\Omega$

$$Q = \frac{1}{2\pi fCR} = \frac{1}{2(3.14)100{,}000(0.000000005)0.1}$$
$$= \frac{1}{6.28(0.005)0.1} = \frac{1}{0.00314} = 318.5$$

For an Inductor

(9–13)
$$Q = \frac{2\pi fL}{R}$$

where L = inductance (in henrys)
f = frequency (in cycles per second)
R = resistance (in ohms)

Example: What is the 1,000-cps Q of a 20-h choke coil having a resistance of 250 ohms?

(*a*) $f = 1{,}000$ cps

(*b*) $L = 20$ h

(*c*) $R = 250\Omega$

$$Q = \frac{2\pi fL}{R} = \frac{2(3.14)1{,}000(20)}{250} = \frac{6.28(20{,}000)}{250}$$
$$= \frac{125{,}600}{250} = 502.4$$

Equations (9–11), (9–12), and (9–13) show that Q varies with frequency. At a given frequency, Q varies directly with inductance and inversely with capacitance. What is regarded as a high Q value at one frequency may be low or medium at another frequency.

The Q value is a convenient number, since it expresses directly the quality of a capacitor or inductor. A high-Q component is a good one. This means simply that X (which is desired) is high, with respect to R (which is undesired and serves only to waste power).

SUMMARY OF CHAPTER 9

Alternating current and voltage are in phase only in a circuit containing pure resistance. Current **Lags** voltage by 90 deg in a **Pure Inductive Circuit** and **Leads** voltage by 90 deg in a **Pure Capacitive Circuit.**

Impedance (Z) is the net opposition offered by reactance and resistance in combination in an a-c circuit. Impedance is the vector sum of resistance and reactance; that is, $Z = \sqrt{R^2 + X^2}$. Impedance is expressed in ohms.

Ohm's Law for Alternating Current is similar to Ohm's law for **Direct Current,** with the difference that impedance is substituted for resistance in each formula: $I = E/Z$, $Z = E/I$, $E = IZ$.

The **Net Reactance** of an inductor and capacitor in combination is the difference of their individual reactances: $X = X_L - X_c$.

The **Resonant Frequency** is that frequency at which $X_L = X_c$. In a series-resonant circuit, maximum current flows at resonance; in a parallel-resonant circuit, maximum voltage appears across the circuit at resonance. The resonant frequency $f_r = 1/(2\pi\sqrt{LC})$.

Phase Angle (ϕ) expresses the time difference between current and voltage in a-c circuits. $\tan \phi = X/R$.

The **Time Constant** (t) of an R-C circuit equals RC sec. For an R-L circuit, $t = L/R$ sec.

The product of volts and amperes gives **Power** (in watts) in a d-c circuit. In an a-c circuit, EI does not give the true power but the *volt-amperes.* This value must be multiplied by the cosine of the phase angle ($\cos \phi$), known as the *power factor,* to obtain true a-c power. Thus: $P = EI \cos \phi$. Power factor varies from zero (for a theoretical pure reactance) to 1 (for a pure resistance).

Figure of Merit (Q) expresses the quality of a capacitor or inductor as a reactor. It is the ratio of reactance to resistance. Thus, $Q = (2\pi fL)/R = 1/(2\pi fCR)$.

QUESTIONS AND EXERCISES

1. Why does capacitive current lead voltage?
2. Why does inductive current lag voltage?
3. Calculate the 60-cps impedance of a 2½-h, 5-ohm inductor.

4. Calculate the 1,000-cps impedance of a circuit having a 15-μfd capacitor connected in series with a 10-h choke coil.

5. What is the impedance of a circuit in which 5 ma flow when 75 v are applied?

6. Calculate the net reactance at 1 megacycle of a circuit in which a 2-mh inductor is connected in series with a 0.001-μfd capacitor.

7. Calculate the resonant frequency of a circuit containing 10 μh and 50 $\mu\mu$fd.

8. Calculate the time constant of: (*a*) a circuit containing 0.02 μfd and 1 megohm in series; (*b*) a circuit containing 30 h and 10 ohms in series.

9. Calculate the power where voltage = 440 v, current = 15 amp, and phase angle = 10 deg.

10. Calculate the 400-cps Q of: (*a*) a 0.0001-μfd capacitor having 0.01 ohm resistance; (*b*) a 6-h inductor having a resistance of 100 ohms.

Chapter 10

Transformer Action and Transformers

10.1 INDUCTION BETWEEN COILS

When alternating current flows in a coil, an alternating magnetic field is set up around the coil. A counter emf is induced in this coil (*primary coil*), and an emf also will be induced in another coil (*secondary coil*) placed in the field.

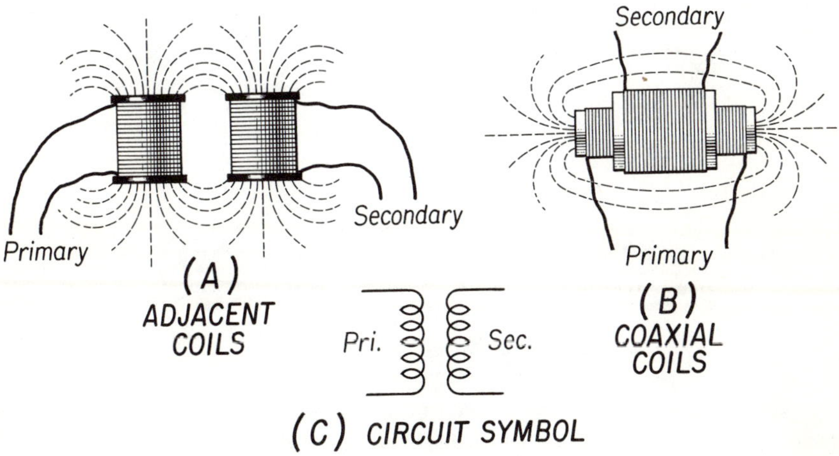

Fig. 10–1 Basic Transformer

The mechanism of induction between coupled coils permits the transfer of electrical energy from primary to secondary without a direct physical connection. The magnetic field is the only connecting link.

The *transformer* is an important electrical device. Most of the world's electric power is transmitted through transformers. Yet it is one of the simplest components. A transformer consists of a primary coil and secondary coil coupled together magnetically for efficient induction. The coils may be supported side-by-side, as shown in Figure 10–1(A), end-to-end, or one coil may be wound on top of the other, as shown in Figure 10–1(B).

These are only a few of several possible arrangements. All constructions are arranged best to utilize the induction between coils. Some transformers have several secondary windings (coils); others have several primaries as well. The simplest possible arrangement is one primary and one secondary, although occasionally the primary and secondary are combined ingeniously into a single winding. (See Section 10.8.)

10.2 MUTUAL INDUCTANCE

The phenomenon of mutual inductance is the quantitative basis of transformer action. The mutual inductance, M, between two coils is 1 h when a current change at the rate of 1 amp per second in one coil induces an emf of 1 v across the other coil. (See Sections 8.5 and 8.8, Chapter 8, for a discussion of mutual inductance.)

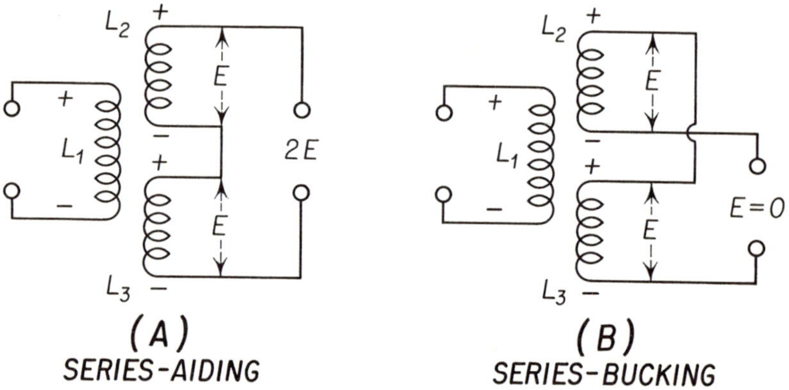

Fig. 10–2 Mutual Inductance Effects

Mutual inductance may have positive or negative effects. This is shown by the illustrations of transformer action in Figure 10–2. Here, a transformer is shown with two secondaries, L_2 and L_3. An alternating emf is applied to the primary, L_1. At one instant, the top of the primary is positive and the bottom negative, as shown. Assume that the top of each secondary is positive at this instant and that an equal voltage, E, is induced across each secondary.

If the secondaries are connected in the manner shown in Figure 10–2(A), so that the negative terminal of L_2 goes to the positive terminal of L_3, the secondary voltages will add in series and the total output voltage will be $2E$, the sum of the two secondary emf's. The secondaries are said to be connected in *series-aiding.*

If the two secondaries are connected in the manner shown in Figure 10–2(B), so that the positive terminal of L_2 goes to the positive terminal of L_3, the secondary voltages will buck each other and the total output voltage will be zero if the two voltages are equal, or their difference if they are unequal. The secondaries in this instance are said to be connected in *series-bucking* or *series-opposition.*

10.3 COUPLING

Coils or windings are said to be *coupled* when mutual inductance exists between them. The extent of the coupling depends upon the amount of flux linkage; that is, the number of lines of force linking the turns of the separate coils. When the flux linkage is great (M high), the coupling is referred to as *close* or *tight.* When flux linkage is slight (M low), the coupling is called *loose.*

Coupling may be tightened by placing coils close together and parallel. It may be loosened by separating the coils or situating them at right angles.

Tight coupling is employed in transformers intended for the transmission of electric power. Loose coupling is used in some r-f transformers.

10.4 EFFECT OF CORE

Since a core of magnetic material acts to concentrate the magnetic flux of a coil, a core will tighten the coupling between the windings of a transformer. In the air-core transformers illustrated by Figure 10–1, flux concentration cannot reach a high value, since many lines of force pass into the air and do not link the coils.

The number of magnetic lines of force per unit area, called the *flux density,* is greater in the core material than in air. This accounts for the ability of the core to concentrate the flux. Good core materials have low *reluctance.* (That is, they offer little opposition to magnetic flux, just as a low resistance offers little opposition to current flow.)

The effectiveness of the core material is expressed as the ratio of the flux in a closed magnetic path of the material to the flux which would be set up in an air path, having the same dimensions, by the same *magnetomotive force* (force setting up the magnetic flux). This ratio, called the *permeability* of the material, is greater than 1, since the permeability of air is unity.

Figure 10–3 shows several constructions employed in core-type transformers. In Figure 10–3(A), the primary and secondary coils are wound on the two separate legs of the core. In Figure 10–3(B), a

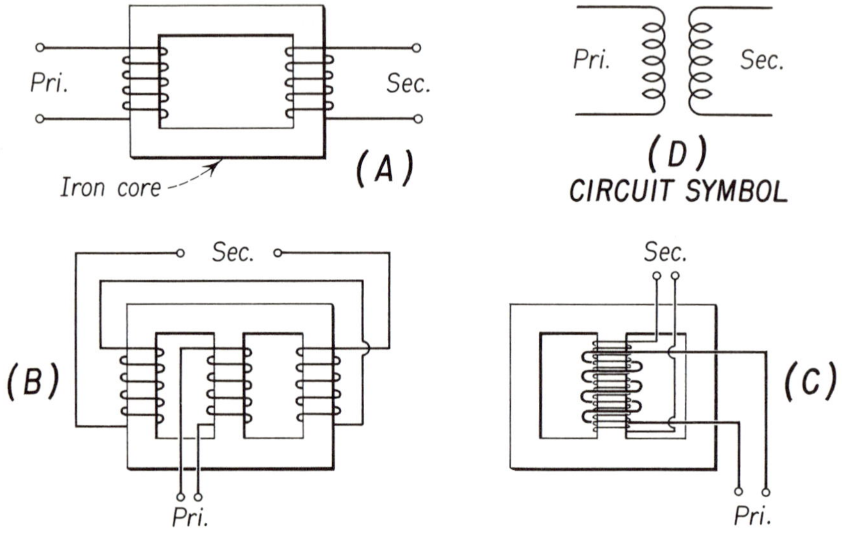

Fig. 10–3 Core-Type Transformers

three-legged core is employed, with the primary wound on the center leg and the secondary wound as two separate coils, one on each of the other legs, connected in series-aiding. In Figure 10–3(C), the secondary is wound on top of the primary on the center leg of the core.

10.5 COEFFICIENT OF COUPLING

In order to determine how well two coils are coupled, it is necessary that the degree of coupling be expressible. We are enabled to do this by the *coefficient of coupling* (k) which takes into consideration the self-inductance of the two coils and the mutual inductance between them.

(10–1) $$k = M/\sqrt{L_1 L_2}$$

where M = mutual inductance (in henrys)
L_1 = self-inductance of primary (in henrys)
L_2 = self-inductance of secondary (in henrys)

The coefficient of coupling is expressed as a figure. The unit has no name, such as the ohm, farad, etc. The coefficient of coupling can never exceed 1.

10.6 STEP-UP TRANSFORMER

The voltage induced in a coil by an alternating magnetic field is proportional to the number of turns in the coil. Conversely, the strength of a field is proportional to the number of turns in the coil carrying the alternating current which sets up the field, and to the a-c voltage applied to the coil.

From this, it is seen that an emf applied to a primary coil with a given number of turns will induce an emf of the same value across a secondary coil with the same number of turns. A rule is: *the secondary voltage* (E_s) *equals the primary voltage* (E_p) *multiplied by the ratio of the secondary turns* (N_s) *to the primary turns* (N_p).

(10–2) $$E_s = E_p(N_s/N_p)$$

Example: The primary of a certain transformer has 230 turns; the secondary 1,150 turns. 115 v are applied to the primary. What is the value of the secondary voltage?

$$E_s = 115(1{,}150/230) = 115(5) = 575 \text{ v}$$

Note in this example that the secondary voltage is five times higher than the primary voltage, because there are five times as many turns in the secondary winding as in the primary. $N_s/N_p = 5$. The term N_s/N_p is called the *turns ratio*. When the turns ratio is greater than 1, as in the preceding example, the transformer is of the *step-up* type. Step-up transformers are employed to obtain higher voltages from low-voltage a-c sources.

In all transformers, the primary ampere-turns equal the secondary ampere-turns; that is, $I_pN_p = I_sN_s$. From this expression, we find that the primary current (I_p) equals the available secondary current multiplied by the turns ratio.

(10–3) $$I_p = I_s(N_s/N_p)$$

Also, the available secondary current $I_s = I_p(N_p/N_s)$. From this, it is seen that the available secondary current in a *step-up* transformer must be less than the primary current. In fact, it is equal to the primary current multiplied by the reciprocal of the turns ratio.

10.7 STEP-DOWN TRANSFORMER

Further application of Equation (10–2) to the transformer having less secondary turns than primary turns shows that N_s/N_p in this case will be less than 1, and the secondary voltage, E_s, will be less than the primary voltage, E_p.

A transformer in which N_s/N_p is less than 1 is known as a *step-down* transformer and is used to obtain a low voltage from a higher-voltage a-c source. Application of Equation (10–3) shows that the available secondary current in the step-down transformer is higher than the primary current.

10.8 AUTOTRANSFORMER

For space-saving and simplicity, the primary and secondary sometimes are combined in a single tapped winding on a core, as shown in Figure 10–4. This arrangement is called an *autotransformer.*

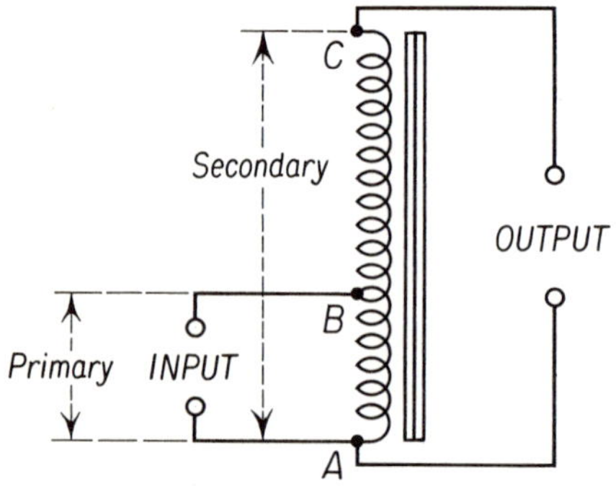

Fig. 10–4 Autotransformer

A few turns at the start of the coil are included between *A* and *B*, while a large number of turns is included between *B* and *C*. When an a-c voltage is applied to *A* and *B* as the primary, the magnetic flux surrounding coil section *AB* links the turns between *A* and *C* and induces a higher voltage across the latter proportional to the turns ratio *AC*/*AB*.

This is step-up action. If, conversely, an a-c voltage is applied to terminals *A* and *C* as the primary, a lower voltage is induced across *A* and *B* as the secondary because of the step-down turns ratio *AB*/*AC*, and the autotransformer acts as a step-down transformer.

A disadvantage of the autotransformer is its inability to provide complete isolation between primary and secondary circuits, since the secondary is a part of the primary, and vice versa.

10.9 SPARK COIL

Figure 10–5 illustrates the construction of a special type of transformer used to obtain a high a-c voltage from a low d-c voltage. This device is called a *spark coil,* from the fact that its earliest practical use was the production of sparks for automobile ignition and for pioneer radio telegraphy.

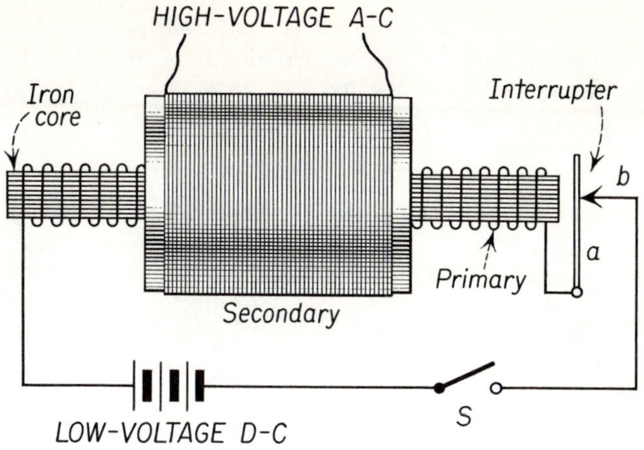

Fig. 10–5 Spark Coil

The core of the spark coil is a long rod made up of a bundle of soft-iron wires insulated from each other to reduce eddy-current losses. (See Section 8.8, Chapter 8.) The primary coil, consisting of relatively few turns of heavy, insulated wire, is wound on the core. The secondary consists of many turns of fine, insulated wire and is wound in a number of layers on top of the primary.

The primary is connected in series with a d-c source (such as a battery), Switch S, and a special buzzer (*a*, *b*) called an *interrupter*. (See Section 5.14, Chapter 5, for a discussion of buzzer action.)

When Switch S is closed, direct current flows through the primary, magnetizing the core. The expanding flux cuts the turns of the secondary and induces a voltage across the latter proportional to the primary-to-secondary turns ratio. The magnetism of the core attracts Spring *a* of the interrupter, thereby breaking the primary circuit. At this instant, the primary current is stopped, and the lines of force collapse toward the core, causing the secondary voltage to reverse polarity. With the cessation of primary current, however, the core loses its magnetism, and Spring *a* returns to Contact *b*, once again closing the primary circuit.

This make-and-break action is repeated at a regular rate determined by the tension of Spring *a* and its distance from the core and Contact *b*. In this way, a high a-c voltage is obtained at the secondary and its frequency may be varied by adjusting the buzzing rate of the interrupter.

10.10 VIBRATOR-TYPE TRANSFORMER

Another device similar to the spark coil is shown in Figure 10–6. This is the *vibrator-type* transformer, also used to convert a low voltage direct current into high voltage alternating current.

The primary of this transformer is *center-tapped;* that is, a tap is taken at the center turn of the entire winding, thereby dividing the primary into two equal sections, *AB* and *BC*. For voltage step-up, the secondary-to-primary turns ratio is made high.

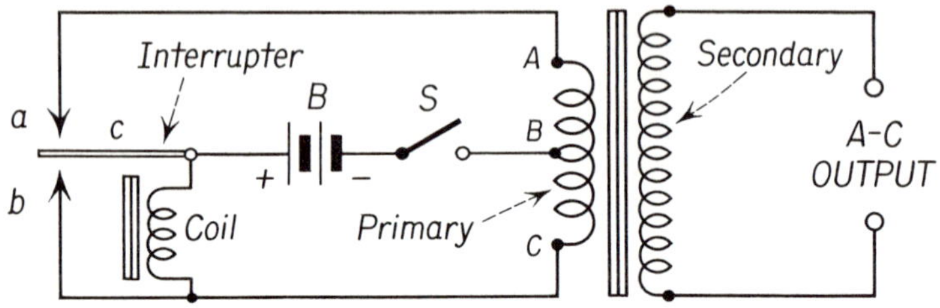

Fig. 10–6 Vibrator-Type Transformer

A buzzer-type interrupter is connected so as to switch current from Battery *B* first through the top half (*AB*) of the primary and then through the bottom half (*BC*). This interrupter consists of the magnet coil, contact spring *c* (often called the *vibrating reed*), and stationary contacts *a* and *b*.

When Switch *S* is closed, current flows from Battery *B*, through the magnet coil and the lower half of the primary (from *B* to *C*). The energized magnet coil attracts Spring *C* which, on touching Contact *b*, short-circuits the coil and destroys its magnetism. Without the magnetic attraction, Spring *c* is released from the core of the coil and springs against Contact *a*. Current then flows from Battery *B*, through the upper half of the primary (from *A* to *B*). But now, the magnet coil again is energized, since the short circuit has been removed, and the spring is drawn into contact with *b*.

As the spring thus vibrates back and forth between Contacts *a* and *b*, current flows first from *A* to *B* and then from *C* to *B*, through the primary winding of the transformer. This results in an alternating magnetic field which links the turns of the secondary winding and induces a high a-c voltage across the latter.

Vibrator-type transformers are widely used in portable equipment, such as automobile radios, radio transmitters, and d-c–a-c converters.

10.11 USE OF TRANSFORMER IN IMPEDANCE MATCHING

It was stated in Section 10.6 that the secondary ampere-turns (I_sN_s) of a transformer equal the primary ampere-turns (I_pN_p). In a theoretically perfect transformer without losses, the secondary volt-amperes (E_sI_s) equal the primary volt-amperes (E_pI_p).

The primary impedance $Z_p = E_p/I_p$ (from Ohm's law for alternating current, $Z = E/I$). Likewise, the secondary impedance $Z_s = E_s/I_s$. Figure 10–7 shows transformer impedance relationships. Here, the primary is supplied with 1 amp at 1 v from an a-c generator having an internal impedance (resistance) of 1 ohm. The transformer has a

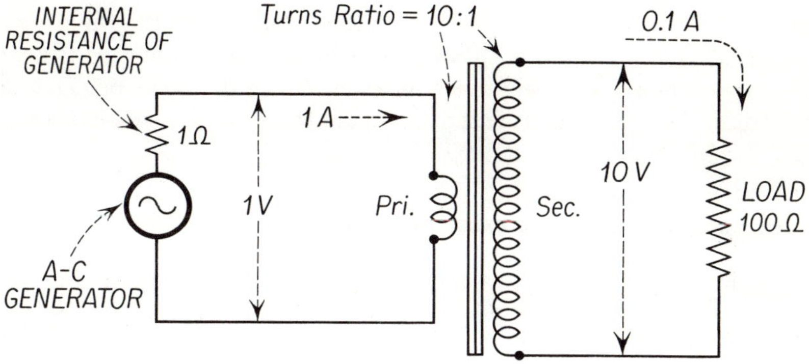

Fig. 10–7 Transformer Impedance Relationships

step-up turns ratio of 10 to 1. The primary volt-amperes = 1 × 1 = 1. Because of the 10:1 turns ratio, the secondary voltage = 10 × 1 = 10 v. In order to keep the secondary volt-amperes equal to the primary volt-amperes and thus to preserve the 10-v secondary potential, the secondary current cannot exceed 0.1 amp. (Secondary va = 10 × 0.1 = 1.) Thus, the load impedance connected to the secondary equals 100 ohms.

An inspection of the values in Figure 10–7 shows that the turns ratio is 10:1, but that the impedance ratio is 100:1. The impedance ratio is the square of the turns ratio.

(10–4) $$Z_s/Z_p = (N_s/N_p)^2$$

Or, the turns ratio is the square root of the impedance ratio.

(10–5) $$N_s/N_p = \sqrt{Z_s/Z_p}$$

Example: What transformer turns ratio will be required to provide an impedance ratio of 25:1?

(*a*) $Z_s/Z_p = 25$

(*b*) $N_s/N_p = \sqrt{Z_s/Z_p} = \sqrt{25} = 5$;

that is, 5:1.

In Figure 10–7, the transformer is said to match the 100-ohm load to the 1-ohm generator. *Impedance matching* between generator and load is important because maximum power is delivered only when a load impedance equals the internal impedance of a generator. If the two impedances are not equal, a transformer with the proper impedance ratio (square of the turns ratio) must be inserted between the generator and load.

When a given load impedance (Z_L) is connected to the secondary terminals of a transformer, the impedance "looking" into the primary equals $Z_L(N_p/N_s)$. This is called the *reflected impedance* of the load. Conversely, when an impedance (Z_p) is connected to the primary of a transformer, the reflected impedance viewed from the secondary terminals equals $Z_p(N_s/N_p)$.

10.12 MULTIWINDING TRANSFORMER

Practical transformers are not limited to a single primary and single secondary winding. Several coils often are wound on one core to supply different voltages and currents.

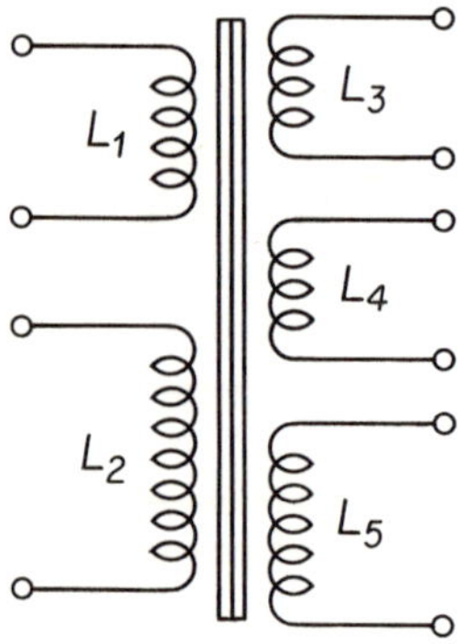

Fig. 10–8 Multiwinding Transformer

Figure 10–8 shows an example of such a multiwinding transformer. Here, L_1, L_2, L_3, L_4, and L_5 are separate windings. One of these coils may be used as the primary winding and the others as secondaries.

10.13 EQUIVALENT CIRCUIT OF PRACTICAL TRANSFORMER

A theoretically perfect, or *ideal*, transformer would be represented by a primary winding, secondary winding, and core, just as shown in the preceding illustrations. However, a practical transformer is not so simple. The windings have resistance, the core has losses, and not all of the flux links both coils. Although every effort is made in good design to eliminate these factors, a small amount of each unavoidably remains in the finished transformer.

Plate 10–1 Large Oil Filled Station-Type Transformer. This outdoor-type transformer is employed to step down 12,000 volts from a high-voltage transmission line to 240 and 120 volts for consumer use. (Courtesy, General Electric Co.)

Figure 10–9 shows the equivalent circuit of a practical transformer. R_c represents the core losses, which are in phase with the applied voltage and therefore may be shown as a resistance. A small amount of flux (in this case, called "leakage flux") does not link the two coils, and, since it causes self-induction, it gives rise to *leakage inductance*

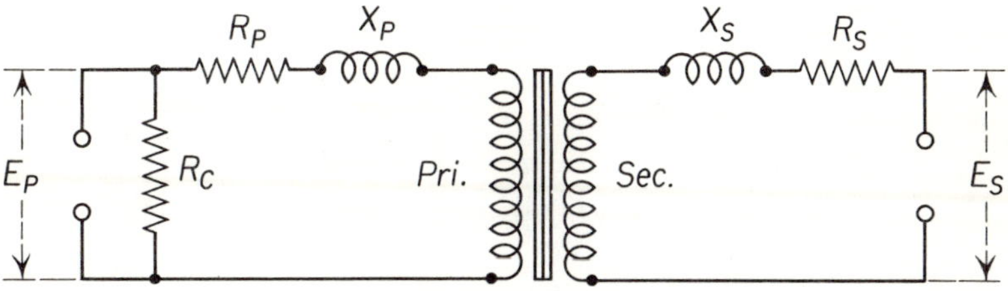

Fig. 10–9 Equivalent Circuit of Transformer

in both primary and secondary circuits. These effects are shown by the *leakage reactances*, X_p and X_s. The effect of these reactances is to reduce, via voltage drops, the voltage actually across the primary and (E_s) at the secondary output terminals. Voltage E_s thus is lower than the turns ratio would indicate, because of the various losses. Further voltage drops are caused by the resistance (R_p) of the primary winding and the resistance (R_s) of the secondary winding.

Distributed capacitances are not shown in Figure 10–9 but do appear in parallel with the windings in a practical transformer. They act to resonate the primary and secondary at frequencies determined by the values of distributed capacitance and the inductance of the windings.

The ideal transformer would transmit to a circuit connected to its secondary terminals the full amount of power fed into its primary winding. Owing to the losses indicated in Figure 10–9, however, the output power is somewhat less than the input power. The ratio of output to input in any device is called efficiency (n). Thus

$$n = 100(P_o/P_i) \tag{10–6}$$

where n = efficiency (in per cent)
P_o = power output
P_i = power input

Example: What is the efficiency of a transformer delivering 200 w output as a result of 230 w input?

$$n = 100(200/230) = 100(0.8695) = 86.95\%$$

10.14 SATURABLE REACTOR

Figure 10–10 shows a special type of transformer in which the principle of magnetic saturation is employed to control the flow of an alternating current. This device is called a *saturable reactor.*

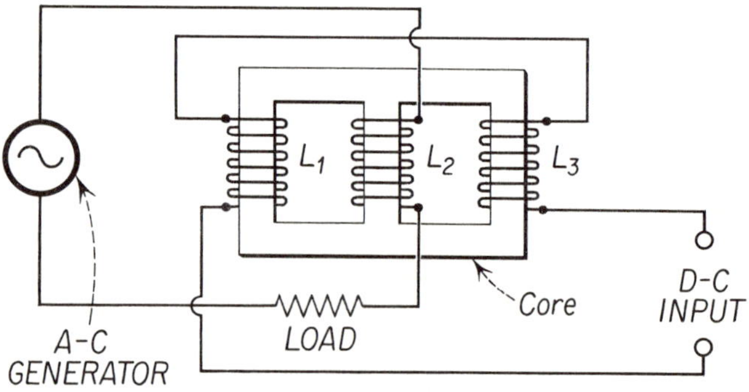

Fig. 10–10 Saturable Reactor

A three-legged core is employed. The *a-c winding* (L_2) is wound on the center leg and is connected in series with an a-c generator and load. The amount of current passing into the load thus depends upon the reactance of L_2. The *control winding* is split into two sections; one half (L_1) is wound on one outside leg, and the other half (L_3) on the other outside leg. L_1 and L_3 are connected in series-bucking so as to remove from the D-C INPUT terminals any a-c voltage induced in L_1 and L_3 by alternating current in L_2.

Direct current flowing through L_1 and L_3 in series tends to saturate the core. This lowers the inductance and reactance of L_2 proportionately by reducing the total flux in the core. As a result, more current is able to flow from the generator and through the load. At high d-c control current levels, the core becomes completely saturated, the reactance of L_2 disappears, and the load receives maximum current from the generator. With zero direct current through L_1 and L_3, winding L_2 has maximum reactance, and the load current is near zero.

Commercial saturable reactors provide a wide range of smooth control of alternating current.

10.15 MAGNETIC AMPLIFIER

When the core of the saturable reactor is made from a special high-permeability alloy having a hysteresis loop with steep sides, it will saturate with quite small values of control current. This allows large values of load current to be controlled with small, often tiny, levels of control current. This is a condition of *amplification.*

Commercial magnetic amplifiers operate on this principle, although some varieties use somewhat more complicated circuits than the one shown in Figure 10–10. Magnetic amplifiers have come into wide civilian and military use since the end of World War II.

10.16 RADIO-FREQUENCY TRANSFORMERS

All of the transformers illustrated up to this point in this chapter have been of the iron-core type, except Figures 10–1 and 10–2. The iron-cored transformer is satisfactory for use only at power and audio frequencies. (High-quality, alloy-cored transformers designed for high-fidelity sound reproduction are usable up to approximately 100 kc.)

At radio frequencies, transformers must either be air-cored or must utilize special core materials suitable for operation at these high frequencies. Such materials include powdered iron, permalloy dust, and ferrites. Cores for some special, low radio-frequency (r-f) applications are made from *magnetostrictive* metals (for example, Monel metal, nickel, Nichrome, Invar, etc.). Magnetostriction is the name of a property which is evidenced by an expansion or contraction of the material when it is magnetized, and conversely by a permeability change when the material is stretched or compressed.

Radio-frequency transformers are wound with zigzagged turns (in the multilayered types) or with spaced turns (in the single-layer types) to reduce distributed capacitance.

10.17 PEAKING TRANSFORMER

The *peaking transformer* is a special unit used in certain industrial electronic systems in which a wave form consisting of sharp, narrow spikes or peaks is required. The core of the peaking transformer contains very little iron. Indeed the core is so thin that it would be called skimpy if the transformer were to be used for conventional purposes.

Because there is so little iron, the core saturates easily. When the primary winding of the peaking transformer is operated from a sine-wave a-c voltage, saturation occurs shortly after the primary current wave has left zero. The flux pattern therefore is flat-topped and shows no change in amplitude over a large part of each half-cycle. The only *change* in flux takes place during the short time interval when the current is reversing, and this rapid change causes the peaks or spikes in the secondary voltage wave form.

SUMMARY OF CHAPTER 10

A **Transformer** consists basically of a primary coil and secondary coil closely coupled. Alternating current flowing through the primary induces an alternating emf across the secondary. Air-core transformers are efficient only at radio frequencies. Iron-cored transformers are employed at power and audio frequencies.

Transformer action depends upon **Induction.** Its extent is determined by the **Mutual Inductance,** M, between primary and secondary windings. Coupling is *close* or *tight* when M is high and *loose* when M is low. The **Coefficient of Coupling,** $k = M/\sqrt{L_1 L_2}$. The maximum possible value of k is 1.

The **Secondary Voltage,** $E_s = E_p(N_s/N_p)$, where E_p is the primary voltage, N_s the number of secondary turns, and N_p the number of primary turns. The factor N_s/N_p is called the **Turns Ratio** of the transformer. N_s is higher than N_p in a **Step-up** transformer; N_s is lower than N_p in a **Step-down** transformer.

Primary Current $I_p = I_s(N_s/N_p)$, where I_s is the *secondary current,* and N_s/N_p is the turns ratio.

The **Primary Volt-Amperes** (E_pI_p) equals the **Secondary Volt-Amperes** (E_sI_s). In an ideal transformer, the **Secondary** (or **Output**) **Power** (P_o) also would equal the **Primary** (or **Input**) **Power** (P_i). Because of losses, however, this is not true in a practical transformer, P_o always being somewhat less than P_i. Thus, **Transformer Efficiency,** n, (in per cent $= 100(P_o/P_i)$.

Impedance Matching is essential in all circuits, since maximum power is delivered to a load only when the impedance of the load equals the internal impedance of the generator. When this condition is fulfilled, the generator and load impedances are *matched.* When the load impedance (Z_L) has some value other than that of the generator (Z_g), a transformer must be inserted between generator and load to match impedances. The **Impedance Ratio** provided by the transformer equals the square of the turns ratio, $Z_s/Z_p = (N_s/N_p)^2$.

The **Saturable Reactor** utilizes the ability of a current, usually direct current, flowing through one winding to saturate the core and thus to vary the inductance and reactance of a second winding. The **Magnetic Amplifier** is a special adaptation of the saturable reactor in which a large alternating current flowing in one winding and through a load is controlled by means of a much smaller current flowing in the control winding.

QUESTIONS AND EXERCISES

1. How does a core intensify transformer action?
2. Define: (*a*) loose coupling; (*b*) tight coupling.
3. The primary coil of a certain transformer has a self-inductance of 100 mh. The secondary coil has an inductance of ½ h. The mutual inductance between the two windings is 1/10 h. What is the coefficient of coupling?
4. A transformer primary has 58 turns; the secondary 485 turns. What is the secondary voltage when 2¼ v are applied to the primary?
5. In Example 4 above, what is the maximum secondary current when the primary current is ¼ amp?
6. If the primary voltage = 100 v and the primary current = 2 amp, what is the turns ratio of a transformer when the secondary voltage = 25 v?
7. What turns ratio is needed in a transformer to match 100 ohms primary impedance to 3 ohms secondary impedance?
8. What impedance ratio is provided by a transformer having a 5:1 turns ratio?
9. What is the efficiency of a 2:1 step-up transformer in which the output current is 10 per cent less than one-quarter of the input current?
10. Explain *leakage reactance* in transformers.

Chapter **11**

Generators and Motors

11.1 RUDIMENTARY A-C GENERATOR

The elementary a-c generator consists of a single loop of wire rotating in a magnetic field between two pole pieces and provided with a pair of slip rings. Two brushes collect the alternating current at the rings.

The operation of a basic generator of this type is described in Sections 6.3 and 6.4, Chapter 6. The reader is referred to those sections for a review of the fundamental principles.

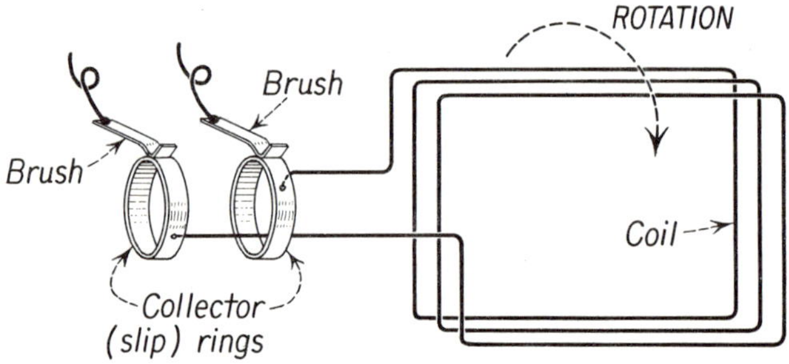

Fig. 11–1 Multiturn Armature Coil

11.2 PRACTICAL A-C GENERATORS

While the simple single-coil generator serves to demonstrate the production of alternating current by an electromagnetic machine, it has little practical value because its output is so low.

A practical generator employs many turns of wire in its armature, as illustrated in Figure 11–1, to increase the total voltage. (The voltage is the sum of the voltages developed by the individual turns.) Also, these turns or loops are wound around a laminated iron core to concentrate the magnetic flux. The core is attached to the rotating shaft. Collectively, the armature consists of core, turns of wire, and slip rings.

An a-c generator properly is called an *alternator.* Figure 11–2 shows details of a two-pole *magneto-type* alternator. Figure 11–2(A) shows the shape of the armature core and general arrangement of the

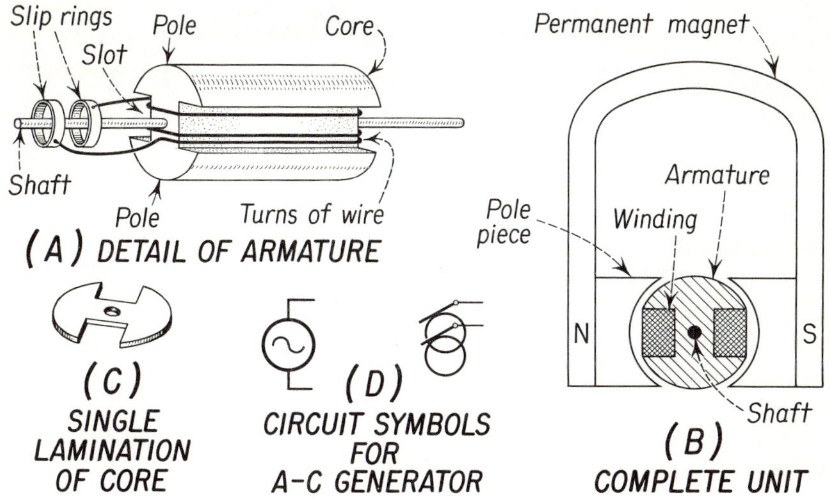

Fig. 11–2 Simple 2-Pole Magneto-Type Alternating-Current Generator

armature employed in this type of machine. While only three turns are shown in this figure, the core slots actually are wound full of insulated wire, as may be seen in the armature cross section in Figure 11–2(B).

Fig. 11–3 Four-Pole Armature

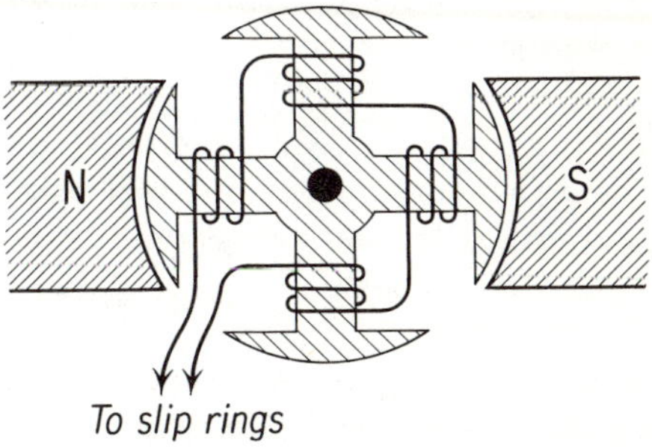

A permanent magnet supplies the field in this type of alternator and is responsible for the name *magneto.* The armature is mounted so as to rotate between curved pole pieces attached to the poles of this magnet. [See Figure 11–2(B).] The air gap between the pole pieces and the armature is very small. The frequency of the generated emf increases with the speed of rotation.

For increased voltage, more than two poles may be cut in the armature. A coil is wound around each pole, as shown in Figure 11–3, and the coils are connected together in series-aiding.

The magneto-type of alternator is employed only for low power output and in small machines. Its size is restricted by the inconvenience of large permanent magnets. The electromagnet is a more feasible field member for machines of practical size. Figure 11–4 shows one way of

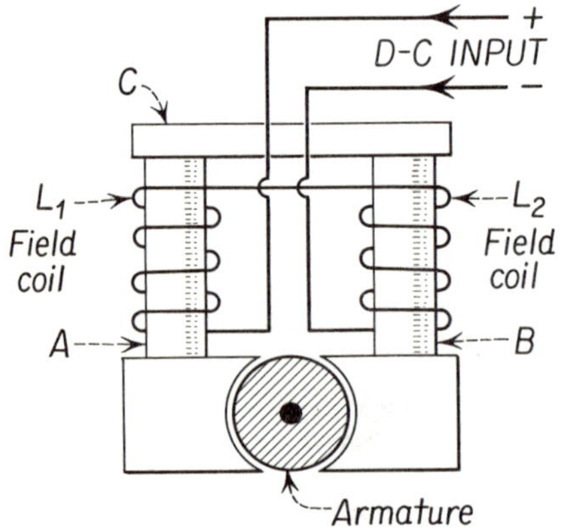

Fig. 11–4 Separately-Excited Alternator

employing electromagnetism. Here, the magnetic field is supplied by an electromagnet consisting of Poles *A* and *B*, Yoke *C*, and Coils L_1 and L_2. The two coils are connected together in series-aiding and to a d-c source, termed the *exciting source.* The technical term for the stationary member of the alternator (collectively, the field coils and poles in Figure 11–4) is the *stator,* and for the rotating member (the armature in Figure 11–4) the *rotor.*

Stationary fields and rotating armatures are employed only in low-powered alternators. An important disadvantage of this arrangement is the requirement that the brushes and slip rings handle high voltages in commercial machines. This often introduces a sparking at the brushes and calls for exceptionally good insulation in the armature assembly. The use of a rotating magnetic field is more feasible in large-size, high-voltage machines. This allows simple connections, without moving parts, to the high-voltage stator, while the slip rings and brushes are confined to the low-voltage d-c excitation circuit of the rotor.

Figure 11–5 shows the basic construction of a rotating-field alternator. The rotor contains a number of poles (cores) around its circumference, each being wound with a coil of wire. The coils are connected in series and to the slip rings. The brushes are connected to a source of direct current through the *field rheostat* by means of which the direct current through the rotor coils may be adjusted. Each rotor pole therefore is magnetized.

The stator contains a number of poles, in this instance matching the number in the rotor. Each pole is wound with a coil of wire and the coils are connected in series with each other and to the a-c output terminals. The rotor is mounted inside the stator ring with a small air gap between the rotor and stator poles.

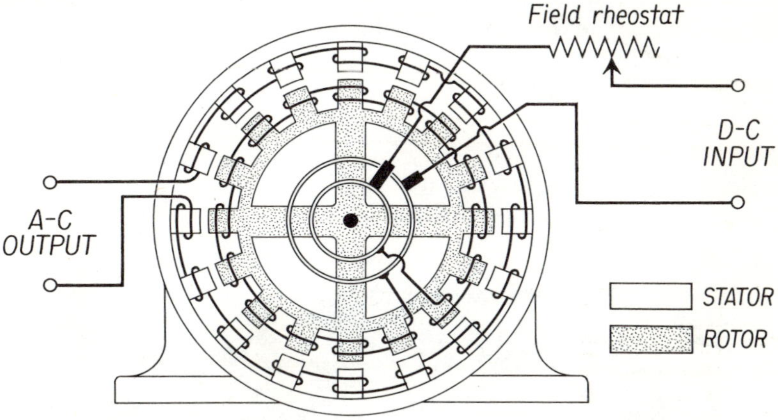

Fig. 11–5 Basic Construction of Rotating-Field Alternator

When the rotor revolves, its magnetized poles move past the stator poles. The lines of force surrounding the rotor poles cut the turns of the stator coils and thus generate an emf in the stator winding. The emf reverses polarity each time a rotor pole reaches the point halfway between two stator poles. The strength of this induced emf is controllable to some extent by adjustment of the field rheostat.

Since the emf reverses polarity at each instant the rotor poles are midway between stator poles, the frequency (f in cycles per second) equals the rotational speed [n in revolutions per second (rps)] multiplied by one-half the number of poles (p).

(11–1) $$f = (np)/2$$

Example: An 8-pole alternator rotates at a speed of 900 rpm. What is its frequency?

$$n = 900/60 = 15 \text{ rps}$$

$$f = (np)/2 = (15 \times 8)/2 = 120/2 = 60 \text{ cps}$$

When speed is expressed in revolutions per minute (rpm)

(11–2) $$f = (np)/120$$

From Equation (11–1), it is seen that an advantage of the multipolar generator is its ability to generate a given frequency at lower rotational speeds. Relatively slow-moving generators, such as those turned by water wheels, thus are able to generate commercial frequencies while spinning at low rates. For example, a two-pole generator would have to rotate at 3,600 rpm to generate the commercial frequency of 60 cps, while a 32-pole machine can accomplish this at only 225 rpm.

The rotor of an alternator must be turned mechanically. Depending upon the size of the alternator, this force may be (1) manual, (2) wind, (3) water, (4) steam, (5) electricity, (6) internal combustion engine, or (7) mercury or semisolids (in atomic generators).

Direct current for the field winding occasionally is supplied by storage batteries. More often, however, it is furnished by a small d-c generator (*exciter*) driven by the alternator shaft. The direct current for some alternators is obtained from the same windings that produce the alternating current.

Plate 11–1 Outdoor-Type Hydroelectric Generator in the Lower Owens Gorge Generating Plant in California. (Courtesy, City of Los Angeles Department of Water and Power)

11.3 RUDIMENTARY D-C GENERATOR

Figure 11–6 shows one of the earliest mechanical generators of direct current, the *disc dynamo* of Michael Faraday. This device is found today only in school science laboratories.

In the Faraday dynamo, which was the forerunner of modern d-c generators, a copper disc is rotated between the poles of a permanent magnet. Free electrons in the disc move with the disc, and this electron motion sets up a magnetic field which strengthens the field in which the disc rotates.

Fig. 11–6 Faraday's Disc Dynamo

The intensification is focused in the region between the electrons and the rim of the disc and pushes the electrons toward the center of the disc. This causes the edge of the disc to acquire a positive charge (deficiency of electrons) and the center a negative charge. A small wire (brush) making sliding contact with the edge of the spinning disc and a similar wire contacting the center will deliver a d-c voltage, and electrons will flow through an external closed circuit connected to these contacts.

11.4 FUNCTION OF COMMUTATOR IN D-C GENERATOR

The simple loop rotating in a uniform magnetic field [Figure 6–2(B), Chapter 6] or a simple coil rotating in the field (Figure 11–1) can produce only an alternating emf. The reason for this is that each half of the loop moves past a north pole and south pole in succession, the polarity of the emf reversing as the poles are exchanged.

A d-c voltage can be produced only if some means is provided for automatically switching one output terminal always to that half of the loop passing the north pole, and the other output terminal always to the half passing the south pole. Such an automatic switching mechanism, operated by the rotating armature shaft, is shown in Figure 11–7(A). Instead of the pair of slip rings employed in the a-c generator, a single ring has been split in half. This split ring is called a *commutator,* and each of its halves is called a *bar, section,* or *segment.* As the shaft rotates, the upper brush is always in connection, through the uppermost commutator bar, to the portion of the loop passing the north pole, while the lower brush is connected, through the lowermost commutator bar, to the portion passing the south pole.

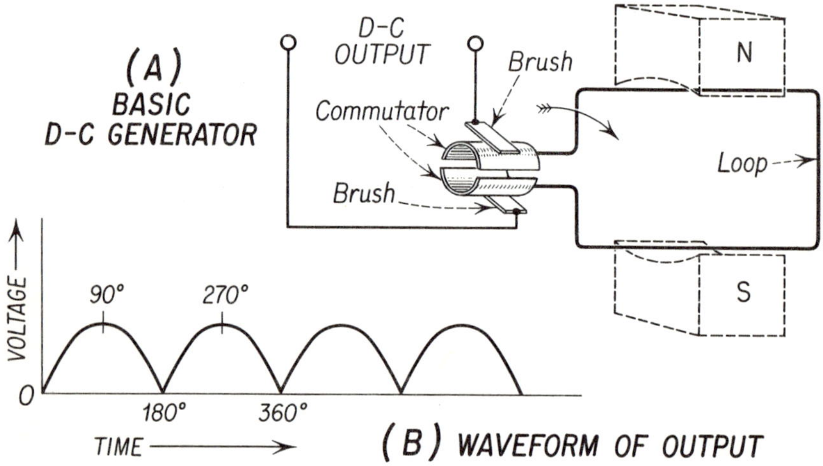

Fig. 11–7 Action of Commutator in Direct-Current Generator

As the loops of the coil pass through points of different magnetic intensity from zero to maximum, the generated emf increases from zero to maximum. During a 360-deg rotation (one complete revolution), the voltage describes two complete humps, as shown in Figure 11–7(B). Thus, the output of the d-c generator is not a steady direct current, like the output of a battery, but rises and falls periodically between zero and maximum—*but never reversing polarity.* This is called a *pulsating direct current.* The wavelike pattern is termed *commutator ripple.* The faster the armature revolves, the higher is the frequency of this ripple.

The single-coil armature serves only to demonstrate the principle of the d-c generator and commutation. Very little power could be produced with such an arrangement, and the ripple [as shown in Figure 11–7(B)] would be rather severe. Instead of a single loop, a coil of many turns might be used. However, this would serve only to raise the voltage. The ripple would not be improved.

Figure 11–8(A) shows how coils and commutator bars are added to the armature. The commercial armature is wound, as a number of separate coils, on an iron or steel core. The purpose of the core is to concentrate the magnetic flux as well as to provide mechanical support for the coils. For simplicity and clarity, the core is shown as a ring in Figure 11–8(A), but the coils of commercial armatures are wound in the slots of a *drum-type* core [Figure 8–11(B)] made by stacking together a number of disc-type laminations of the shape shown.

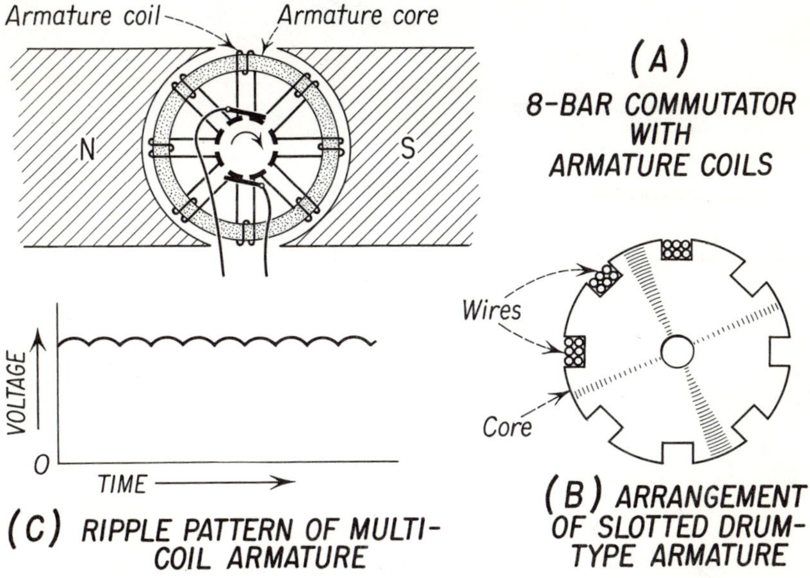

Fig. 11–8 Multisection Commutation

Figure 11–8(A) shows that the armature coils are connected to adjacent commutator bars, each bar being common to two coils. Some coil is always at the point in the magnetic field where the induced emf is maximum, and some other coil at zero or minimum. Still other coils are at intermediate voltage points. The instantaneous emf appearing at the brushes is the sum of the voltages in the separate coils, since all of the coils are connected in series between the two brushes. The result is a nearly steady direct current, resembling the wave shape in Figure 11–8(C). The ripple intensity decreases as the number of coils and commutator bars increases. (The ripple frequency increases.)

With a multiloop armature, many coils per unit time pass through the point of maximum magnetic intensity. The emf accordingly does not have the opportunity to fall completely to zero at the end of each pulsation, as in Figure 11–7(B), but drops only to a slightly lower value as the next coil comes into the maximum field. [See Figure 11–8(C).] This results in a lower-amplitude ripple and a higher average emf.

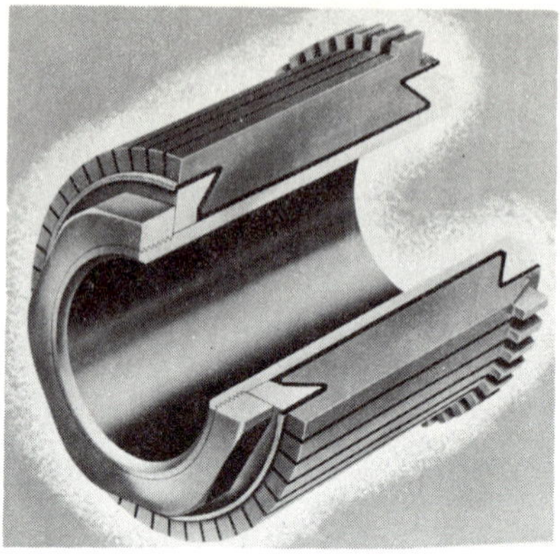

Plate 11–2 Cutaway View of Multibar Commutator. The copper segments are spaced apart by insulating separators. (Courtesy, Nippert Electric Products Co.)

11.5 SERIES-WOUND D-C GENERATOR

In the previous examples, it has been assumed that the magnetic field was produced by a permanent magnet. This is not true in practical machines.

The fields are self-contained and electromagnetic in d-c generators. Figure 11–9 illustrates one type of machine in which the field coils are

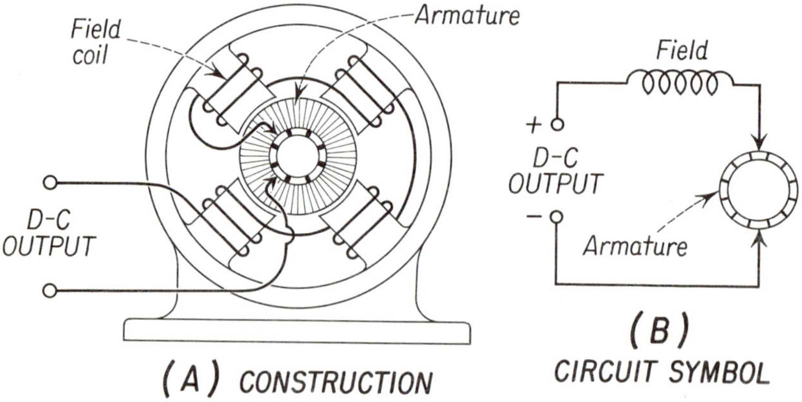

Fig. 11–9 Series-Wound Direct-Current Generator

connected in series with the armature. In this *series-wound* d-c generator, the field coils are connected in series with each other and in series with the armature and output terminals. Figure 11–9(A) shows four field poles for illustration, but various machines employ different numbers of poles.

Under open-circuit conditions (that is, when no load is connected to the d-c output terminals), the output voltage of this generator is quite low and is due to the effect of residual magnetism in the structure. When an external load is connected, the load current flows through the field and armature coils, and the magnetic field is proportional to this current. As the load resistance is lowered, the current and voltage increase over a certain range, and then the voltage decreases as magnetic saturation is reached.

Series-wound d-c generators are not widely used at present. They once were employed extensively in series arc lighting, now obsolete.

11.6 SHUNT-WOUND D-C GENERATOR

Figure 11–10 shows the circuit of a *shunt-wound* d-c generator. As the name implies, the field coil and armature are connected in parallel, or shunt in this type of machine.

When the shunt-wound generator first is started up, the slight residual magnetism in the field poles causes a current to be induced in the armature windings. A portion of this current flows through the field coil and increases the magnetism. This, in turn, causes a larger

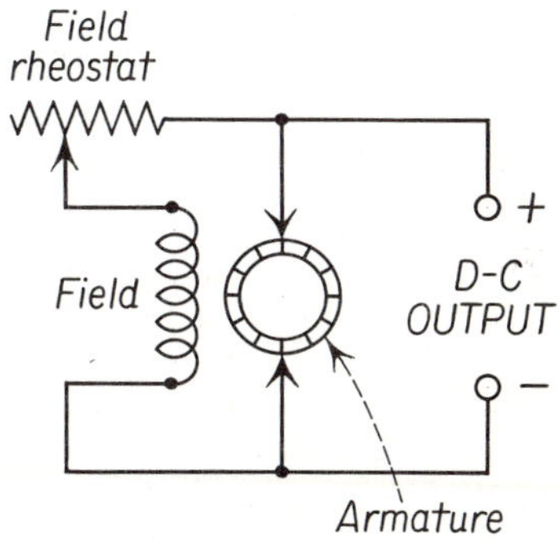

Fig. 11–10 Shunt-Wound Direct-Current Generator

current to be induced in the armature and a larger current to flow through the field coils. This action causes a build-up of the voltage until the point of magnetic saturation is reached. This point, or somewhat below it, represents the normal output voltage of the generator.

The *field rheostat* serves for adjustment of the field current to desired levels and, by its consequent control of the magnetization level, acts as an output-voltage control. The field (*exciting*) current is a small percentage of the total generator current.

The shunt-wound generator has relatively poor *voltage regulation.* By this is meant that its terminal voltage drops rather sharply as the load current is increased. This characteristic has caused the shunt-wound generator to be superseded in many applications by the *compound-wound* machine.

11.7 COMPOUND-WOUND D-C GENERATOR

Figure 11–11 shows the circuit of a *compound-wound* d-c generator. Both series and shunt fields are employed in this type of machine. Increased load current boosts the magnetization due to the series winding, since this current must pass through the series coils. This, in turn, raises the generated voltage which would tend to slump if the shunt field alone were used. This improved voltage regulation has brought the compound-wound generator into widespread use.

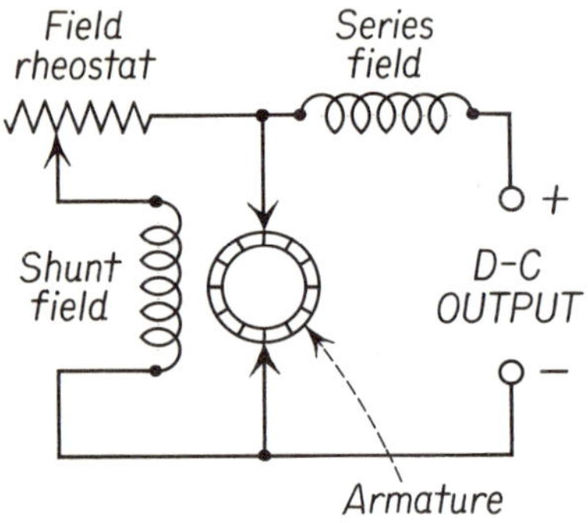

Fig. 11–11 Compound-Wound Direct-Current Generator

The series and shunt field coils are wound on the same poles. (See Figure 11–12.) When the sizes of the coils are so chosen with respect to each other that the voltage is maintained constant for all current loads within a rated range of values, the generator is said to be *flat-compounded.* In some applications, particularly where line voltage drop is to be overcome in order to stabilize the voltage at a point remote from the generator, the series coils may be proportioned so that the full-load voltage exceeds the no-load voltage. The generator then is said to be

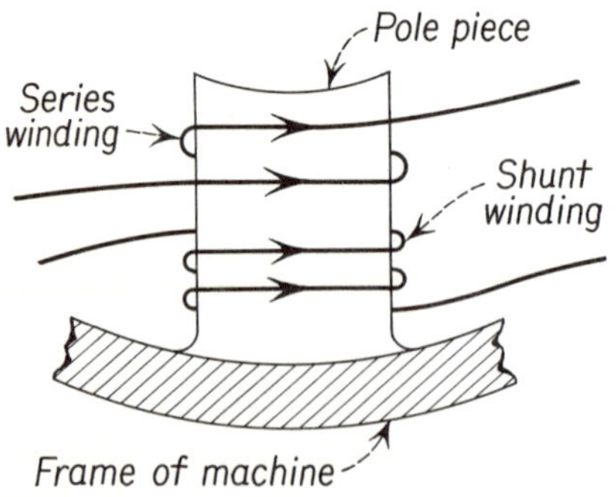

Fig. 11–12 Compound Field Winding

overcompensated. In an *undercompensated* machine, the series-field magnetization is too weak to maintain the output voltage constant under varying current loads.

The series field usually is wound in the same direction as the shunt field, as shown in Figure 11–12, in order to aid the latter. This arrangement is termed a *cumulative compound* winding. In a *differential compound* winding, the two fields are wound in opposite directions and oppose each other. As the load current increases, in this case, the series-field magnetization increases but counteracts the shunt-field magnetization. This eventually changes the direction of the magnetization and reverses the polarity of the generated emf.

The amount of compounding may be adjusted by means of a rheostat connected in parallel with the series coil. Compounding is maximum when the series-field current is highest.

Placement of Brushes. The point at which the brushes make contact with the commutator is very important. Figure 11–13 shows how the tip of a brush rests on the commutator.

The brush contacts a single bar only when it is directly over the center of the bar. At other times, it contacts two bars simultaneously. This latter condition is shown in Figure 11–13 where the brush is seen

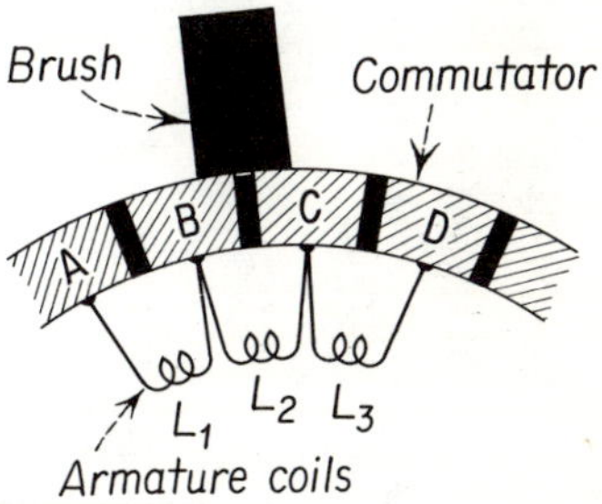

Fig. 11–13 Brush and Commutator Relationship

contacting bars *B* and *C*. In this position of the commutator, the brush thus short-circuits armature coil L_2. As the armature rotates in a clockwise direction, Bar *C* will move away from under the brush, and Bar *B* will be in contact. Later, the brush will straddle *A* and *B* and short-circuit coil L_1.

If a coil has an induced voltage (owing to its cutting of lines of force) at the time the brush makes contact with the commutator bars connected to this coil, short-circuit current will flow through the brush. In Figure 11–13, for example, short-circuit current will flow from L_2 to *C*, through the brush, to *B*, and back to L_2 if the coil is cutting lines of force. If the induced voltage is high, the current may be sufficient to damage the coil and brush.

When a commutator bar moves from under the brush (as when *C* moves clockwise), a spark will accompany the breaking of contact between the brush and bar if the coil concerned has a high induced voltage. This sparking eventually will damage the commutator bars and brushes.

In order to prevent these two bad effects—coil short-circuiting and commutator sparking—the brushes must be so oriented on the commutator that coils are switched when they are passing through zero or minimum field. This point of zero field, under actual operating conditions, is called *machine neutral.*

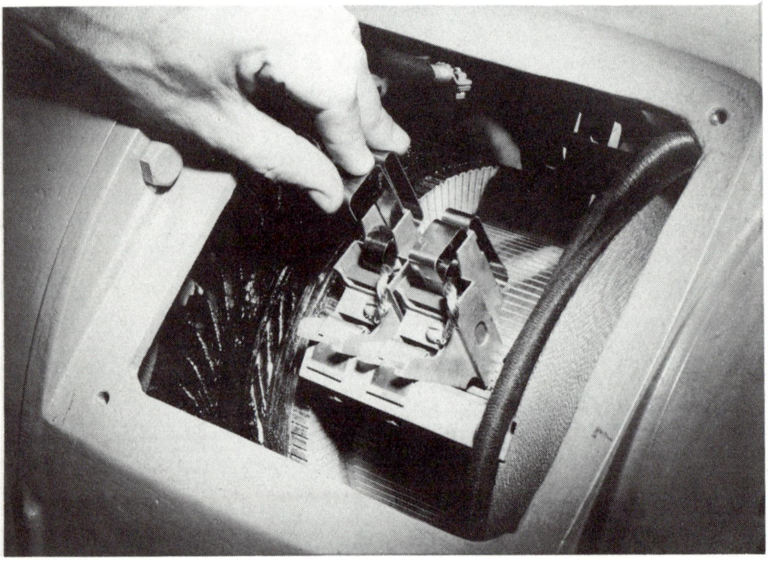

Plate 11–3 Closeup of Commutator, Brush Holder, and Armature of Direct-Current Motor. Spring is being disengaged by hand for replacement of brush. (Courtesy, General Electric Co.)

Actually, machine neutral does not coincide with the *neutral axis* of the generator. The neutral axis is a line running perpendicular to the lines of force and halfway between facing field poles. But the armature coils are cutting no lines at this position only when no current is being drawn from the generator. When current flows, the magnetization of the armature core causes a distortion of the field, so that the coils cut lines at the neutral axis. The point at which no lines are cut is slightly ahead of the neutral axis and is the machine neutral.

Thus, in Figure 11–14, the neutral axis is the line *ab*, but lines are cut here when current is drawn, owing to distortion of the field by the armature. The brushes therefore are placed 180 deg apart along the axis *cd* (the machine neutral) which, for clockwise rotation, is *ahead* of the neutral axis. Here, no lines (or very few lines) are cut, and there is no sparking damage or short-circuit damage.

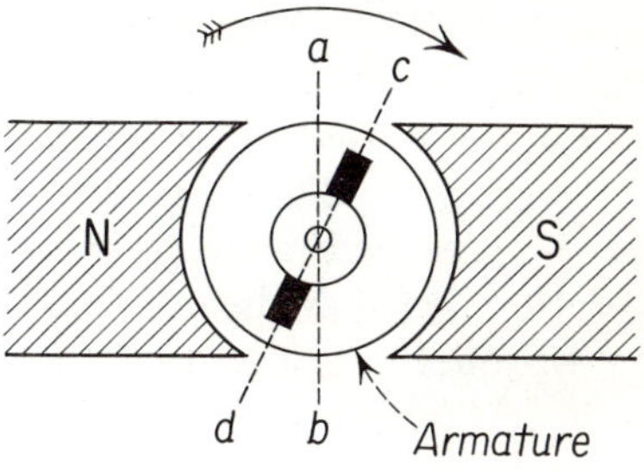

Fig. 11–14 Brush Placement in Direct-Current Generator

Voltage Regulation. Terminal voltage in the ideal theoretical generator remains constant whether full rated current or no current is drawn from the machine. *Voltage regulation* (VR) is a figure (expressed as a percentage) which shows the extent to which the voltage falls between no-load and full-load current drains.

(11–3) $$\mathrm{VR} = 100(E_1 - E_2)/E_2$$

where VR = voltage regulation (in per cent)
E_1 = no-load voltage (in volts)
E_2 = full-load voltage (in volts)

Example: A generator has an open-circuit terminal voltage of 100 v. This emf drops to 95 v under full current load. What is the voltage regulation of the generator?

$$\mathrm{VR} = 100(100 - 95)/95 = (100 \times 5)/95 = 500/95 = 5.26\%$$

11.8 VOLTAGE CONTROL IN D-C GENERATORS

Two general methods are available for varying the output voltage of a d-c generator: (1) changing the speed of the prime mover—water wheel, steam engine, gas engine, etc.—that turns the generator, and (2) varying the field magnetization. The latter is accomplished by means of a rheostat in series with the field winding. Decreasing the resistance raises the field current and increases the magnetization.

The control of field current is not practical in series-wound generators, since the field current is also the load current here. In a compound-wound machine, the rheostat would be connected in series with the shunt field.

11.9 SERIES-WOUND D-C MOTOR

When current is passed through a wire placed in a uniform magnetic field, the reaction between the uniform field and the field set up around the wire by the current flow will cause the wire to move. This effect is the basis of d-c motor operation.

Figure 11–15 illustrates the phenomenon. The uniform magnetic field is shown in Figure 11–15(A), and the field surrounding the current-carrying wire in Figure 11–15(B). When the wire is placed in the field, the latter becomes distorted in the manner shown in Figure 11–15(C). The reason for this is that the circular lines around the wire [Figure 11–15(B)] flow in one direction above the wire and in the other direction

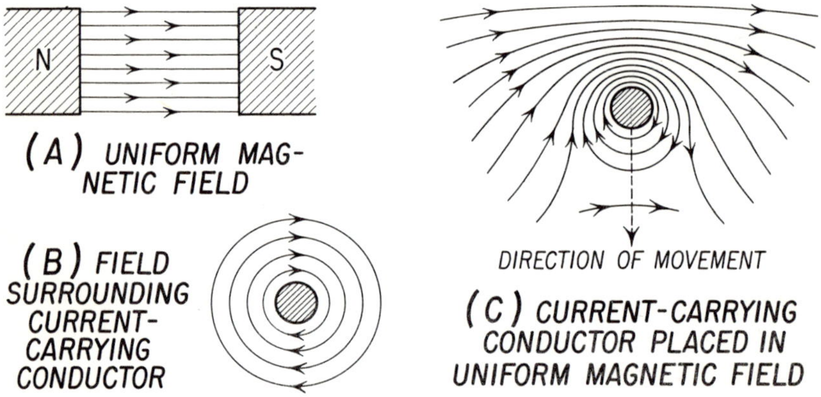

Fig. 11–15 Motion of Current-Carrying Conductor in Uniform Magnetic Field

below it. When the wire is placed in the uniform field; the lines above the wire are in the same direction as the lines of the uniform field and therefore strengthen the latter, while the lines below the wire are in the opposite direction and tend to cancel the lines in this region. The result is the crowding of lines above and thinning of lines below the wire, pushing the wire downward.

How this principle is utilized in the motor is depicted in Figure 11–16. The conductor has been bent into a single-turn loop or coil in Figure 11–16(A) and mounted on a shaft to rotate between curved magnetic pole pieces. A commutator and brushes are provided so that, as the coil rotates, one side of the loop always is connected to the positive terminal of Battery *B*, and the other side to the negative terminal. Current flowing through the coil is controllable by Rheostat *R*. Figure 11–16(B) shows how current through the coil causes crowding of lines of force above the positive half of the coil and below the negative half. [Compare with Figure 11–15(C).] This crowding of lines imparts a

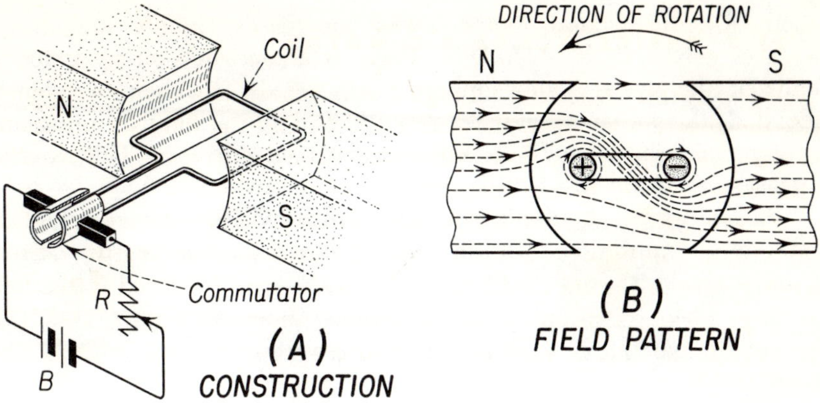

Fig. 11–16 Direct-Current Motor with Single-Loop Armature

torque (turning force) which causes the coil to rotate counterclockwise. The rotational speed is governed by the current through the coil and is controllable by means of Rheostat *R*.

The construction in Figure 11–16(A) is seen to bear a striking resemblance to the single-coil d-c generator [Figure 11–7(A)]. As in the generator, the best performance cannot be obtained with a single coil. For better efficiency, a slotted, drum-type armature [Figure 11–8(B)] containing a number of coils is used. Modern motors also contain more than two field poles in many instances. Construction of motor and generator are similar.

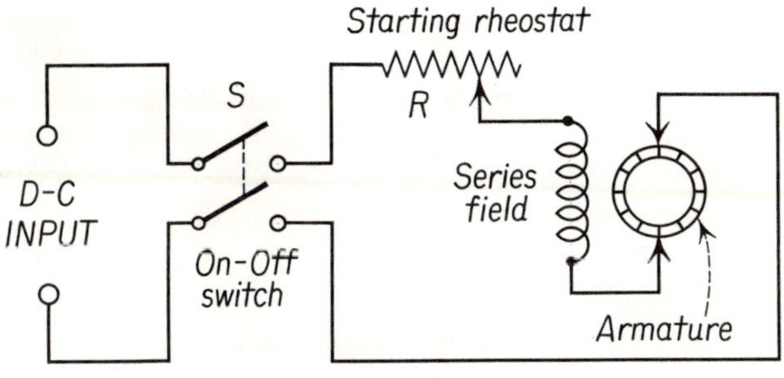

Fig. 11–17 Series-Wound Direct-Current Motor

As in generators, the armature and field of a d-c motor may be connected in series. Figure 11–17 shows the circuit of a series-wound d-c motor. The rheostat is advisable for starting the motor, since excessive current may be drawn when the armature is at rest. At the start, all of the resistance of the rheostat is in the circuit. As the resistance then gradually is cut out, the motor gains speed. At full speed, all of the resistance may be cut out since the running current is not excessive.

In a series-wound motor, speed decreases rapidly and torque increases as the shaft is loaded. There is a tendency for an unloaded motor of this type to develop excessive speeds. (When the shaft is unloaded, the motor accelerates rapidly, and the armature builds up a counter emf which gradually reduces the field current. This, in turn, increases the speed still further, and the motor soon runs away.)

In motors, as in d-c generators, brush position is important for sparkless commutation. In the motor, however, the brushes are set *behind* the neutral axis.

11.10 SHUNT-WOUND D-C MOTOR

Figure 11–18 shows the circuit of a shunt-wound motor. Note the similarity to the shunt-wound d-c generator.

The starting rheostat, *R*, initially is set to its high-resistance point. When the Switch, S, is closed, full current flows through the shunt field winding, and the field magnetization is maximum. The armature cannot

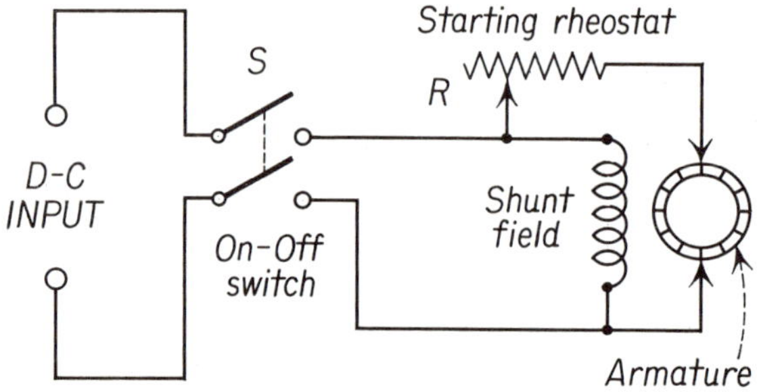

Fig. 11–18 Shunt-Wound Direct-Current Motor

receive full current, however, until the rheostat has been advanced to its lowest-resistance setting. This allows the full torque of the motor to be obtained at starting.

The speed of a shunt-wound motor is very nearly constant from no load to full load. The torque increases with speed approximately in direct proportion.

11.11 COMPOUND-WOUND D-C MOTOR

Like the compound-wound generator, the compound-wound motor utilizes the separate advantages of series and shunt field windings. Two types of compound winding are shown in Figure 11–19.

When the series and shunt fields are wound in the directions shown in Figure 11–19(A), the ampere-turns of the series winding aid those of the shunt winding. This is the *cumulative compound* motor. This motor has the high starting torque of a series motor but cannot run away in speed because the field is sustained by the shunt winding.

When the series and shunt fields are wound in the directions shown in Figure 11–19(B), the ampere-turns of the series winding buck, or oppose, those of the shunt winding. This is the *differential compound*

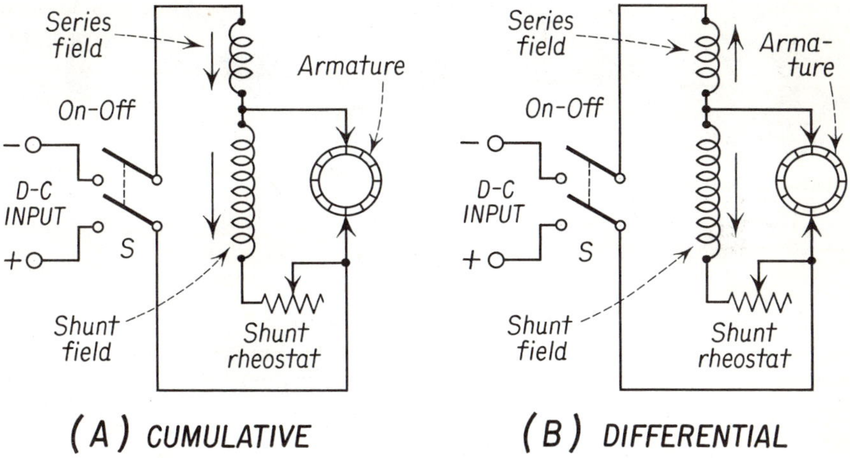

Fig. 11–19 Compound-Wound Direct-Current Motors

motor. When a load is applied to the shaft of this motor, current increases through the series winding and boosts the latter's field strength. This increase, bucking the shunt field, decreases the net field strength and permits increased armature current. The latter, in turn, produces more torque to compensate for the load. The result is constant speed under varying load.

11.12 INDUCTION A-C MOTOR

An *induction* motor consists basically of a stationary field coil wound around poles inside the frame of the machine, and a revolving rotor. The rotor resembles somewhat the armatures in motors of other types. The field and rotor sometimes are referred to as *primary* and *secondary*, or as *stator* and *armature*.

In this type of motor, the magnetic field is caused to rotate electrically by the applied alternating current, and it pulls the rotor around with it. There is no electrical connection between the field coils and the rotor. Induction motors are widely used because of their simplicity and economy.

Figure 11–20 serves to illustrate the principle of the induction motor. The rotor is a simple bar of iron here and is mounted on a shaft equidistant between north and south magnetic poles. The poles rotate clockwise, and the rotor keeps aligned with them. In Figure 11–20(A), the axis of the poles is vertical, so the rotor assumes the vertical position.

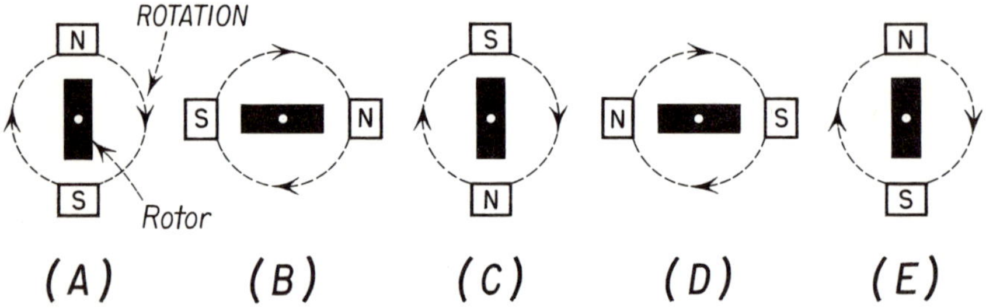

Fig. 11–20 Representation of Rotating Field and Rotor

At (*B*), the poles and rotor have turned through 90 deg to the horizontal position. At (*C*), they are vertical but upside down with respect to the starting position. At (*D*), three-quarters of a revolution (a rotation of 270 deg) has been completed; and at (*E*), a full rotation has been made. While in this illustration the poles themselves move around the circle,

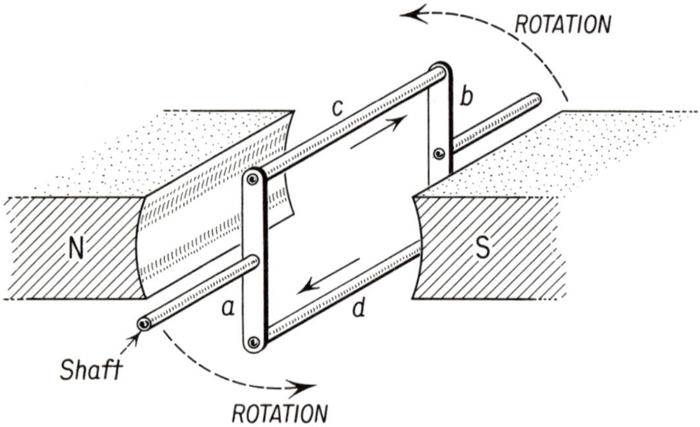

Fig. 11–21 Principle of Induction Motor

the actual physical poles do not move in an induction motor; they remain stationary, but their field is rotated electrically. How this is accomplished is explained later in the text.

Figure 11–21 illustrates the *basic* arrangement of the rotor of the induction motor. Here, *c* and *d* are straight rod- or bar-type copper conductors, while *a* and *b* are heavy short-circuiting end bars. The magnetic poles, *N* and S, rotate in a counterclockwise direction, as indicated by the curved arrows. Actually, as mentioned before, it is not the pole pieces that are rotating but the magnetic field itself. As the field rotates, its lines cut conductors *c* and *d* and induce a current shown by the straight arrows. This current sets up a magnetic field around the conductor. With current flowing in the direction shown, a north pole is induced on the conductor adjacent to the rotating south pole, and a south pole on the conductor adjacent to the rotating north pole. The rotor accordingly follows the revolving field because the latter attracts its poles.

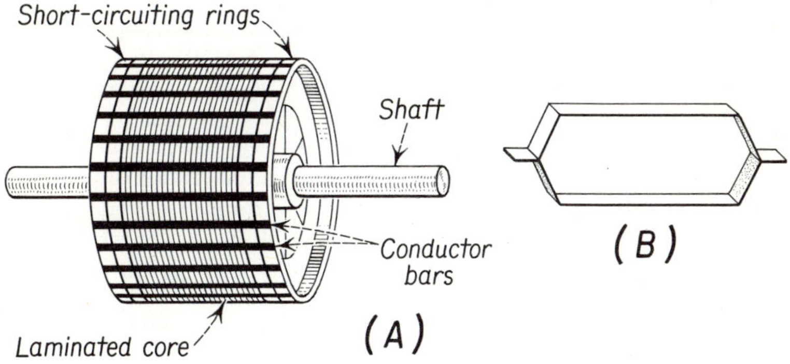

Fig. 11–22 Construction of Squirrel-Cage Rotor

The rotor of a commercial induction motor employs many more conductors than the two shown in Figure 11–21. One familiar type of rotor is built on a laminated, cylindrical iron core containing many slots around its periphery and resembling the core of a drum-type armature. Straight rods or bars of copper are pressed into the slots, and their ends are connected to short-circuiting rings. The bars usually are not insulated, since their resistance is much lower than that of the surrounding core material. This type of construction is shown in Figure 11–22(A). Because this arrangement resembles the exercise wheel found in pet cages, it is called a *squirrel-cage rotor.*

Instead of copper bars, numerous flat coils made from copper strip sometimes are used in squirrel-cage rotors. One such coil, cut and stretched out from a single copper strip, is shown in Figure 11–22(B). Since such a loop is a continuous coil, no short-circuiting rings are required in a rotor using these coils.

Since the field flux must cut across the conductors in the rotor in order to induce currents therein, the field must travel faster than the rotor is turning. If the rotor rotation were *synchronous* with that of the field (that is, at the same speed), there would be no inducing of current and therefore no induction motor action. The difference between synchronous speed and actual rotor speed is called the *slip* of the motor. Slip (s) is expressed in per cent. Thus

$$s = 100(S_s - S_r)/S_s \tag{11-4}$$

where s = slip (in per cent)
S_r = rotor speed
S_s = synchronous speed

Example: The synchronous speed of a 60-cps induction motor would be 1,800 rpm. (Synchronous speed S_s = 60 times frequency/number of *pairs* of poles.) The rotor speed is 1,750 rpm. What is the slip?

$$s = 100(1{,}800 - 1{,}750)/1{,}800 = 100(50)/1{,}800$$
$$= 5{,}000/1{,}800 = 2.78\%$$

When a simple induction motor is connected to a common *single-phase* a-c line, it is unable to start because the field is not rotating. The motor may be started, however, by giving its shaft an initial twist. This causes the rotor conductors to cut lines of force and to set up their own

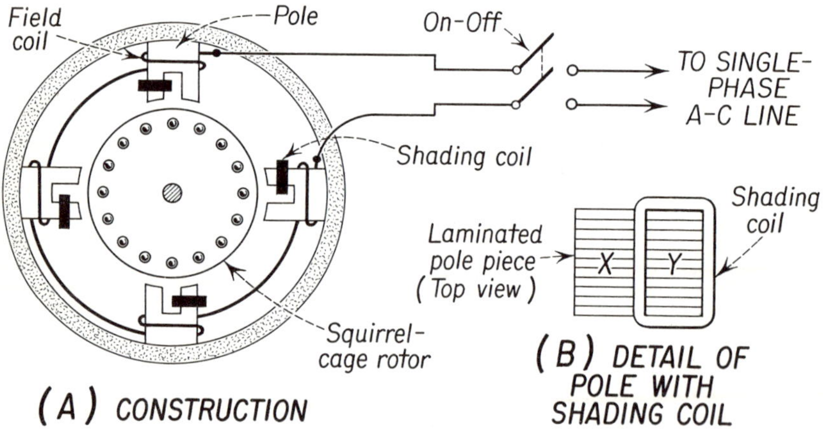

Fig. 11–23 Shaded-Pole Induction Motor

field which then interacts with the regular field to maintain rotation. Starting a motor in this way is inconvenient and not very practical in most instances. Several electrical methods of self-starting accordingly have been devised.

Figure 11–23 shows one type of self-starting, single-phase induction motor. In this type, a short-circuited copper-bar ring, called a *shading coil,* encircles a portion of the tip of each field pole [Region *Y* in Figure 11–23(B)]. Alternating flux in the field pole induces a current in the shading coil. This current opposes the build-up of flux in Region *Y* of the pole. (See Lenz's law, Section 8.3, Chapter 8.) Flux in Region *X*, the unshaded part of the pole, thus builds up ahead of that in *Y*. This momentary flux shift approximates a rotating field and provides the torque that starts the rotor turning. Once the rotor is under rotation, the cutting of lines of force by its conductors provides true induction-motor action.

The shading-coil method of starting is not practical for large motors. It is to be found principally in light-duty appliance motors, such as those employed in clocks, fans, and instruments.

The so-called *universal* motor, found in many appliances and electrically driven hand tools, may be operated from alternating or direct current and in the former application is classed as an induction motor. This type is identical with the series-wound d-c motor (Figure 11–17) except that the field structure and armature core *must* be laminated, to prevent eddy currents. The armature windings and commutator are the same as in a d-c machine. On alternating current, current in both field and armature reverses each time the applied voltage changes polarity. With the reversal occurring in both windings in this manner, the direction of rotation of the armature remains unchanged.

11.13 REPULSION A-C MOTOR

The rotor of a straight *repulsion* motor is wound like the armature of a d-c motor. That is, its coils are connected to commutator bars.

The magnetic flux of the field coils induces current which is somewhat more than 90 deg lagging in the armature. The curves in Figure 11–24 illustrate this condition. The torque at any instant is the sum of the instantaneous flux and current values, and it is positive from 90 to 180 deg and negative from 180 to 270 deg. The positive half-cycle of torque is the smaller, and this force is regarded as attractive, while the negative half-cycle is the larger and is regarded as repulsive.

The repelling force due to the negative torque is utilized to drive the armature of the repulsion motor. This is accomplished by placing brushes in such a position on the commutator that the armature coils are closed by them only when the coils are positioned so as to receive the effect of this repelling action. Coils whose commutator bars are not in contact with the brushes are open-circuited at this time.

Figure 11–25 shows the basic arrangement of the repulsion motor. The *energy brushes, a* and *b*, short-circuit the armature coils in which the torque-producing current is induced. Brushes *c* and *d*, called *compen-*

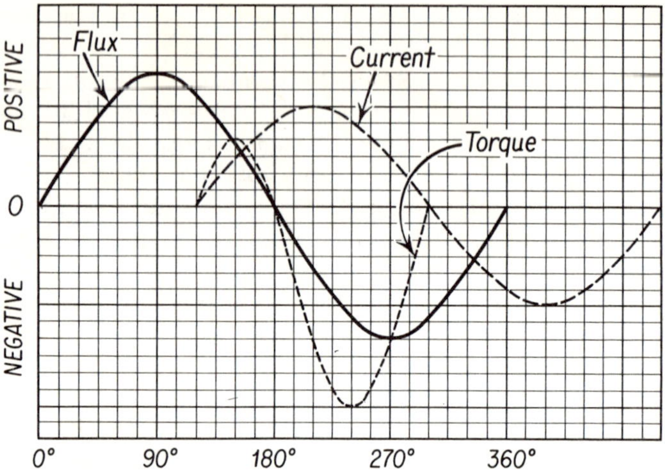

Fig. 11–24 Operating Curves of Repulsion Motor

sating brushes, are oriented 90 deg from the energy brushes and are connected to the armature winding as shown. The field flux induces a current in this *compensating winding* which serves to stabilize the motor speed.

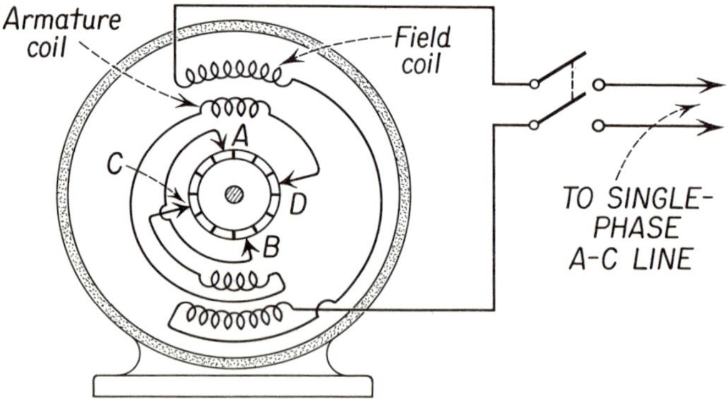

Fig. 11–25 Repulsion Motor

11.14 SPLIT-PHASE A-C MOTOR

In the *split-phase* motor, two separate field windings are provided. One is the *starting winding,* the other the *running winding.* The former is in the circuit only during the interval required to start the motor and bring it up to normal running speed.

Figure 11–26 shows the basic construction of a typical split-phase motor for use with single-phase a-c supply. The two windings are

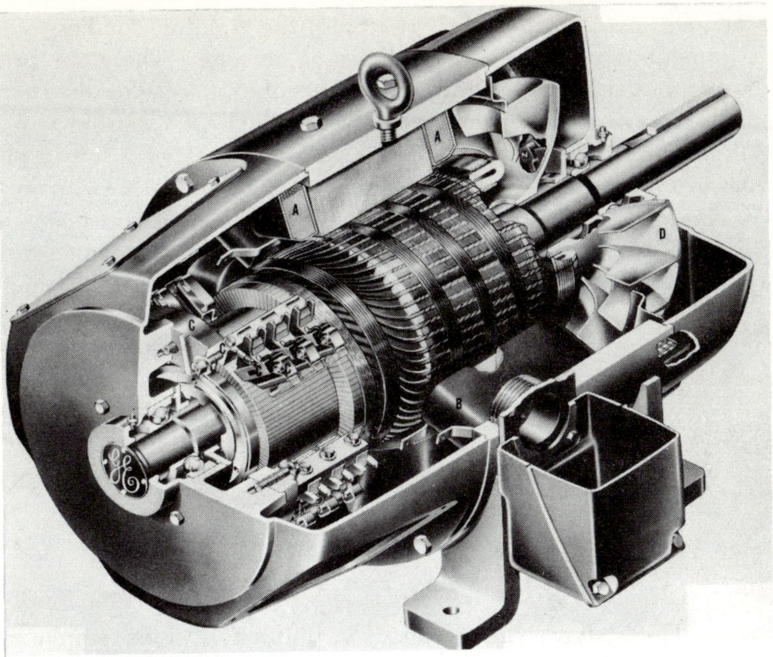

Plate 11–4 Cutaway View of Modern Alternating-Current Motor. The slotted, drum-type armature and the commutator are easily visible. Other parts in view are (A) and (B) field coils, (C) brush rigging, and (D) cooling fan. (Courtesy, General Electric Co.)

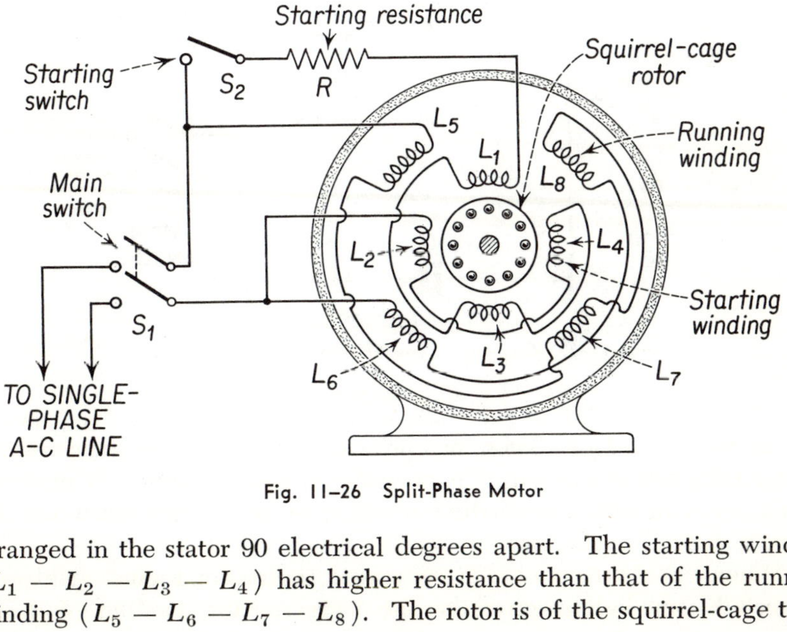

Fig. 11–26 Split-Phase Motor

arranged in the stator 90 electrical degrees apart. The starting winding ($L_1 - L_2 - L_3 - L_4$) has higher resistance than that of the running winding ($L_5 - L_6 - L_7 - L_8$). The rotor is of the squirrel-cage type.

When the main switch (S_1) is closed, the running winding is connected immediately to the line and passes current. The rotor cannot turn, however, since the resulting flux is not rotating. When the starting switch (S_2) then is closed, current flows through the starting winding and is limited by the starting resistance, R. This current is out of phase with that in the running winding, because of the relative positions of the two windings on the stator and the differences in their resistances and inductances. This causes a rotating field which starts the rotor turning. When the rotor approaches synchronous speed, the starting winding no longer is needed, and the starting switch, S_2, is opened to remove this winding from the circuit. The motor then runs as an induction motor. In practical split-phase motors, S_2 is a centrifugal switch which opens automatically when the motor attains running speed.

Split-phase starting is limited to small (fractional-horsepower) motors. Split-phase motors draw large starting current and have low starting torque.

11.15 CAPACITOR-TYPE A-C MOTOR

Figure 11–27 shows the circuit of an induction motor employing a high-capacitance electrolytic capacitor (C) and an auxiliary field winding for starting.

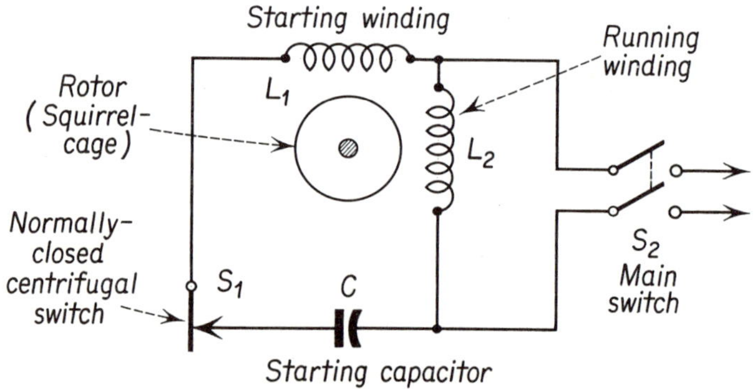

Fig. 11–27 Capacitor-Type Alternating-Current Motor

The stator of this motor contains a starting winding (L_1) and a running winding (L_2). These windings are situated 90 electrical degrees apart on the stator structure. A normally closed, centrifugally operated switch, S_1, connects the starting winding to the single-phase a-c line through starting capacitor C.

When the main switch (S_2) is closed, the running winding (L_2) is connected immediately across the line. The starting winding (L_1) also is connected to the line but receives current through the capacitor (C). This current accordingly is out of phase with that through L_2. A rotating field accordingly is produced and the rotor is caused to turn. When the rotor reaches 75 to 80 per cent of synchronous speed, centrifugal action causes the automatic switch (S_1) to open and disconnect the starting winding and capacitor from the line. The machine then continues to run as a squirrel-cage induction motor.

Capacitor-start induction motors are characterized by high starting torque.

11.16 SYNCHRONOUS MOTOR

The *synchronous* motor is so called because it runs in synchronism (that is, at the same speed) with the a-c generator supplying power to it. It runs at this constant speed under all loads within its rated range. It is the most efficient a-c motor.

The synchronous motor strongly resembles an a-c generator, or alternator. In this motor, the rotor winding is connected to a pair of slip rings; the field winding to the a-c line. Direct current, usually from a d-c generator, must be supplied to the rotor. In addition to its winding, the rotor also contains the short-ciruited bars of a squirrel-cage rotor.

The direct current is removed from the rotor when the motor is to be started. When alternating current is applied, the machine starts up like any other squirrel-cage induction motor. When the rotor is turning nearly at synchronous speed, the direct current is applied, and the rotor synchronizes with the revolving flux of the stator at full synchronous speed.

The conventional synchronous motor of the type described in the preceding paragraphs is not usually built in smaller than 5 hp size, because of its d-c requirement. Small-sized synchronous motors are made to run without d-c excitation, however. Examples of such fractional-horsepower motors are the ones used in electric clocks, timers, sequence switches, and the chart-driving mechanisms of electric recorders.

Figure 11–28 shows the basic construction of one type of self-starting, synchronous clock motor (Telechron). In this example, the rotor is composed of several hardened steel rings, each with two crossbars and spaced parallel to each other along the shaft. The ring-and-bar assembly often is enclosed in an aluminum shell. The rotor turns between split pole pieces, one section of each of the latter carrying a shading coil. A multiturn exciting (field) coil is mounted on the yoke of the field magnet.

Rotating action of the flux produced by the field and shading coils starts the rotor turning. (The starting torque is heightened by energy consumed in the rotor as a consequence of the high hysteresis loss in the hardened steel.) At a rotor speed near synchronism, the flux follows the low-reluctance path through the crossbars of the rotor. The rotor then becomes magnetized permanently and attains synchronous speed (3,600 rpm when the supply frequency is 60 cps).

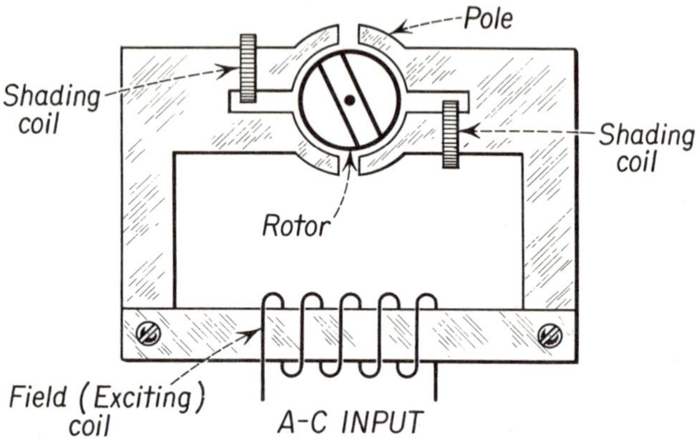

Fig. 11–28 Synchronous Motor without Direct Current

11.17 MOTOR GENERATORS

A motor and generator may be coupled together mechanically so that the motor turns the generator. This usually is accomplished by coupling together the shafts of the two machines, as shown in Figure 11–29, although a generator sometimes is belt-driven by a motor.

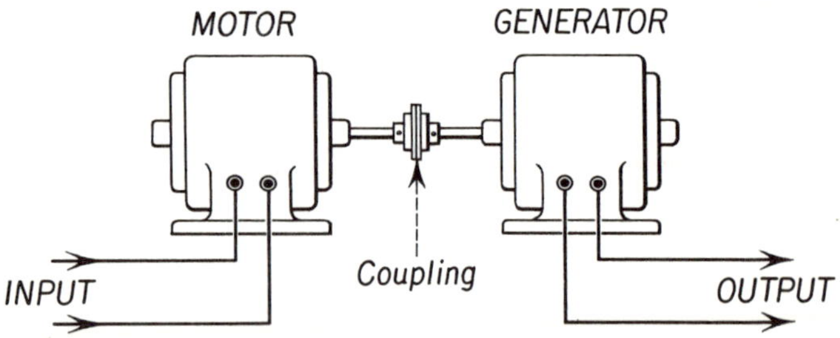

Fig. 11–29 Basic Arrangement of Motor-Generator

The two-machine combination, called a *motor generator,* is convenient for producing a desired type of power from an available type. The possible combinations are:

(*a*) d-c motor–a-c generator
(*b*) d-c motor–d-c generator
(*c*) a-c motor–d-c generator
(*d*) a-c motor–a-c generator

Thus, alternating current may be provided where ordinarily only direct current is available, and vice versa. Or alternating current or direct current of one voltage and/or current level may be converted into alter-

Plate 11–5 Commercial Motor-Generator Set (Courtesy, Electric Specialty Co.)

nating current or direct current at another level. Also, alternating current of one frequency or phase may be converted into alternating current of another frequency or phase.

11.18 DYNAMOTORS AND ROTARY CONVERTERS

The *dynamotor* is a special form of motor generator in which the motor and generator are combined in a single machine. A common field is employed, and the generator armature coils are wound on the same core as the motor armature coils. Separate commutators (or slip rings) and brushes are provided for motor and generator.

Figure 11–30 shows this arrangement. Here, a shunt field is employed. The motor is operated with direct current and the generator delivers direct current. This arrangement is widely used in small machines to obtain high-voltage direct current from a low-voltage d-c source. An example is the so-called radio-type dynamotor operated from a storage battery and delivering high-voltage direct current to portable electronic equipment.

Although d-c–d-c dynamotors are fairly common, d-c–a-c machines are commercially available. In this type, there is a motor commutator on one end of the armature shaft and slip rings on the other end.

The *rotary converter* resembles the dynamotor in construction. However, a single armature winding serves both the motor and generator sections of this machine. A commutator is mounted on one end of the armature shaft and slip rings on the other. [See Figure 11–31(A).] Taps are taken on the armature coil at points 180 deg apart and connected to corresponding commutator bars. The armature coil is connected also to the slip rings.

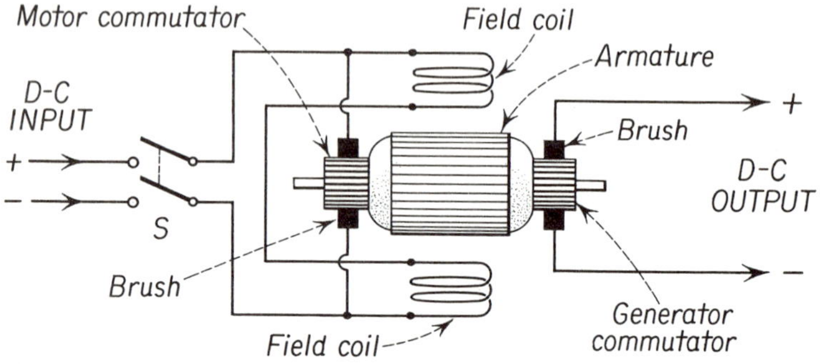

Fig. 11–30 Dynamotor

Figure 11–31(B) illustrates the operating principle of the rotary converter. Suppose direct current is supplied to the input to drive the motor. When the armature is in the position *AB* shown, the slip rings receive the full d-c voltage. When the armature rotates one-quarter revo-

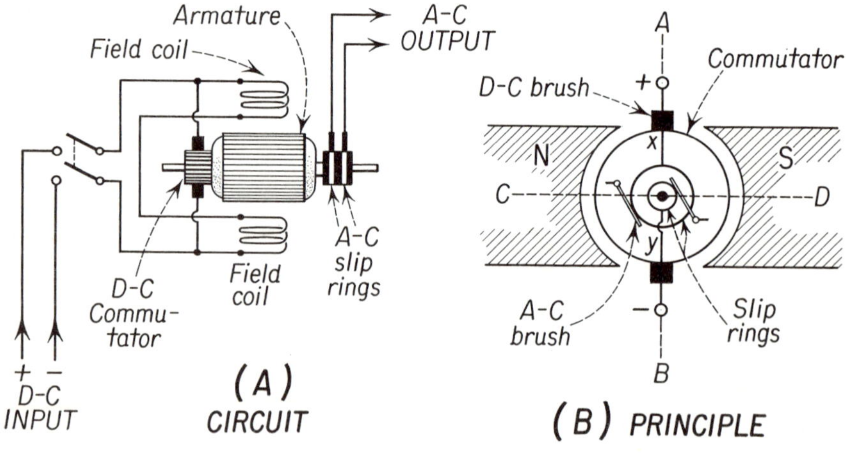

Fig. 11–31 Rotary Converter

lution to *CD*, the voltage present at the a-c brushes is zero. One-quarter revolution farther, the output voltage again is equal to the d-c voltage, but the output polarity has reversed. At three-quarter revolution, the voltage is zero; and at the completion of one revolution, it again is equal to the d-c voltage. Thus, in one complete revolution, the output voltage at the a-c brushes passes through one complete a-c cycle.

Instead of running on direct current, as just described, alternating current may be applied to the slip rings, and the machine will run as a synchronous a-c motor and deliver d-c output at its commutator. Because of this possible mode of operation, the rotary converter also is called a *synchronous converter.*

Because the armature windings of the rotary converter are common to both motor and generator, the output voltage varies with input voltage. In some installations, some difficulty might be experienced also because of the lack of isolation between input and output circuits of the rotary converter, especially where operation of the machine sets up radio interference or where separate grounds are required for the a-c and d-c circuits.

SUMMARY OF CHAPTER 11

An **A-C Generator** or **Alternator** is composed of a field structure and armature, respectively called *stator* and *rotor,* or *primary* and *secondary.* Unless the field poles are permanent magnets, the coils of either the stator or rotor are excited with direct current from an external source.

In an alternator, either the field assembly or the armature may rotate. In the rotating-field type, the rotor contains a number of poles around which are wound coils connected to a d-c source through slip rings. This keeps the brushes in the low-voltage circuit.

The **Frequency** of an alternator, $f = (np)/2$ when rotational speed, n, is expressed in revolutions per second. $f = (np)/120$ when n is in revolutions per minute.

A **Commutator** added to the rotating armature converts an alternator into a d-c generator. Direct current generated by this arrangement is not smooth and steady like the output of a battery but rises and falls rhythmically between zero and maximum *in one direction* like one-half of an a-c wave. The characteristic wave pattern is termed *commutator ripple.* Ripple amplitude is decreased, that is, the direct current is smoothed, by increasing the number of armature coils and commutator segments, or bars.

The **Field Winding** of the d-c generator often requires no external excitation; that is, the field current is derived from operation of the generator itself. (External excitation sometimes is used.) Depending upon the arrangement of field coils and armature, d-c generators are classified as **Series-, Shunt-,** or **Compound-wound.** In a **Cumulative Compound** winding, the two field windings are wound to aid each other magnetically; in the **Differential Compound** winding, they oppose.

Voltage Regulation refers to the change of generator output voltage between no-load and full-load operation. VR (in per cent) equals $100(E_1 - E_2)/E_2$.

The **D-C Motor** utilizes a practical application of Lenz's law—namely, a current-carrying conductor is displaced in a magnetic field. The current-carrying coils of an armature in a uniform magnetic field intensify this effect and cause rotation of the armature core supporting them.

D-C Motors, like generators, are of the **Series-, Shunt-,** and **Compound-wound** types.

The **Induction A-C Motor** has a stationary field assembly and revolving rotor. The rotor consists of short-circuited bars or coils on a laminated iron core. The bar type is called a *squirrel-cage* rotor. The magnetic field in the induction motor rotates, pulling the rotor around with it. A straight induction motor, when operated from single-phase alternating current, is not self-starting, since the field is not rotating when the motor is standing still.

The **Rotating Field** in an induction motor cuts the rotor bars, inducing currents in them. These currents set up fields which react with the main field to make the rotor turn. Rotor and field cannot revolve at the same speed, since no bars would then be cut by the field flux. The difference between field (synchronous) speed and actual rotor speed is called *slip* (s). $s = 100(S_s - S_r)/S_s$ per cent.

Self-starting in the induction motor may be secured by means of shading coils on the tips of the field poles. The shading coil causes an initial twist of the flux, and this starts the rotor turning.

The **Universal A-C Motor** has a commutator and is similar to the series-wound d-c motor, except for a laminated field and armature core. On alternating current, this motor rotates in one direction because *both* field and armature currents reverse when the a-c cycle reverses.

The **Repulsion Motor** is self-starting. It has an armature and commutator like a d-c motor. However, the brushes are oriented to such a position that armature coils are short-circuited by them and provide true induction-motor action only when the rotor is in such a position that the short-circuited coil can utilize the repulsive phase of the torque produced by the field.

The **Split-Phase Motor** uses a *starting winding* separate from its *running winding.* Its rotor is of the squirrel-cage type. Currents in the two windings are out of phase when the power first is switched on, and they produce a rotating field which starts the rotor turning. When synchronous speed is approached, the starting winding is disconnected by a centrifugal switch.

The **Capacitor-Type Motor** also employs separate starting and running windings. The starting winding is connected to the power line through a high-capacitance electrolytic capacitor. The current in the starting winding is out of phase with that in the running winding and produces the initial field rotation necessary to start the rotor. When the rotor approaches synchronous speed, a centrifugal switch opens and disconnects the capacitor and starting winding from the circuit.

The **Synchronous Motor** is so called because it rotates in step (synchronism) with the alternator supplying power to it. The rotor of this motor contains armature coils to which direct current is applied through slip rings, and also the short-circuited bars of a squirrel-cage rotor. Alternating current is applied to the field winding.

A **Motor Generator** consists of a motor and generator coupled together mechanically so that the motor drives the generator. Motor-generator sets are popular for obtaining high-voltage direct current from low-voltage direct current. They are used also in all other a-c and d-c combinations.

The **Dynamotor** is a compact form of motor generator in which the armature coils of the motor and generator are wound on the same core. Separate commutators (or commutator and slip rings) are provided. The dynamotor has the same shape, and often size as well, as a single motor or generator.

The **Rotary Converter** is a special form of dynamotor in which a single set of armature coils is common to both the d-c and a-c sections of the machine. The rotary converter may be driven from direct current to produce alternating current, or vice versa. In the latter condition, it runs as a synchronous motor.

QUESTIONS AND EXERCISES

1. Describe the operation of a simple magneto-type alternator.
2. Draw a simple rotating-field alternator, showing field poles, field coils, rotor poles and coils, and slip rings.
3. Calculate the frequency of a two-pole alternator rotating at a speed of 1,800 rpm.
4. Explain commutator action in a d-c generator.
5. Draw simple circuit diagrams of the following d-c generators: (*a*) series-wound; (*b*) shunt-wound; (*c*) cumulative compound-wound; (*d*) differential compound-wound.
6. Why must the brushes in a d-c generator be placed *off* the neutral axis?

7. A certain generator delivers a no-load voltage of 110 v. Its full-load voltage is 107 v. Calculate its voltage regulation.

8. Draw simple circuit diagrams of the following d-c motors: (*a*) series-wound; (*b*) shunt-wound; (*c*) cumulative compound-wound; (*d*) differential compound-wound.

9. Explain what is meant by *slip* in an induction motor.

10. Calculate the slip percentage in a four-pole, 60-cps induction motor rotating at 1,700 rpm.

11. Why does not the rotor of a universal motor oscillate back and forth on alternating current?

12. Explain the operation of a repulsion motor.

13. Explain the operation of a split-phase motor.

14. Explain the operation of a capacitor-start motor.

15. Explain the operation of a synchronous motor.

16. Describe the differences between: (*a*) motor generator; (*b*) dynamotor; (*c*) rotary converter.

Chapter **12**

Polyphase A-C Systems

12.1 ADVANTAGES

Up to this point, the a-c systems discussed have all been *single-phase*. That is, only a single alternating voltage has been involved. In *polyphase* systems, however, two, three, or more voltages are utilized, and these emf's differ in phase.

Polyphase systems offer many advantages over single-phase systems. Some of these are: (1) Polyphase motors run more smoothly than single-phase types, because polyphase power is more uniform, not pulsating like single-phase power. (2) Polyphase motors are more efficient. Most very large motors are of the polyphase type. (3) These motors are simpler. A rotating field, for example, is obtained easily with polyphase alternating current, and motors are inherently self-starting when not single-phase. (4) Less copper (in lines) is needed in a three-phase system to transmit a given amount of power over the same distance than in single-phase transmission. (5) Polyphase motors and generators have a higher power rating than single-phase motors of comparative physical size. (6) Small two-phase motors may readily be controlled in speed and direction by vacuum tubes.

12.2 TWO-PHASE

If the coils in an alternator are wound on separate rotor poles spaced 90 deg apart, as in Figure 12–1(A), the voltage E_1, across one coil (1–2) which is in line with the field poles and cutting lines of force at the maximum rate, will be at its maximum value; while voltage E_2 across the other coil (3–4) will be at zero, since no lines of force are being cut. The phase relations between these two voltages, E_1 and E_2, are shown in Figure 12–1(D).

As the rotor revolves, E_1 and E_2 will alternate, always 90 deg out of phase, as shown by the graph. This is the principle of the two-phase generator. Figure 12–1(E) shows a two-phase generator in which the field, instead of the rotor coils, rotates.

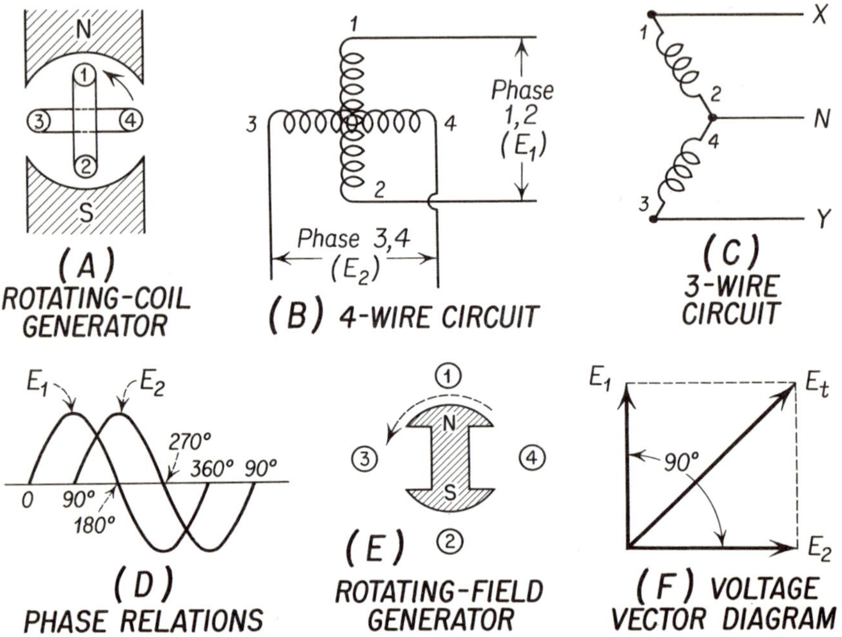

Fig. 12–1 Generation of 2-Phase Voltages

Figures 12–1(B) and (C) show the two separate generator coils in the 90-deg physical relationship. Single-phase voltages are available separately across each phase (1–2 or 3–4) in Figure 12–1(B). This would necessitate connecting Coil 1–2 to one pair of slip rings and Coil 3–4 to another pair. This seldom is done in practice. Instead, the coils are connected together at one point, as shown in Figure 12–1(C), to form a three-wire circuit. Wires *X*, *Y*, and *N* (neutral) are connected to each of the three slip rings.

It is apparent from Figure 12–1(D) that, although voltages E_1 and E_2 on each side of neutral (N) are equal, the total voltage (E_t) between X and Y is not the sum of E_1 and E_2 because of phase relations. Instead, this voltage is the vector sum of E_1 and E_2. $E_t = \sqrt{E_1{}^2 + E_2{}^2}$. [See vector diagram, Figure 12–1(F).] Since $E_1 = E_2$ and the phase angle is 90 deg, from geometry $E_t = \sqrt{2}$ times E_1 or E_2. The square root of 2 is 1.414, so

(12–1) $$E_t = 1.414E_1 = 1.414E_2$$

Example: A certain two-phase alternator delivers 115 v across each phase. What is its phase-to-phase voltage?

$$E_t = 1.414(115) = 162.5 \text{ v}$$

12.3 THREE-PHASE

Throughout industry, three-phase distribution of electric power is the most frequently encountered of all polyphase systems. In this

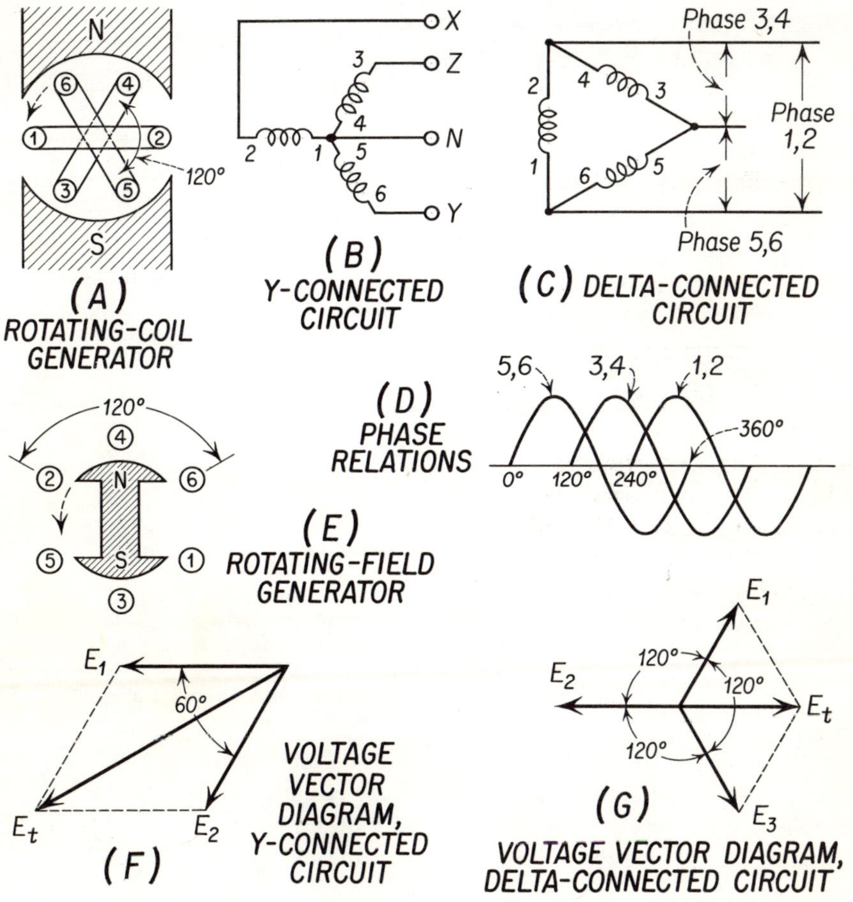

Fig. 12–2 Generation of 3-Phase Voltages

system, three coils are employed with an angle of 120 deg between coils. Figure 12–2(A) shows this arrangement. Figure 12–2(E) shows the alternative arrangement of a rotating field, with stationary conductors or coils spaced 120 deg apart.

In the three-phase generator, each coil voltage is 120 deg out of phase with the other. Figure 12–2(D) shows this relationship. When the corresponding ends of all three coils (phases) are returned to a common (neutral) connection, *N* [as in Figure 12–2(B)], four slip rings are required for connections *N*, *X*, *Y*, and Z. This is the *Y*, or *star*, connection. When all unlike terminals of the coils are connected to form a closed loop or *delta*, as in Figure 12–2(C), the number of slip rings may be reduced to three.

In the three-phase system, voltage is available across each phase. Thus, in Figure 12–2(B), *N* is the common or neutral wire. Voltage E_{1-2} is available between wires *X* and *N*, E_{3-4} from Z to *N*, and E_{5-6} from *Y* to *N*. In a symmetrical circuit, each phase is loaded externally by the same amount.

In the Y-connected circuit, the coils are spaced apart by 120 deg [Figures 12–2(A) and (E)], and their voltages are 120 deg out of phase. It is clear from Figure 12–2(A), however, that connecting *adjacent* ends of coils together for the Y circuit actually places these coils and their phases 60 deg apart. The voltage between any two phases in this circuit is the vector sum of these phases. Assuming the phase voltages to be equal, the vector sum of two such voltages displaced by 60 deg equals $\sqrt{3}$ times either of the voltages. The square root of 3 is 1.732. Therefore, the voltage, E_t, between any two phases in the Y-connected three-phase circuit is

$$E_t = 1.732E_1 = 1.732E_2 = 1.732E_3 \tag{12–2}$$

Example: If a three-phase generator delivers 110 v per phase, what is the voltage between any two of its phases?

$$E_t = 1.732(110) = 190.5 \text{ v}$$

The vector diagram in Figure 12–2(F) shows these voltage relationships.

The vector diagram in Figure 12–2(G) shows the voltage relationships in the delta-connected three-phase circuit. Here, the voltage E_t between any two phases is equal and opposite to the voltage in the third phase. Thus, the voltage between phases E_1 and E_3 is equal to the third phase, E_2.

12.4 DELTA AND Y CONNECTIONS

Figures 12–2(B) and (C), respectively, show Y and delta connections of three-phase generator coils. Figure 12–3 shows some of the connections possible between three-phase alternators and loads.

In Figure 12–3(A), a Y-connected alternator is connected through a four-wire line to a Y-connected load (Z_1, Z_2, Z_3). The load may be balanced ($Z_1 = Z_2 = Z_3$) or it may be unbalanced. In the balanced system, currents I_1, I_3, and I_4 are equal. Under these conditions, no current flows through the neutral wire; that is, $I_2 = 0$. When the load is unbalanced, the phase currents will be unequal, and current will flow in the neutral. An unbalanced *three-wire* Y system may be connected by omitting the neutral wire.

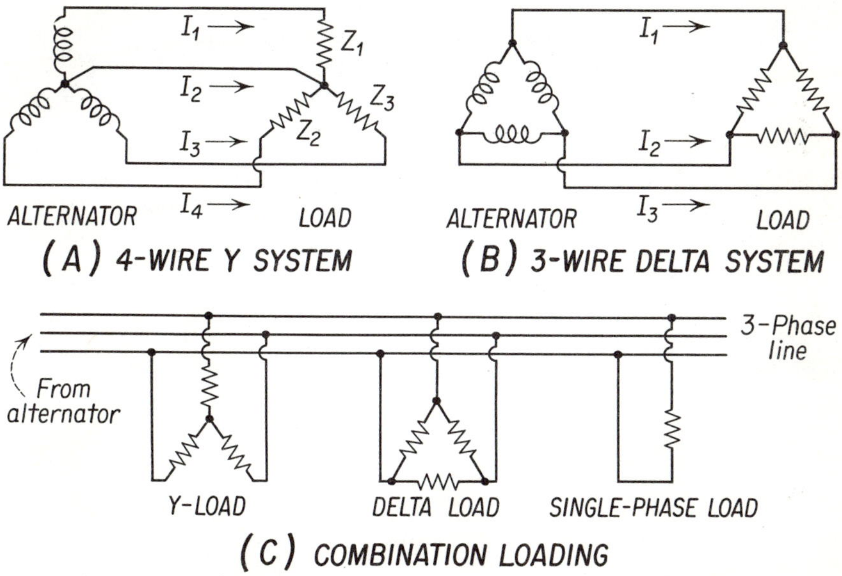

Fig. 12–3 Three-Phase Load Circuits

Figure 12–3(B) shows the connection of a delta-connected alternator to a delta-connected load through a three-wire line. The delta load may be balanced (that is, all load elements equal and currents displaced by 120 deg) or the system may be unbalanced. Three identical impedances are found in such three-phase devices as motors, hence the latter constitute balanced loads. In the balanced three-phase delta system, the line current, I_1, is equal to $1.732I_p$ (where I_p is either one of the phase currents).

In a practical power system, several loads of different types often are connected to a three-phase line. Thus, in Figure 12–3(C), a Y-load, delta-load, and single-phase load are shown connected appropriately to a three-phase line.

Total power (P) in a balanced three-phase system is equal to $\sqrt{3}$ times the power ($EI \cos \phi$) calculated for one of the phases:

$$P = 1.732EI \cos \phi \tag{12–3}$$

12.5 GENERATOR CIRCUITRY

In Figures 12–1(A) and (E) and 12–2(A) and (E), only a few coils are shown in the explanatory generator diagrams. In actual practice, however, coils are wound on the entire armature structure of the alternator.

This type of construction is illustrated by Figure 12–4, showing a three-phase alternator. All coils included in opposite 60-deg arcs (A–A', B–B', and C–C') are connected in series with each other to form a *phase*

Fig. 12–4 Multicoil 3-Phase Generator

group. The time phase of the resultant emf in each of these groups is the same as that of the center coil (marked black in Figure 12–4) in the group. The three groups are connected either delta or Y, as explained previously.

In a two-phase alternator, the coils are divided into two groups displaced by 90 deg. These two groups then are connected together according to Figure 12–1(B) or (C).

Polyphase circuits having more than three phases are possible and practical. Both six-phase and twelve-phase power are found in industry, but to a lesser extent than three-phase. In polyphase alternators for generating this power, it is necessary to space the windings appropriately into time-displaced phase groups.

12.6 MOTOR CIRCUITRY

One of the great advantages of polyphase alternating currents is the ease with which a rotating magnetic field may be obtained with them. The reader will recall that it was pointed out in the discussion of a-c motors in Chapter 11 that single-phase alternating current ordinarily will not produce a rotating field in a motor at rest, and some

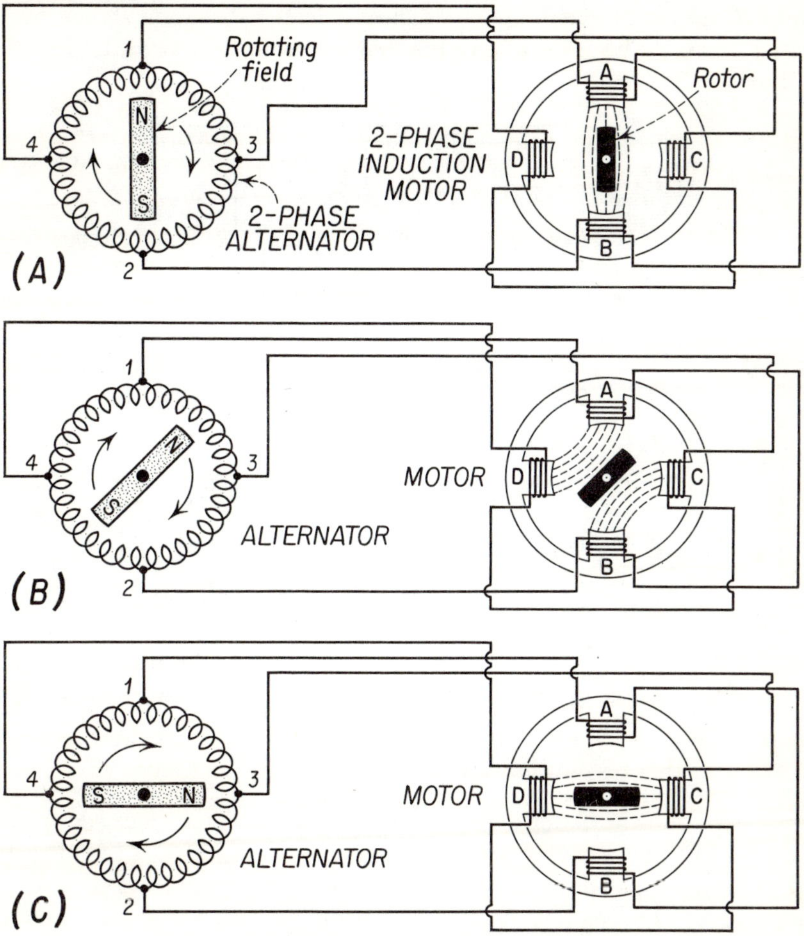

Fig. 12–5 Two-Phase Induction Motor Operation

provision must be made to start the machine. In polyphase motors, the field rotates from the beginning, starting as well as running the motor.

Figure 12–5 illustrates two-phase induction-motor operation. In the alternator, the revolving field is shown as a rotating magnet. In the motor, the rotor is shown as a simple iron bar mounted on a shaft so as to turn between the field poles. During one phase of voltage from the alternator, motor poles *A* and *B* are magnetized, and during the next phase, poles *C* and *D* are magnetized. Because the current is 90 deg out of phase in the two halves of the circuit, the magnetization of one pair of poles will be zero when that of the other pair is maximum.

When the alternator field is cutting stator turns in the position indicated by the north and south poles in Figure 12–5(A), motor poles *A* and *B* are magnetized, and their flux is in the direction shown by the dashed lines. This flux aligns the bar rotor in the vertical direction, as shown.

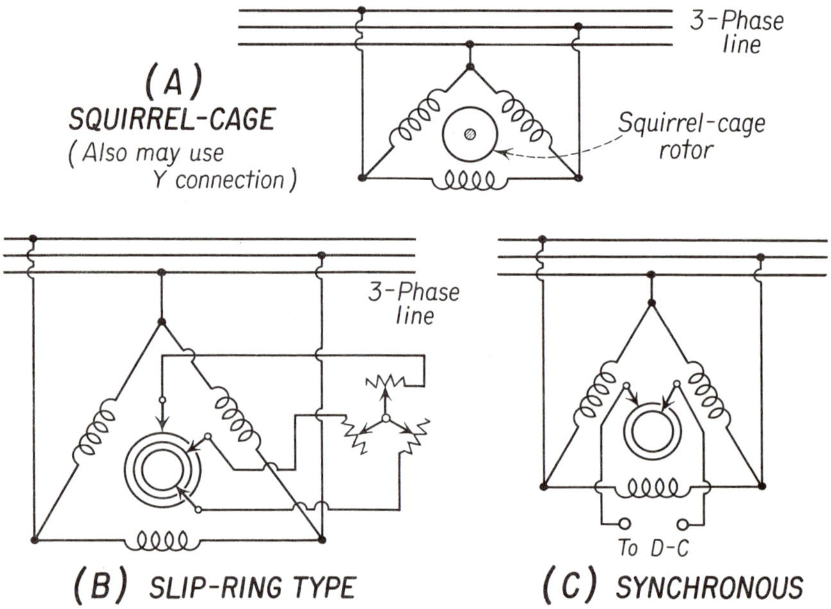

Fig. 12–6 Some 3-Phase Motor Circuits

When the alternator field has advanced halfway between phases, as shown in Figure 12–5(B), the phase currents are in the same direction and are equal, and the lines of force in the motor pass from *A* to *D* and from *B* to *C*. This causes the rotor to turn to the position shown, between Poles *A* and *C*, and *B* and *D*.

When the alternator field has advanced to the next phase [Figure 12–5(C)], the flux in the motor passes between Poles *C* and *D*, and the rotor is turned to the horizontal position between these poles.

Next, the alternator field would be halfway between 3 and 2, and the motor rotor halfway between poles *C* and *B*. As the alternator passes through succeeding phases, the magnetic field of the motor rotates in synchronism and the motor follows the rotation to complete a revolution.

In the practical induction motor, the squirrel-cage rotor revolves in the same manner as the simple bar shown in Figure 12–5. While this example illustrates action of a two-phase motor, a three-phase motor is operated in a similar manner from a three-phase alternator. The three-phase motor has more poles, but its magnetic field rotates in a similar manner between poles, pulling the rotor around with it.

Figure 12–6 shows several three-phase a-c motor circuits. While these circuits are representative, it must be remembered that they are not the only ones.

SUMMARY OF CHAPTER 12

Polyphase Alternating Current offers several advantages over single-phase: smooth, efficient motor operation and simpler motor construction; higher power rating in alternators of a given physical size; and less copper to transmit a given amount of power the same distance as by single-phase alternating current.

In a **Two-Phase Alternator,** coils are spaced so as to be displaced 90 time-degrees. The two voltages generated by a two-phase alternator are 90 deg out of phase. Two-phase alternator coils may be connected for three-wire or four-wire output.

In a two-phase system, with equal phase voltages, the **Total Voltage** between the two phases is $\sqrt{2}$ times one phase voltage. Thus, $E_t = 1.414E_1 = 1.414E_2$.

In a **Three-Phase System,** the three voltages are time-displaced by 120 deg. The coils of a three-phase alternator may be connected Y (or *star*) for four-wire output, or *delta* for three-wire output.

The **Voltage Between Phases** in a three-phase Y-connected system with equal phase voltages is $\sqrt{3}$ times either of the phase voltages: $E_t = 1.732E_1 = 1.732E_2 = 1.732E_3$. In a delta-connected system, the voltage between two equal-voltage phases is equal and opposite to the voltage in the third phase.

Polyphase Loads may be connected to the three-phase line as either Y or delta circuits. The load is **Balanced** when each of its three elements is equal to the others.

Power in a balanced three-phase circuit is $\sqrt{3}$ times the power measured or calculated for one phase: $P = 1.732EI \cos \phi$.

Polyphase Alternating Current Higher Than Three-Phase is practicable. Six- and twelve-phase currents occasionally are encountered in industry and in some equipment.

Rotating Magnetic Fields, such as are required in induction motors, are obtained comparatively easily with polyphase current.

QUESTIONS AND EXERCISES

1. Name three advantages of polyphase alternating current.
2. How many degrees phase difference is found in a two-phase system?
3. How many degrees phase difference is found in a three-phase system?
4. Draw: (*a*) three-wire, two-phase output circuit showing alternator coils; (*b*) four-wire, two-phase output.
5. Calculate the voltage between the two phases of a two-phase system in which each phase voltage is 115 v.
6. Draw the simple circuit of; (*a*) three-phase, Y-connected generator; (*b*) three-phase, delta-connected generator.
7. Calculate the voltage between two phases in a Y-connected, three-phase system in which each phase voltage is 230 v.
8. Calculate the voltage between two phases in a delta-connected three-phase system in which each phase voltage is 440 v.
9. Calculate the power (in watts) in a balanced three-phase system in which voltage = 115 v, current = 5 amp, and power factor = 95 per cent.
10. Draw connected to a three-phase line the following resistance loads: (*a*) delta; (*b*) Y.

Chapter 13

Rectifiers

13.1 VACUUM-TUBE TYPE

A *rectifier* is a device, other than a motor generator, dynamotor, or rotary converter, for changing alternating current into direct current. This is an important device, since most electric power is supplied as alternating current and must be converted into direct current by the user when direct current is required. The process whereby direct current is obtained from alternating current is called *rectification.*

The vacuum tube is a widely used type of rectifier. Figure 13–1(A) shows the basic construction of one type of vacuum-tube rectifier. A metal plate (sometimes cylindrical in shape, as shown, and at other times box-shaped) is mounted inside an evacuated bulb. A filament, consisting of a loop or coil of wire or strip, is supported in the space inside the plate but not in contact with the latter. Because this tube has only two electrodes, a filament and a plate (sometimes called the *anode*), it is termed a *two-element tube.* It also is called a *diode.*

When the filament is connected to a voltage source, such as Battery B_2 in Figure 13–1(C), the filament is heated by the current flowing through it. It does not burn up because the air has been removed from the bulb. However, the heat does cause free electrons to be emitted by the atoms of the filament, and they surround the filament in a sort of cloud, as shown by the dots. These electrons are bits of negative electricity. (See Section 1.8, Chapter 1.) If a second battery, B_1, is connected between the filament and plate with its positive terminal connected to the plate, the electrons will be attracted to the positively charged plate and will flow to it. This movement constitutes an electric current which passes from Battery B_1, up the filament, across the filament-plate gap, and through the plate lead. This flow is termed the *plate current* or *anode current.*

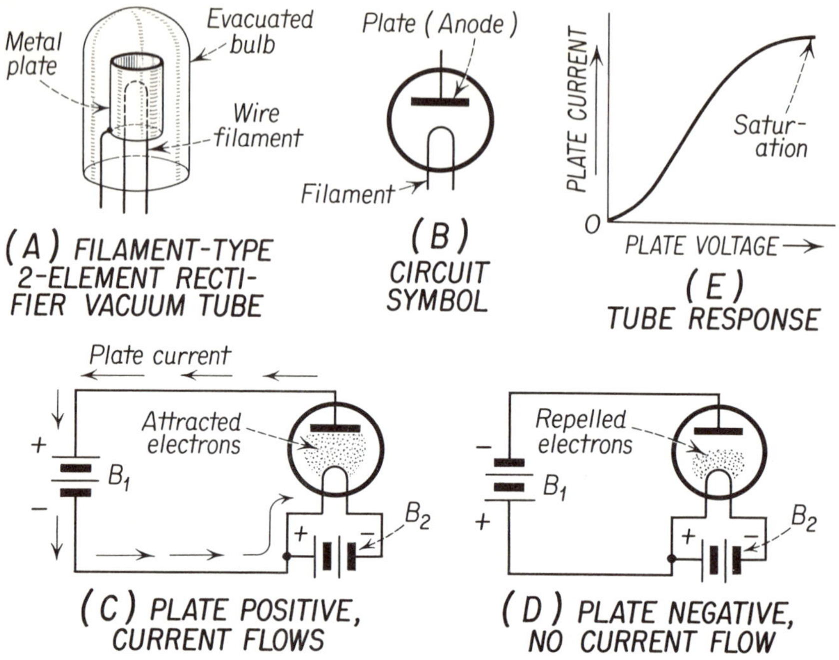

Fig. 13–1 Two-Element Vacuum Tube

Battery B_2 serves only to heat the filament, and its current does not contribute directly to the plate current. As the plate potential is increased, by raising the voltage of Battery B_1, the plate current increases, as shown by the curve in Figure 13–1(E), until a point is reached after which further increases of plate voltage give little or no increase in plate current. This point is termed *saturation,* and it indicates that all of the available electrons are being drawn to the plate.

If Battery B_1 is reversed in the circuit, as shown in Figure 13–1(D), so that its *negative* terminal is connected to the plate, the electrons will be repelled by the negatively charged plate, and no plate current can flow.

This action accounts for the ability of the two-electrode tube to act as a rectifier. When its plate is positive with respect to its filament, the tube passes current; when it is negative, the current is cut off. The two-element tube thus is a one-way valve. If an a-c voltage is applied between plate and filament, current will flow only when the plate "sees" positive half-cycles. Current is blocked when the negative half-cycle is presented to the plate.

Figure 13–2 shows a simple rectifier circuit embracing this principle. Here, the a-c voltage is applied to the tube through a transformer. One secondary winding, L_3, is employed exclusively to heat the filament. The other secondary, L_2, supplies plate voltage. The a-c wave forms are shown by (*a*) and (*b*). Each time the a-c cycle swings positive, the tube conducts and passes a positive half-cycle of current through

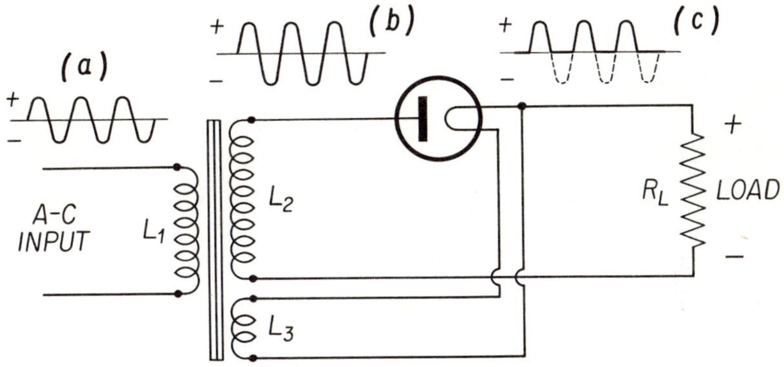

Fig. 13–2 Simple V-T Rectifier Circuit

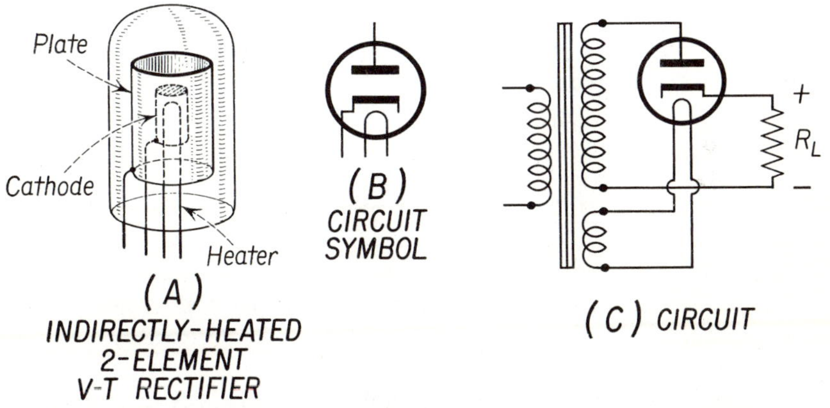

Fig. 13–3 Indirectly Heated (Cathode-Type) Vacuum-Tube Rectifier

the load, R_L. But the negative half-cycles are blocked. The result is a pulsating direct current, as shown by wave form (*c*). Thus, the unaltered output of the rectifier is a ripple current, somewhat similar to the commutator ripple of a d-c generator.

In the tube shown in Figures 13–1 and 13–2, the electrons are emitted directly by the filament. Not all tubes are constructed in this way. Figure 13–3(A) illustrates a rectifier-type tube in which the filament (now called a *heater*) serves only to heat a surrounding *cathode* thimble which emits the electrons. The cathode is coated with a chemical which is an efficient emitter of electrons, and the heater does not make contact with it. The filament-type tube is termed *directly heated,* while the cathode type is called *indirectly heated.*

Figure 13–3(B) shows the circuit symbol for the cathode-type rectifier tube, and Figure 13–3(C) shows the rectifier circuit corresponding to the one shown previously in Figure 13–2. The advantage of the cathode-type tube is complete isolation of the filament and filament-heating voltage from the rectifier circuit.

13.2 GASEOUS-TUBE TYPE

A small amount of suitable gas sometimes is introduced into an electron tube. Electrons from the filament or cathode, traveling toward the plate, strike the gas atoms and knock loose more electrons which

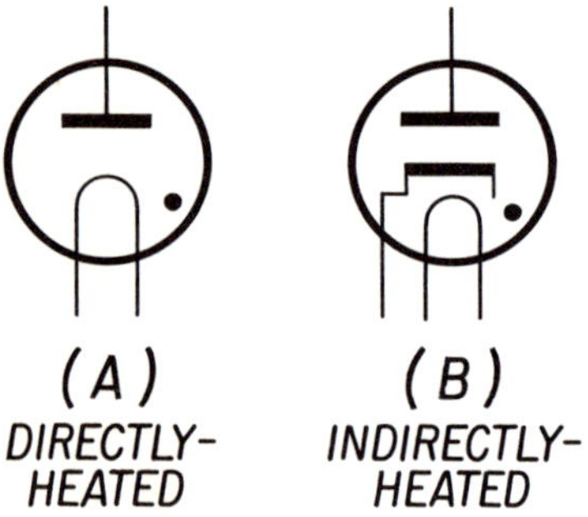

Fig. 13–4 Gaseous-Type Rectifier Tubes

then join the flow and *increase* the plate current. Thus, the gas is ionized by collision. (See Sections 1.9, Chapter 1, and 4.1 and 4.4, Chapter 4.)

Large, heavy-duty rectifier tubes which pass high plate currents often contain mercury vapor. This material ionizes readily and gives rise to heavy current flow.

Figure 13–4 shows the circuit symbols for gas-filled rectifier tubes. Note that the only difference from the vacuum-tube symbol is the large black dot in the lower right-hand region of the tube envelope.

13.3 SEMICONDUCTOR TYPES

Certain crystalline materials, called *semiconductors,* when suitably processed have the peculiar ability to conduct current more readily in one direction than in the other. They are not true valves like the vacuum tube, but act like very low resistances when current flows through them in one direction and like extremely high resistances when the current is in the opposite direction.

If a piece of this material is provided with electrodes and connected correctly in an a-c circuit, it will pass current more or less readily during one half-cycle but will allow only a trace of current to flow during the opposite half-cycle. The material thus serves as a rectifier. The curve in Figure 13–5 shows the typical response of a semiconductor.

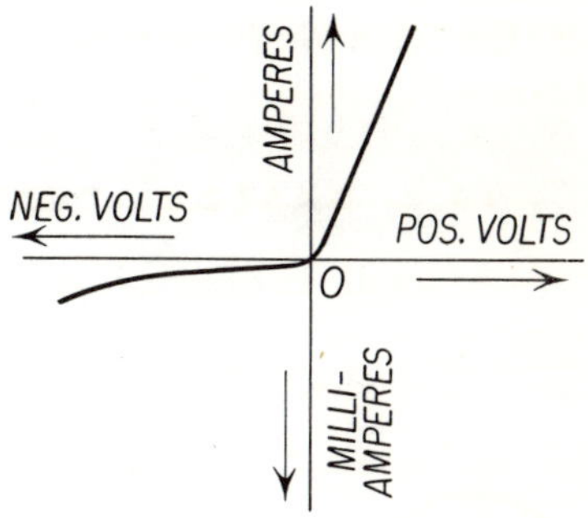

Fig. 13–5 Typical Response of a Semiconductor Rectifier

When the applied voltage is positive, for example, the current flow is in amperes; when the voltage is negative, the current drops to milliamperes. The high-valued current is called *forward current* and is due

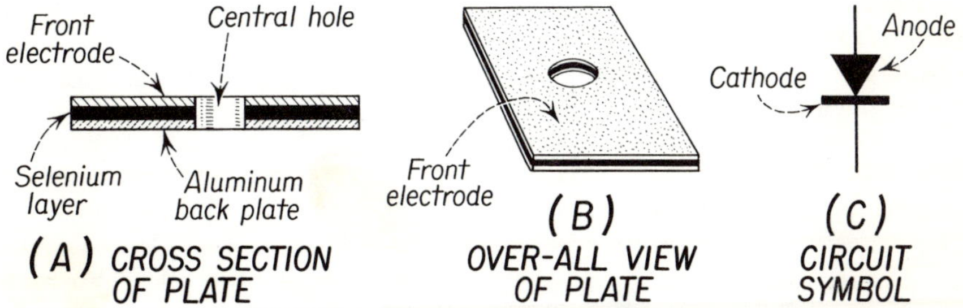

Fig. 13–6 Selenium Rectifier Construction

to the application of *forward voltage.* The weak, opposite current is termed *back* or *reverse current* and is due to the application of *reverse* or *back voltage.*

Figure 13–6 shows the basic construction of a rectifier plate employing *selenium* as the semiconductor. The specially processed selenium is applied evenly to an aluminum plate or disc. This plate is called the *back electrode.* The surface of the selenium layer is sprayed with a low-melting-point alloy, and this metal layer is called the *front electrode.* The rectifier plate thus consists of a sandwich with the selenium between two electrode plates. Flat, spring spiders or lugs make press contact with the front and back electrodes.

When a source of emf is connected to the plate, large values of current flow when the positive terminal is connected to the back electrode and the negative terminal to the front electrode, but only a small amount of current flows when the polarity is reversed. The back electrode, accordingly, has been termed the anode, and it corresponds to the plate of the tube. The front electrode is the cathode, and it corresponds to the filament or cathode of the tube.

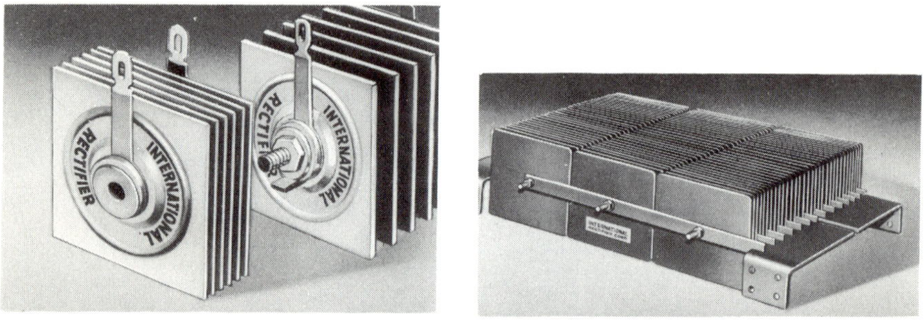

Plate 13-1(A) Small Selenium Rectifiers for Applications under 1 Ampere. (B) Large, Heavy-Current, Three-Phase Selenium Rectifier. (Courtesy, International Rectifier Corp.)

The larger the area of a rectifier plate, the more current can be handled. In order to withstand higher applied voltage than a single plate can handle, several rectifier plates may be connected in series simply by stacking them together (anode to cathode) on an insulating screw passed through the central holes of the plates. Stacked rectifiers are shown in the accompanying photographs.

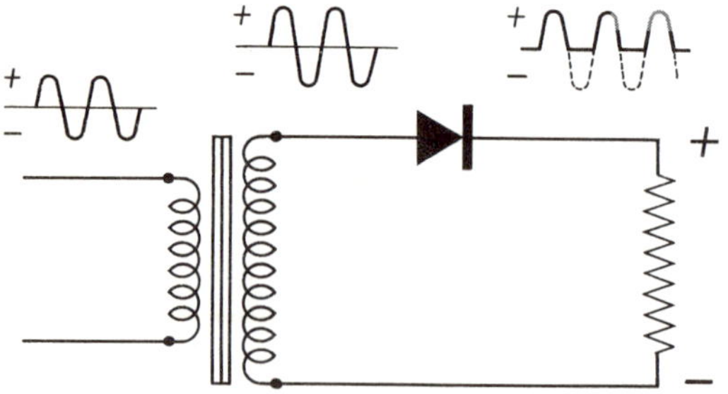

Fig. 13–7 Semiconductor Rectifier Circuit

Figure 13–6(C) shows the circuit symbol for the semiconductor rectifier. This character also serves as a general-purpose rectifier symbol which, for simplicity, often appears in circuit diagrams to represent a tube rectifier. Figure 13–7 shows a simple, semiconductor rectifier circuit. Note that this is similar to the tube circuit (Figure 13–2) except for the absence of the filament and filament-heating winding.

Other kinds of semiconductor rectifiers employ copper oxide, germanium, magnesium-copper sulphide, silicon, or titanium dioxide as the semiconductor material. Not all are of the plate type shown for selenium rectifiers. An advantage of all semiconductor rectifiers, regardless of type, is the absence of a filament. Power is required to heat the filament of a tube, and this reduces the over-all efficiency of the unit. Semiconductor rectifiers are simpler than tubes, as are also the transformers which need no filament windings. However, the voltage ratings of semiconductor rectifiers do not suit them to all applications in which tubes are employed.

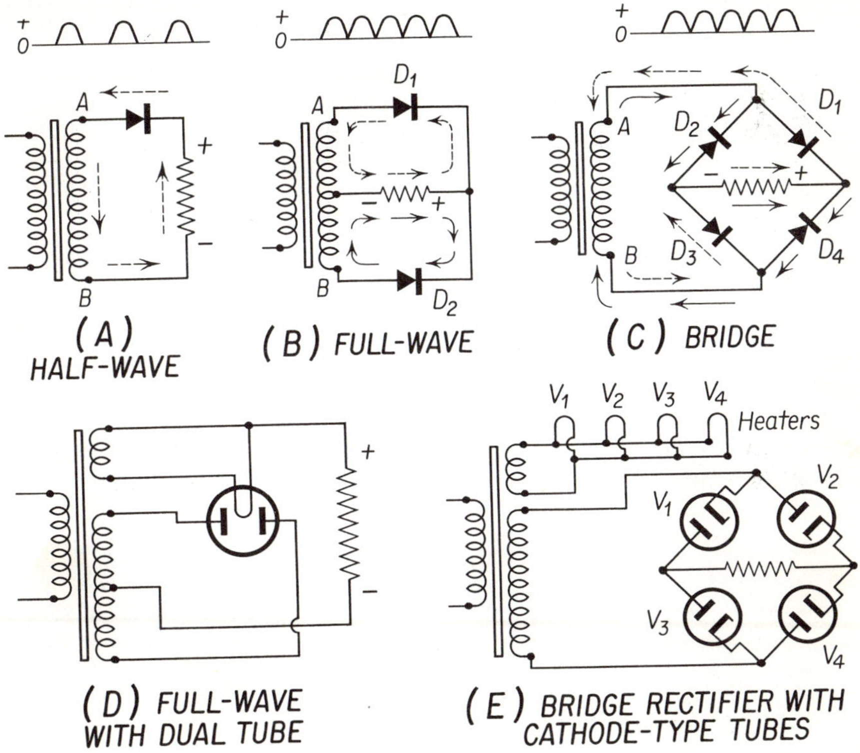

Fig. 13–8 Single-Phase Rectifier Circuits

13.4 HALF-WAVE CIRCUIT

Figure 13–8(A) shows the simplest rectifier circuit. The symbol represents either a tube or a semiconductor rectifier.

When Terminal *A* of the transformer secondary swings positive, electrons pass through the rectifier, load resistor, and transformer winding in the direction indicated by the dashed arrows. When *B* swings positive, the rectifier blocks current flow. The top of the load resistor thus is positive and the bottom negative only during the intervals when the a-c cycle swings *A* positive and *B* negative. The d-c ripple pattern is shown above the circuit.

This *half-wave* circuit is so called because the d-c output is one half of the a-c wave.

13.5 FULL-WAVE CIRCUIT

Two rectifiers and a center-tapped secondary winding are employed in the circuit of Figure 13–8(B).

In this circuit, when Terminal *A* swings positive with respect to the center tap of the secondary winding, electron current flows through Rectifier D_1, the load resistor, and the upper half of the winding in the direction of the dashed arrows. This makes the right-hand end of the resistor positive and the left-hand end negative. At this time, Terminal *B* is negative, and no current can flow through Rectifier D_2.

When the a-c cycle reverses, Terminal *B* swings positive, and electron current flows through Rectifier D_2, the load resistor *in the same direction as before,* and the lower half of the winding in the direction of the solid arrows. Again, the right-hand end of the load resistor is positive. The wave pattern over the diagram shows that a d-c pulsation is obtained for *each* a-c half-cycle, but this pulse always is of the same polarity.

This circuit is termed a *full-wave* rectifier because it utilizes the entire a-c wave. In all single-phase, full-wave rectifier circuits, the ripple frequency is twice the a-c supply frequency.

Figure 13–8(D) shows a full-wave circuit employing a dual tube (or *twin tube*). This tube actually consists of two separate tubes enclosed in a single bulb for convenience and compactness. A single filament is passed between both plates.

13.6 BRIDGE CIRCUITS

Figure 13–8(C) shows a full-wave circuit which does not require a center-tapped secondary on the transformer.

In this circuit, when Terminal *A* of the secondary winding swings positive, electron current passes from Terminal *B* in the direction of the dashed arrows, through Rectifier D_3, the load resistor, Rectifier D_1, and back to secondary terminal *A*. During this half-cycle, the anodes of Rectifiers D_2 and D_4 are negative, and these units cannot conduct. The right-hand terminal of the load resistor accordingly is positive and the left-hand terminal negative.

When the a-c cycle reverses, Terminal *B* becomes positive. Electron current now flows from Terminal *A* in the direction of the solid arrows, through Rectifier D_2, through the load resistor *in the same direction as before,* Rectifier D_4, and back to secondary terminal *B*.

Current flows through the resistor during each a-c half-cycle but does not reverse. This fulfills the conditions for full-wave rectification.

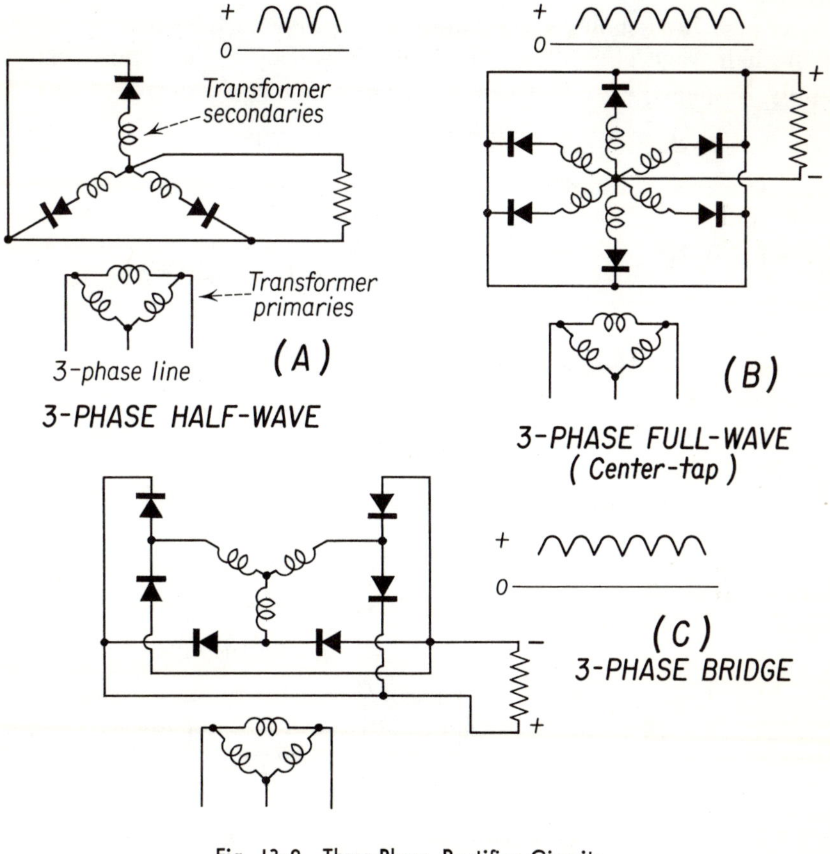

Fig. 13–9 Three-Phase Rectifier Circuits

Figure 13–8(E) shows another bridge rectifier operating in the same way but employing cathode-type tubes. The tubes might have been represented by the rectifier symbols given in Figure 13–8(C), but a separate diagram is given here to show the filament connections. The filaments are wired in parallel and connected to the single filament winding on the transformer. When filament-type tubes are used, however, separate filament windings must be employed for all except Tubes V_2 and V_4 whose filaments may be operated in parallel from one winding. The reason for this is that the filaments are the cathodes in filament-type tubes, and V_2 and V_4 are the only ones with cathodes that may be connected together.

A disadvantage of the bridge rectifier circuit is the fact that there is no common terminal between its input and output circuits. This makes it impossible to ground *both* the input (a-c) and output (d-c) circuits.

Figure 13–9 shows several three-phase rectifier circuits. Note that in the half-wave circuit the d-c ripple frequency is three times the a-c supply frequency. In the full-wave circuit, it is six times.

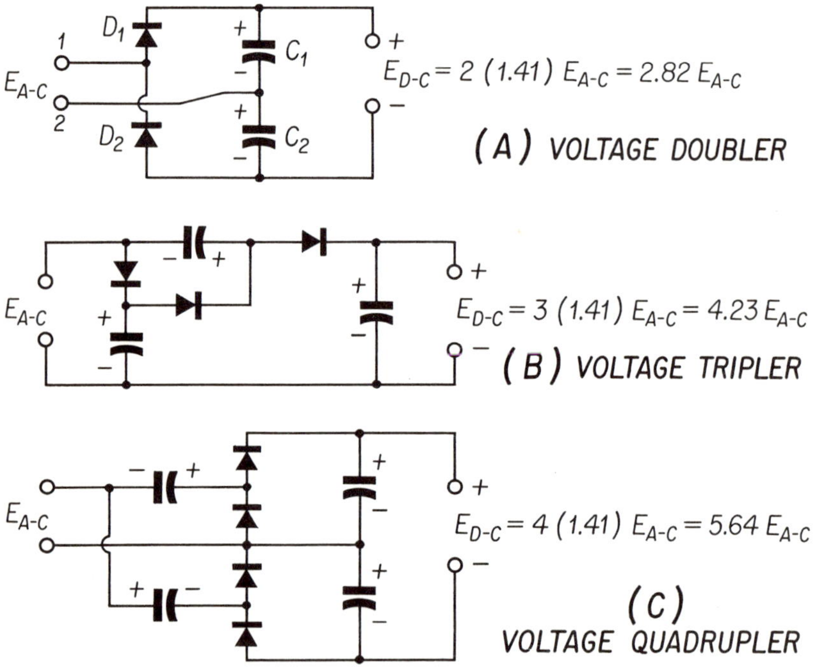

Fig. 13–10 Voltage Multiplier Circuits

13.7 VOLTAGE-MULTIPLIER CIRCUITS

Rectifiers may be combined with capacitors to deliver a d-c output voltage equal to several times the rms value of the applied a-c voltage. Several of these *voltage-multiplier* circuits are given in Figure 13–10.

Figure 13–10(A) shows a *voltage doubler*. In this circuit, when a-c Terminal 1 swings positive, Rectifier D_1 conducts and Capacitor C_1 charges approximately to the peak value of E_{ac} (that is, to $1.414E_{ac}$ rms). During this half-cycle, Rectifier D_2 cannot conduct, since its anode is negative, and Capacitor C_2 is not charged. When Terminal 2 swings positive, Rectifier D_1 is blocked but D_2 conducts, charging Capacitor C_2 approximately to the peak value of E_{ac}. The total d-c voltage now across C_1 and C_2 in series is twice the peak value of E_{ac}, or $2.82E_{ac}$. Capacitances C_1 and C_2 are made high enough to prevent their discharge during reversal of the a-c cycle.

In the *voltage tripler* [Figure 13–10(B)], a doubler and half-wave rectifier have been combined to give a d-c output voltage equal to three times the peak value of the applied a-c voltage. Here, $E_{dc} = 4.23E_{ac}$. Each capacitor has a charged voltage equal to one-third of this d-c output voltage.

The *voltage quadrupler* circuit [Figure 13–10(C)] combines two voltage doublers and has a d-c output voltage equal to approximately four times the peak value of the applied a-c voltage. Here, $E_{dc} = 5.64E_{ac}$.

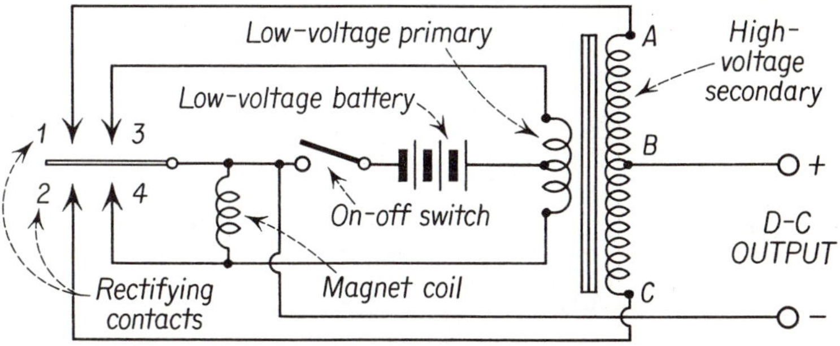

Fig. 13–11 Vibrator-Type Rectifier

13.8 VIBRATOR-TYPE RECTIFIERS

Section 10.10, Chapter 10, describes and Figure 10–6 illustrates a vibrator-type transformer.

It is possible also to employ vibrating contacts to rectify alternating current. Figure 13–11 illustrates a vibrator-type power supply which is provided with an extra pair of contacts (1, 2) for rectification.

As the vibrating reed moves back and forth between Contacts 1 and 2, it connects the bottom d-c output terminal alternately to the top (*A*) and bottom (*C*) of the transformer secondary. Terminal *A* is positive or negative, depending upon whether the reed is at Contact 3 or 4, and the same is true of Terminal *C*. Movement of the reed thus keeps the bottom d-c terminal always negative with respect to the center-tap connection, *B*. Since the same vibrating reed closes the primary circuit, thus initiating the generation of alternating current in the secondary, the switching of the d-c output circuit is in synchronism with the making and breaking of the primary winding in each direction.

Inspection of Figure 13–11 shows that a complete power supply furnishing high-voltage direct current from a low-voltage d-c source is obtained by employing a combination vibrator for both a-c generation and rectification.

13.9 FILTERS

In the preceding sections, it was shown that the output of a rectifier is a pulsating direct current. The ripple varies in frequency, being equal to the a-c supply frequency (f) in single-phase half-wave rectifiers, $2f$ in single-phase full-wave circuits, $3f$ in three-phase half-wave circuits, and $6f$ in three-phase full-wave circuits. The ripple amplitude is lower in three-phase than in single-phase circuits, and the amplitude is lower and frequency higher in higher-order polyphase circuits.

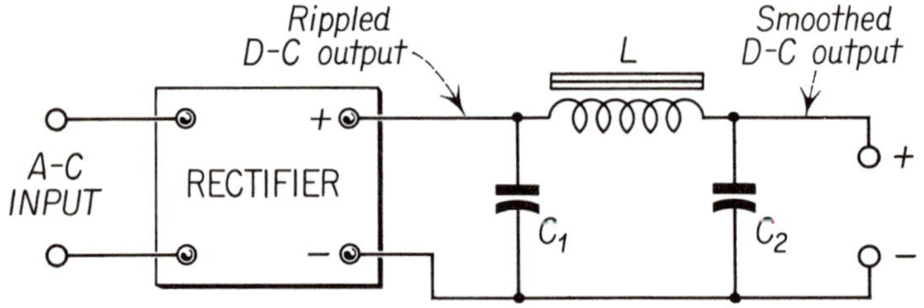

Fig. 13–12 Rectifier Filtering

In practice, the rippled output of a rectifier is smoothed by means of a filter circuit. Filters utilize the reactive properties of capacitors and inductors, usually in combination, to obtain this action.

Figure 13–12 shows a common rectifier filter. Capacitor C_1 cannot pass direct current but short-circuits the ripple component present in the d-c output of the rectifier. The suppression is not complete, however, so the self-induction of Coil L then is utilized to choke out as much as possible of the remaining ripple. If its resistance is kept low, this inductor readily passes the direct current, while opposing the ripple. Here again, however, the suppression is not complete, so Capacitor C_2 is provided to *by-pass* the remainder of the ripple. Comparatively smooth direct current therefore is obtained at the output terminals of the filter.

A small amount of ripple is present even after good filtration; however, it is kept to very low values in high-quality equipment. Filter capacitances are chosen for low reactance at the ripple frequency, while filter inductances are chosen for high reactance and low d-c resistance.

Figure 13–13 shows some of the filter circuits employed for smoothing the output of rectifiers. Several sections are employed in Figure 13–13(A) for additional filter action. The *capacitor-input* class of filter, represented by Figure 13–13(B), provides the highest output voltage for a given input, while the *choke-input* class [Figure 13–13(C)] provides the best voltage regulation. The simplest and most compact of

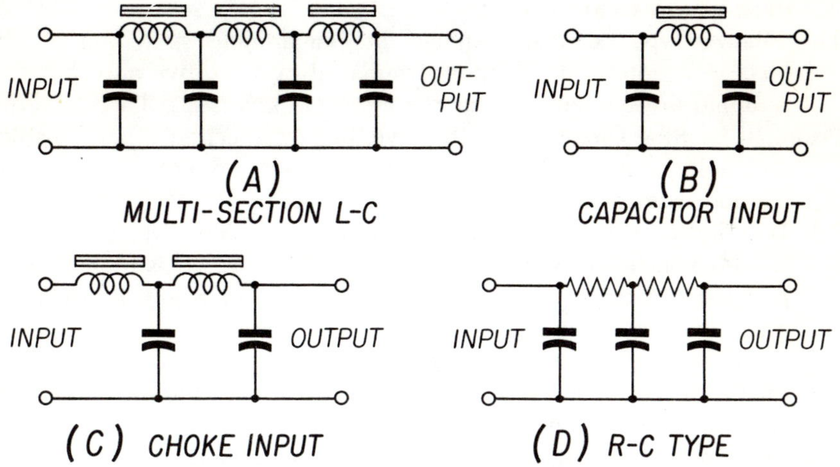

Fig. 13–13 Power Supply Filters

all filters is the *resistance-capacitance* type [Figure 13–13(D)]. However, this type does not have the added choking effect of the inductor, and its filtering action is effective only at low d-c drains. Furthermore, the presence of the resistances in series with the rectifier and load introduces a voltage drop which may be intolerable in some applications.

SUMMARY OF CHAPTER 13

The **Rectifier** is a device for converting alternating current into direct current. Typical rectifiers are vacuum tubes, gaseous tubes, semiconductors, and vibrators. However, these are not the only possible rectifiers. All rectifying devices operate on the basis of one-way conduction.

In the **Two-Element Vacuum Tube,** a hot filament or cathode emits electrons which are attracted by the positively charged plate but repelled when the plate is negative. The *plate current* can flow only when the plate is positive, such as during the positive half-cycle of a-c plate voltage.

In a **Gaseous Tube,** the plate current flow is enhanced by ionization of the gas atoms by colliding electrons.

The **Semiconductor Rectifier** is not a true one-way valve like the vacuum tube. Instead, it conducts current more readily in one direction (anode positive) than in the other (anode negative). The ratio of forward current to reverse current in these rectifiers may reach several hundred thousand to 1. Semiconductor rectifiers (copper oxide, germanium, magnesium-copper sulphide, silicon, and titanium dioxide) have no filaments.

The **Vibrator-Type Rectifier** employs the synchronous switching of contacts to keep one d-c output terminal always positive and the other terminal always negative, as the a-c cycle reverses polarity.

Principal **Rectifier Circuits** are half-wave, full-wave center-tapped, bridge, and voltage multiplier. Both single-phase and polyphase circuits employ these arrangements.

The **D-C Output** of a rectifier is a ripple current. The **Ripple Frequency** is higher in full-wave circuits than in half-wave. It is higher still in polyphase circuits, and here the ripple amplitude is lower. In single-phase systems operated at an a-c supply frequency f, the ripple is equal to f in a half-wave circuit, and $2f$ in full-wave. In a three-phase half-wave circuit, ripple frequency $= 3f$, and in three-phase full-wave circuits, it is $6f$.

The rippled d-c output of a rectifier may be smoothed by passing it through a suitable **Filter.** This device utilizes the reactance of inductors and capacitors to remove the fluctuating component and pass the steady component of the d-c output. Filters may be classed as *single-section, multisection, capacitor-input, choke-input, inductance-capacitance,* and *resistance-capacitance.*

QUESTIONS AND EXERCISES

1. Why does plate current flow in a vacuum-tube rectifier only when the plate is positive?

2. Explain the phenomenon of plate-current saturation in a two-element vacuum tube.

3. Why does higher plate current flow in a gaseous tube?

4. What is the principal difference between a filament-type and cathode-type rectifier tube?

5. (*a*) Describe the structure of a selenium rectifier plate. (*b*) Define *forward current.* (*c*) Define *reverse* ("back") *current.*

6. Draw the following single-phase rectifier circuits: (*a*) half-wave; (*b*) full-wave center-tapped; (*c*) bridge.

7. Draw the following three-phase rectifier circuits: (*a*) half-wave; (*b*) full-wave center-tapped; (*c*) full-wave bridge.

8. Draw the following voltage-multiplier circuits: (*a*) voltage doubler; (*b*) voltage tripler; (*c*) voltage quadrupler.

9. Explain the operation of a vibrator-type rectifier.

10. Draw the following filter circuits: (*a*) two-section *L-C* with choke input; (*b*) single-section *L-C* with capacitor input; (*c*) three-section *R-C* with capacitor input.

Chapter 14

Essentials of Electrical Wiring

The purpose of this chapter is to acquaint the reader with the rudiments of electrical wiring and to familiarize him with some of the materials used in this work. It is not expected that this single chapter will take the place of a workbook. Neither can it possibly cover the subject as fully as a shop manual or textbook of wiring practice. The detailed mechanics of wiring are readily available elsewhere.

The reader of this book who intends to pursue further the subject of industrial electricity must consult additional texts for detailed instruction in the actual wiring of structures. He must also make a careful study of the *National Electrical Code.* (See Reference 1 at the end of this chapter.) The appended reference list contains the titles of several books which will be found helpful to the student of wiring practice.

The student must familiarize himself with all ordinances in his locality governing electrical work and the licensing of electricians. He should attempt no actual wiring, except under the strict supervision of his instructor or a qualified electrician, until he has completed his training and, if required by local regulations, has become duly licensed as an electrician.

14.1 CONDUCTORS

It was explained in Chapter 2 that every circuit consists of one or more generators connected to one or more electrically operated devices by means of suitable wires in such a way that current may pass from the generator, through the devices, and back to the generator. The purpose of all wiring practice is to set up these circuits in an approved and safe manner. Much of the actual work in this direction consists of running and connecting the wiring.

The wires employed in electrical circuits are more properly known as *conductors*. Many types of conductors are used. The one selected for a particular application depends upon the manner in which it must be installed, the environmental conditions under which it is to be operated, and the amount of current it must carry. The provisions of the Code often are the determining factors.

Copper wire is used almost universally because of the good conductivity of this metal. Some conductors consist of a single, solid copper wire, while others have a number of separate wires (*strands*) twisted together to give a larger diameter while at the same time providing more flexibility than a single, solid wire of that diameter. While bare wire is used for some purposes, most conductors are insulated by one or more protective jackets of nonconducting material, usually plastic plus fabric. Asbestos is used also for wires connected to devices that generate heat electrically.

Conductor characteristics and data may be found in standard wire tables. An important characteristic is wire diameter. The larger the diameter, the lower the resistance per unit length. This means that the larger the diameter of the conductor the more current can be carried without overheating. Wire tables show diameter in *mils* (thousandths of an inch) and area in *circular mils*. A circular mil is the area of a 1-mil-diameter circle. American Wire Gauge size numbers for wire extend from No. 0000 (460 mils) to No. 40 (3.145 mils).

For simplicity and convenience in installation, two insulated conductors often are run together in a single *pair* or *cable*. Depending upon type, these wires may be supplied twisted together, run parallel, or molded side by side (example, rubber-covered lamp "cord"). The separate conductors in cables employed in industrial wiring are identified by the color of the insulating jackets—one white for neutral, and other colors depending upon the number of conductors.

14.2 JOINTS AND SPLICES

When conductors are joined together, the junction formed must be tight. A loose connection not only can make and break the circuit intermittently when jostled, but such a connection also introduces harmful resistance into the circuit. The presence of this resistance lowers the efficiency of the circuit. Furthermore, heating results from the I^2R loss in the resistance, and this can lead to a burnout of the conductor at that point and possibly to fire damage to neighboring structures.

When two wires are joined, they must be twisted together (or wound tightly, one around the other) for a sufficient length to guarantee mechanical strength and a large area of intimate contact between the two. Electricians follow standard methods of splicing wires. (In addition to splices, certain other mechanical means for fastening wires together or joining them to terminals also are used in modern wiring practice. These include Stakons, "bugs," wire nuts, and similar devices.)

The first step in making a splice is to remove the insulation from the ends of the wires. (See Figure 14–1.) The insulation first is cut away, as shown in Figure 14–1(A), using a knife and tapering the inner rubber jacket to a cone but being careful not to cut or nick the wire, since cuts will cause the wire to break when bent. (This often is done

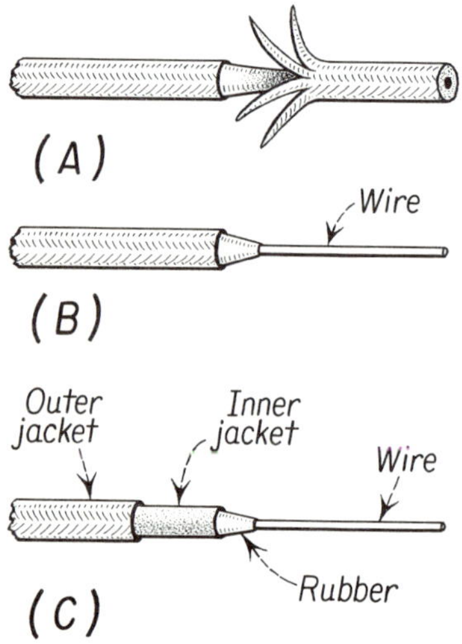

Fig. 14–1 Preparing End of Wire

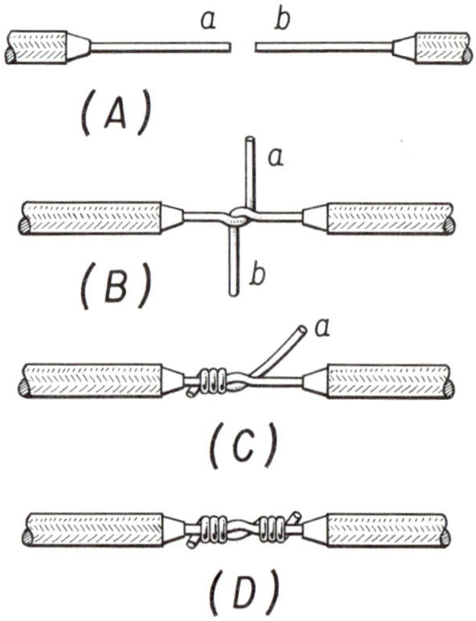

Fig. 14–2 Western Union Splice

with a wire stripper.) The exposed wire [Figure 14–1(B)] must be long enough to allow a complete joint to be made. This will vary from 1 to 3 in., depending upon circumstances. A short piece of the outer jacket of the insulation then is cut away, as in Figure 14–1(C). Finally, the entire length of each exposed wire must be cleaned thoroughly by scraping with a blunt knife blade or with edge of pliers until the wire shines brightly.

Western Union Splice. Figure 14–2 shows the steps in making a Western Union splice. This is a widely used and efficient type of joint. In (A), two conductors, the ends of which have been prepared according to the steps shown in Figure 14–1, are brought together. In (B), they are overlapped, leaving the free ends long enough for twisting around the captive portions. In (C), the free end of *b* is twisted tightly around the captive end of *a*, laying the turns neatly side by side and in contact with each other. The joint is completed in Figure 14–2(D), by similarly twisting the free end of *a* around the captive end of *b*.

Tap Joint. It frequently is required that the end of one conductor be connected to an intermediate point along the length of another. This is called *tapping,* and the splice is termed a tap joint. Figure 14–3 shows the steps in making a tap joint. In (A), the insulation is removed carefully from a length of Conductor 1 to which the connection

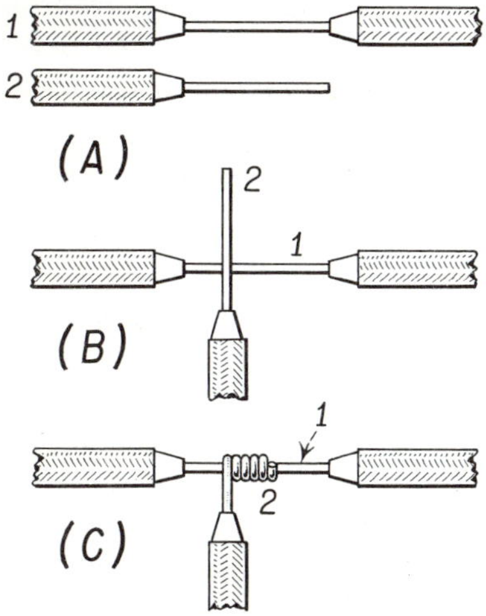

Fig. 14–3 Steps in Making a Tap Joint

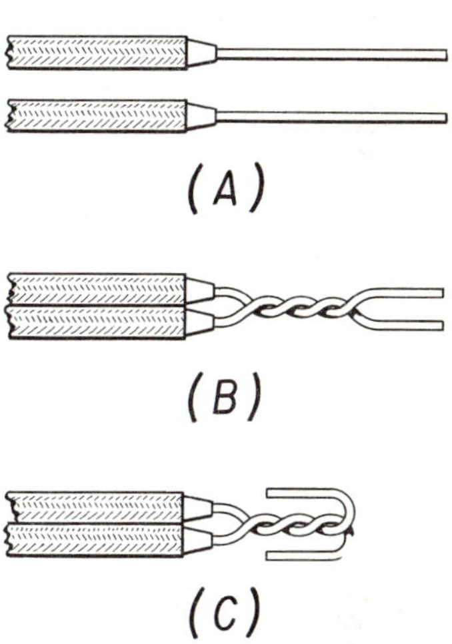

Fig. 14–4 Steps in Making a Rat-Tail Splice

is to be made. Also, the end of Conductor 2 has been prepared according to the instructions given in Figure 14–1. In (B), Conductor 2 is placed in contact with Conductor 1 at right angles. Finally, in (C), the free end of Conductor 2 is twisted tightly around Conductor 1 to complete the splice. The twisted portion should be long enough to ensure mechanical strength and a large-area, intimate electrical contact. The end of Conductor 2 should come up to the insulation on Conductor 1 to help prevent puncturing tape which later may be wrapped around the joint.

Rattail Splice. Sometimes, it is necessary to join the ends of two conductors running parallel in the same direction. This requirement is the opposite of that served by the Western Union splice where the two conductors are in line and, when spliced, become, in effect, a single conductor. Figure 14–4 shows the steps in making a *rattail* splice which serves this purpose. In (A), the ends of the conductors have been prepared in the manner illustrated by Figure 14–1. In (B), the wires are twisted tightly together over a sufficiently long length to ensure mechanical strength and good electrical contact, but leaving two free ends approximately equal in length to that of the twisted joint. In (C), these free ends have been bent back over the joint, and they finally are pressed tightly against the latter.

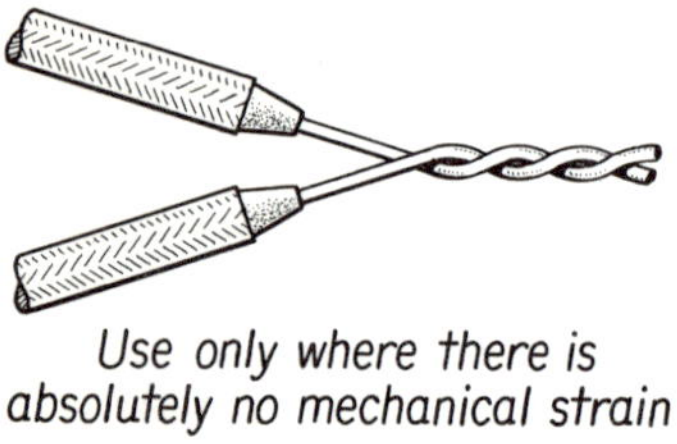

Fig. 14–5 Simple Twist Joint

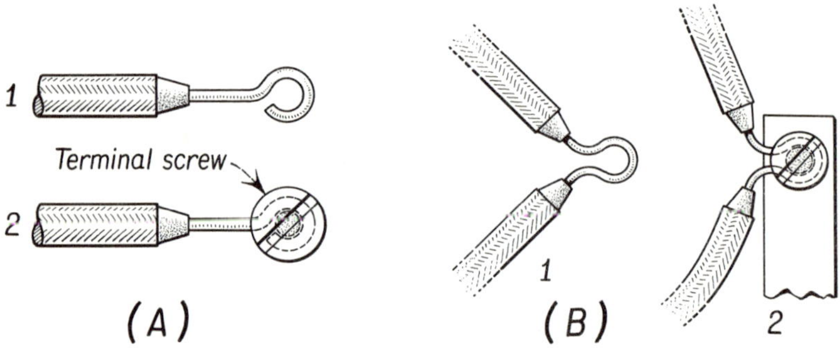

Fig. 14–6 Connection of Conductor to Terminal Screw

Simple Twist Joint. The ends of the wires of two conductors may be twisted together in the simple manner shown in Figure 14–5, without the final bending back of the tips which is a part of the process of making a rattail splice [Figure 14–4(C)]. However, under ordinary circumstances, this type of splice, the *simple twist joint,* should be used only when the connection is temporary. If it must be permanent, there must be absolutely no mechanical strain on the joint, since a pull on either of the conductors will tend to untwist the joint, sometimes even though it has been soldered.

Connecting Conductors to Terminal Screws. Conductors often must be connected to the terminal screws of accessories, wiring devices, and instruments. Figure 14–6 illustrates the proper methods of making such connections.

Figure 14–6(A) shows the ordinary case. In 1, the end of the conductor has been prepared according to the method shown in Figure 14–1, and the wire has been bent into a ring just large enough to clear the threaded shaft of the terminal screw. In 2, the screw has been passed through the ring and driven home to hold the wire tightly. The ring must be placed under the screw in such a way that the ring rotates clockwise from the point at which the wire emerges from the insulation. The reason for this is that the screw turns clockwise and would tend to open the ring, allowing it to slip from under the screw if the ring were turned over on its opposite (counterclockwise) face. Crimp-on screw lugs also are used. They are fastened to the stripped wire by means of a crimping tool.

Figure 14–6(B) shows how a certain part of a conductor may be attached to a terminal screw without breaking the wire. In 1, a short length of insulation has been cut from the conductor, and the wire has been bent into the shape of a ring just large enough in diameter to clear the threaded shaft of the terminal screw. In 2, the screw has been passed through the ring and driven home to provide a tight connection.

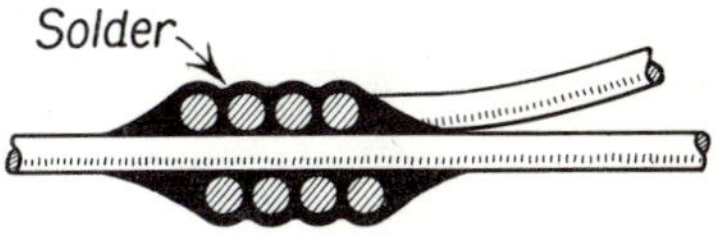

Fig. 14–7 Soldered Joint

Soldering Joints. For additional mechanical strength and protection, all splices should be soldered. Figure 14–7 illustrates the cross section of a soldered joint. This shows the turns of one wire twisted around another, and the solder covering the splice. Enough heat must be applied to the joint during the process to flow the solder into all of the crevices between wires.

When a joint is soldered correctly, it is more efficient electrically as well as mechanically.

Taping Joints. A joint or splice must be insulated as a protection against accidental contact, short circuit, or grounding. The simplest way to do this is to wrap the joint tightly with electrical tape. Rubber tape often is applied first, and this is followed with a wrapping of friction tape. Some electricians use plastic tape.

Figure 14–8 illustrates the taping of a completed joint. Notice that the tape is wound tightly around the joint in one direction [Figure 18–8(A)] to complete the bottom layer and then is wound back to the starting point in the opposite direction [Figure 18–8(B)] to complete the top layer. Start the winding with the tape on the insulation of one wire, proceed to the insulation of the other wire, and end it on top of itself on the insulation of the first wire.

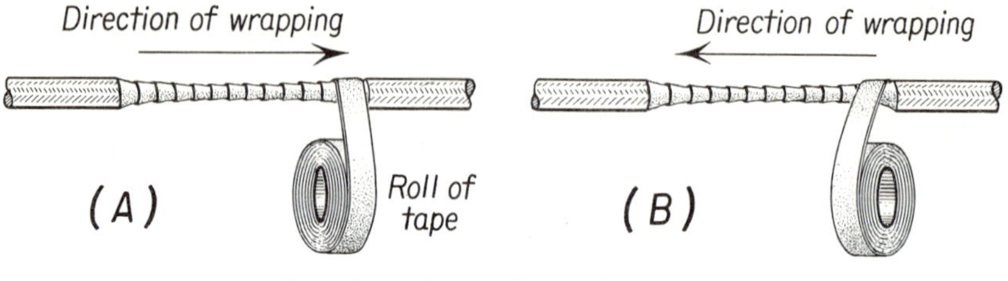

Fig. 14–8 Taping of Completed Joint

14.3 TYING ENDS OF CONDUCTORS

The ends of flexible, insulated conductors should be tied together in a knot for strain relief when the cord containing these conductors passes through a clearance hole in the appliance or fitting to which it is attached. When the knot is made somewhat larger than the clearance hole, any mechanical pull on the cord will be relieved by the knot and

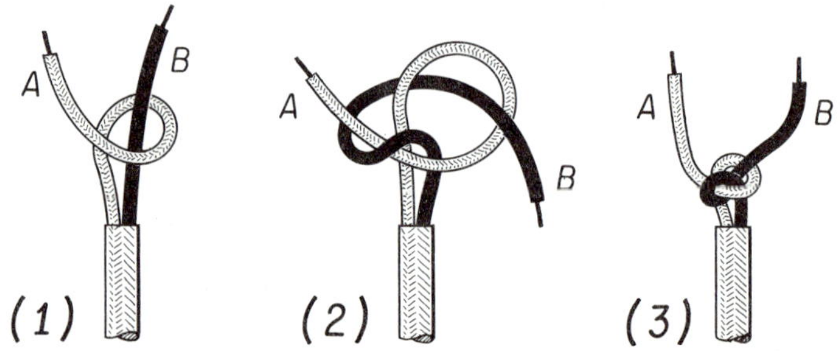

Fig. 14–9 Steps in Making a Strain-Relieving, Underwriters' Knot

thus cannot reach the wires and break them away from their points of connection.

Figure 14–9 shows the steps in making an *Underwriters' knot* which is very effective in this application. These steps are self-explanatory. Notice the use of this knot in the examples in Figures 14–10 and 14–11(C).

14.4 WIRING OF LAMP SOCKET

Figure 14–10 illustrates the wiring of a lamp socket. This example is presented because it is commonplace. However, it is typical of the attachment of a two-wire cord to many other appliances and devices in which the internal connections are made by attaching the wires of a flexible cord to terminal screws and the cord must be provided with a strain-relieving Underwriters' knot.

The first step is to pass the cord through the clearance hole in the cap. The two conductors then are exposed by cutting away a length of the outer insulating jacket. The ends of the wires then are exposed by cutting away a length of the inner insulating jacket of each. The strands of each conductor are cleaned and twisted tightly together. The Underwriters' knot then is tied, the wires are secured under the terminal screws, and the socket is reassembled.

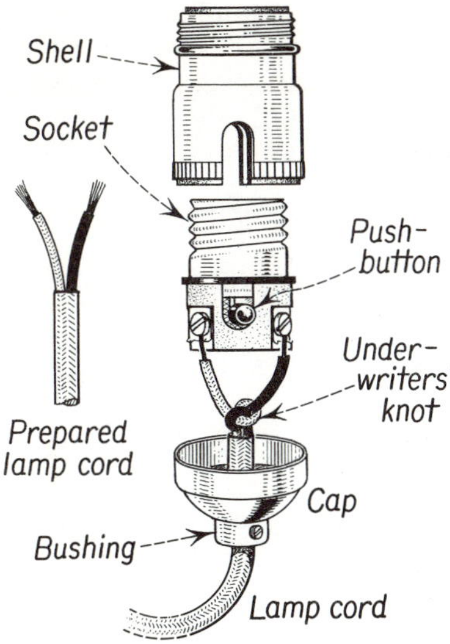

Fig. 14–10 Wiring of Lamp Socket

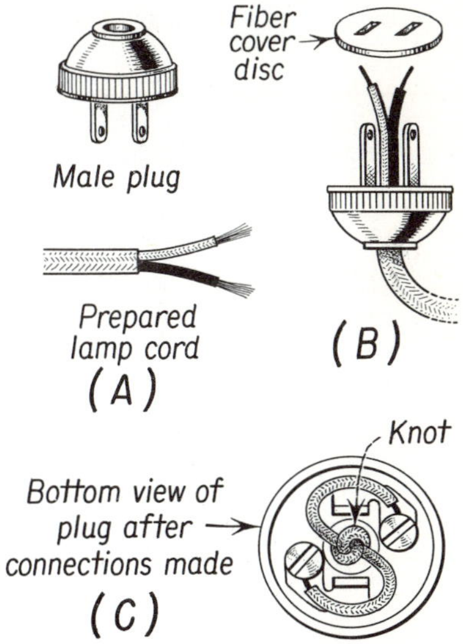

Fig. 14–11 Wiring of Male Plug

14.5 WIRING OF MALE PLUG

Figure 14–11 illustrates the wiring of a male plug. In this instance, flexible lamp cord is prepared and inserted through the clearance hole, and the Underwriters' knot is tied [Figures 14–11(A) and (B)], as explained in the preceding example of the wiring of the lamp socket. After the knot has been tied, each insulated wire is passed around the outside of one of the prongs of the plug, and its wire is fastened to the corresponding terminal screw. This provides additional strain relief.

After the connections have been completed, the prong-clearing fiber cover is pressed into place for protection and insulation.

14.6 BOXES

Metal boxes with access holes (*knockouts* or *pry-outs*) for the entry of cables are provided for switches, junctions, and outlets. These boxes are available in rectangular, round, and octagonal shapes with various dimensions. In modern practice, all connections (splices and joints) are made inside boxes. This is a safety precaution: With wiring connections hidden inside the walls of a building, the danger of fire resulting from overheating in a joint, for example, is lessened when the joint is confined by a protecting box. Electric fixtures are hung from, or attached to boxes installed in the wall or ceiling. This technique, in addition to providing the protection just mentioned, also permits a continuous ground.

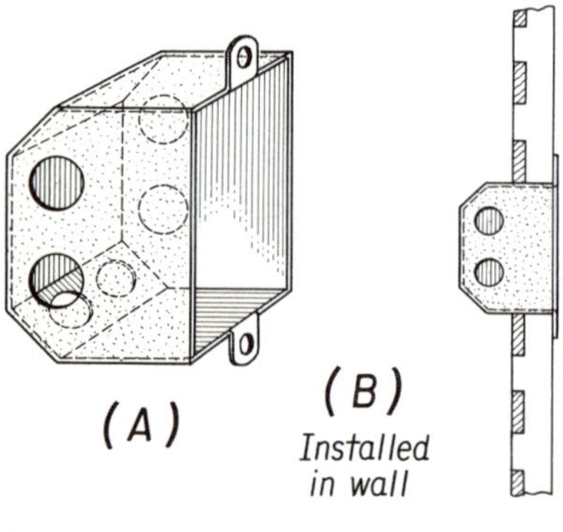

Fig. 14–12 Switch Box

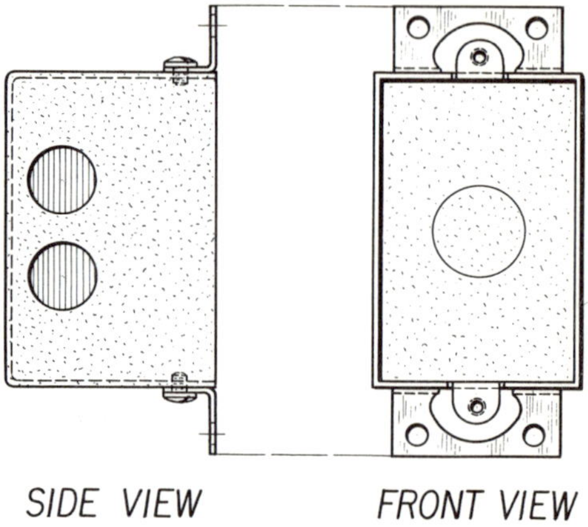

Fig. 14–13 Baseboard Receptacle Box

The student should examine his shop workbook and the catalogues of electrical supply houses for a survey of the numerous types of boxes and fixtures available. Figure 14–12(A) shows the configuration of a common switch box, while Figure 14–13 shows a baseboard receptacle box. The switch mechanism and receptacle unit are fastened to the fronts (right-hand side in the drawings) of their respective boxes after they are installed.

Figure 14–12(B) is an illustrative cross section showing the installation of a switch box in a wall. All boxes must be mounted according to the provisions of the Code. Boxes for ceiling-light fixtures are attached to prescribed hangers.

14.7 KNOB-AND-TUBE WIRING

The knob-and-tube system was the earliest method of electrical wiring. It is still used in some localities but is banned by local ordinance in others. The Code forbids its use in some buildings and locations.

In this system, insulated conductors are supported by porcelain knobs or cleats spaced at regular intervals and are passed through

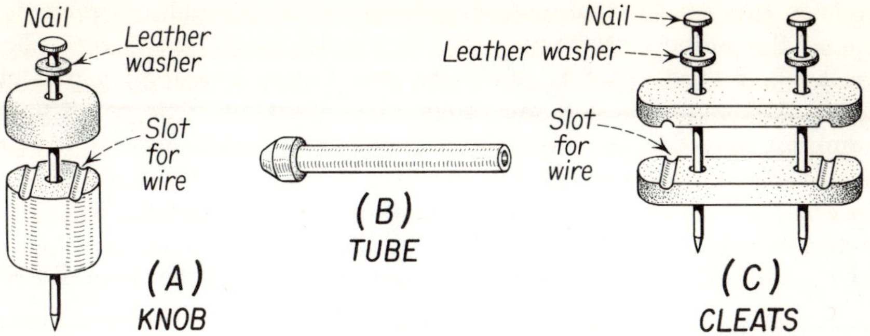

Fig. 14–14 Knob-and-Tube Materials

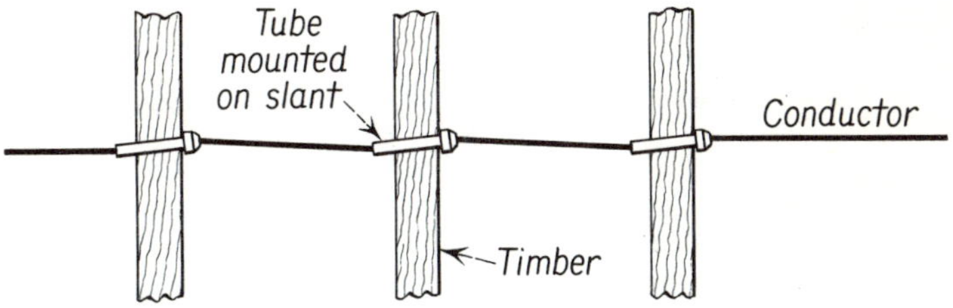

Fig. 14–15 Method of Inserting Tubes in Timbers

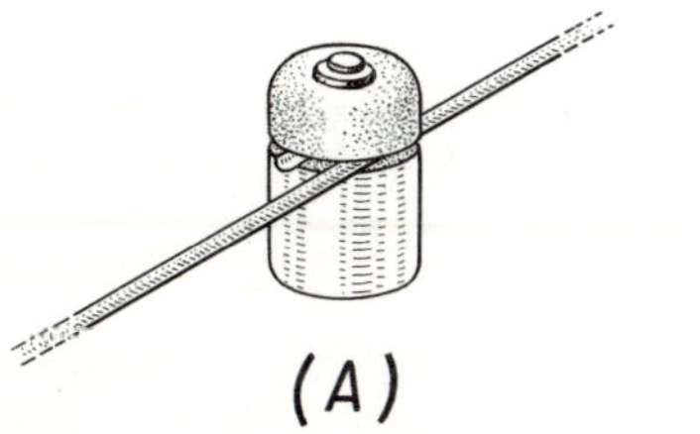

Fig. 14–16 Securing of Conductors by Knob and Cleats

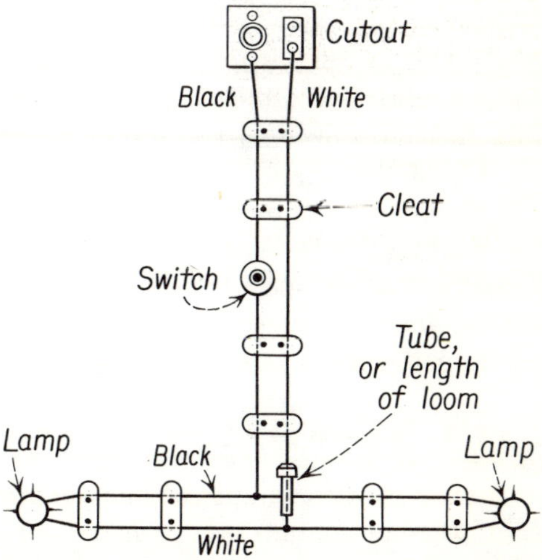

Fig. 14–17 Knob-and-Tube Wiring Installation

porcelain tubes inserted in clearance holes drilled through obstructions, such as the timbers of a building. Figure 14–14(A) shows the split-type knob, Figure 14–14(B) the tube, and Figure 14–14(C) a pair of cleats. Although two slots are provided in the knob, it is permissible to support only one conductor by a knob. The reason for this is that the distance between the slots is less than the spacing required for two wires by the Code. In cleats, the separation between slots is satisfactory for two wires. Figure 14–16(A) shows the method of using a knob, and Figure 14–16(B) the method of using cleats to support conductors.

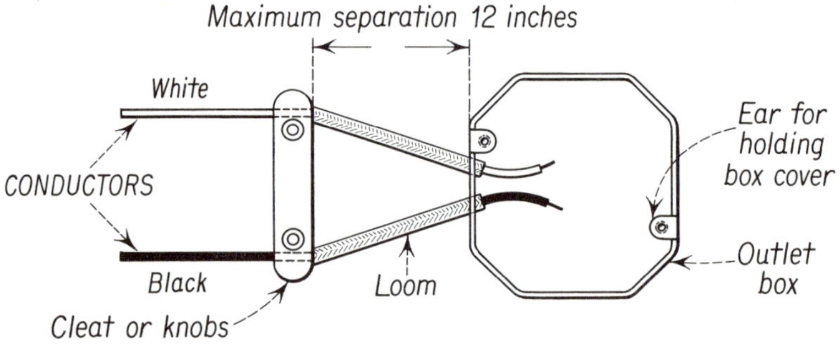

Fig. 14–18 Knob-and-Tube Wiring to Outlet Box

Figure 14–15 illustrates the use of porcelain tubes to pass a conductor through the timbers of a building. The tubes are inserted at an angle, slanting downward from their flanges, so that they will not slip out of their clearance holes.

Figure 14–17 is an illustration of typical knob-and-tube wiring, using two lamps operated from a single switch as an example. Where the wires cross each other, although each is insulated, a tube is slipped over one wire for additional protection. Sometimes, instead of using a tube, a length of *loom* (a kind of flexible, insulating tubing) is slipped over one of the wires.

Figure 14–18 shows the proper method of bringing the conductors in a knob-and-tube system into an outlet box or switch box. The conductors are gripped between cleats spaced not more than 1 ft from the box, and each is covered with a piece of loom extending from the cleat into the box.

14.8 NONMETALLIC SHEATHED CABLE

Romex is one example of moistureproofed and flameproofed flexible cable. It may contain one or more wires. The type shown in Figures 14–19 and 14–20 contains two insulated conductors (each having the rubber plus fabric insulating jackets discussed in Section 14.1). In addition, each conductor is covered by a spiral wrapping of paper tape, and the space between them is filled with a jute cord. This package then is covered with a stout outer braid which forms a nonmetallic sheath. Figure 14–19 shows the general configuration of Romex cable.

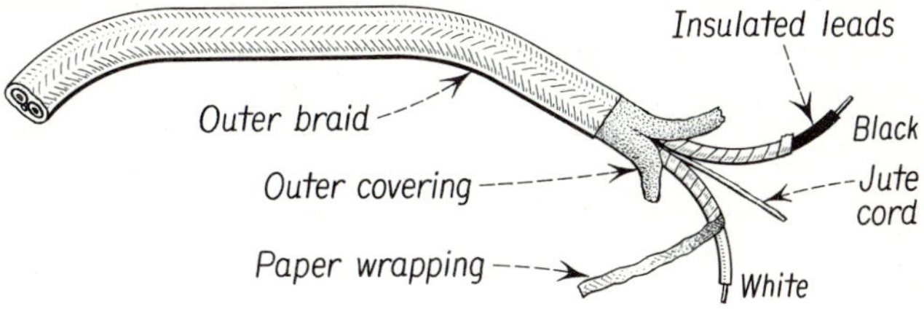

Fig. 14–19 Romex Cable

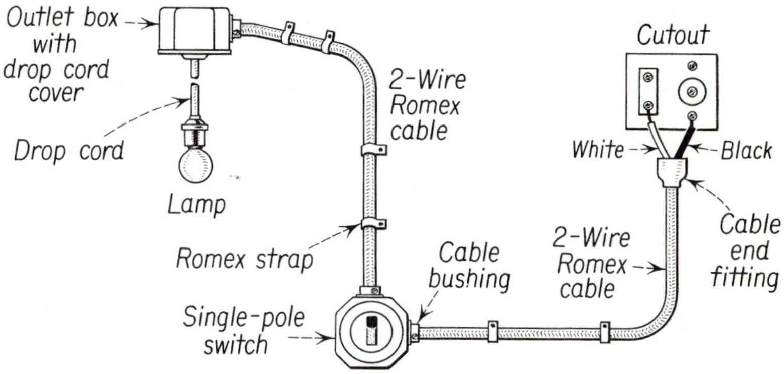

Fig. 14–20 Wiring Installation with Romex Cable

Curved metal straps are supplied for holding the cable to timbers, panels, and other supporting structures. Connectors are provided for fastening the cable solidly at the access holes of boxes.

Figure 14–20 is an example of the use of Romex cable. This is a simple wiring installation including a single switch and lamp, but it is illustrative of the neat manner in which this cable may be run. All taps, junctions, and splices must be made inside metal boxes.

Since Romex has a nonmetallic sheath, its use is permitted only in indoor locations known to be permanently dry. Its outdoor use is not sanctioned.

14.9 METAL RACEWAY

Conductors may be run inside ornamental metal tubing, or *raceway*, when above-surface wiring must be done in finished structures, as on the outside of walls, baseboards, ceilings, workbenches, etc. The conductors are laid in or pulled through the raceway which serves somewhat as a pipe. A well-known trade name for metal raceway is *Wiremold.* Wiremold is supplied in a neutral tan color but may be painted, after installation, to match surrounding surfaces.

Wiremold is obtained in 10-ft lengths [See Figure 14–21(A)] and is cut easily with a hack saw. Separate lengths may be joined together in the manner shown in Figure 14–21(B)–1 and the butt covered as in

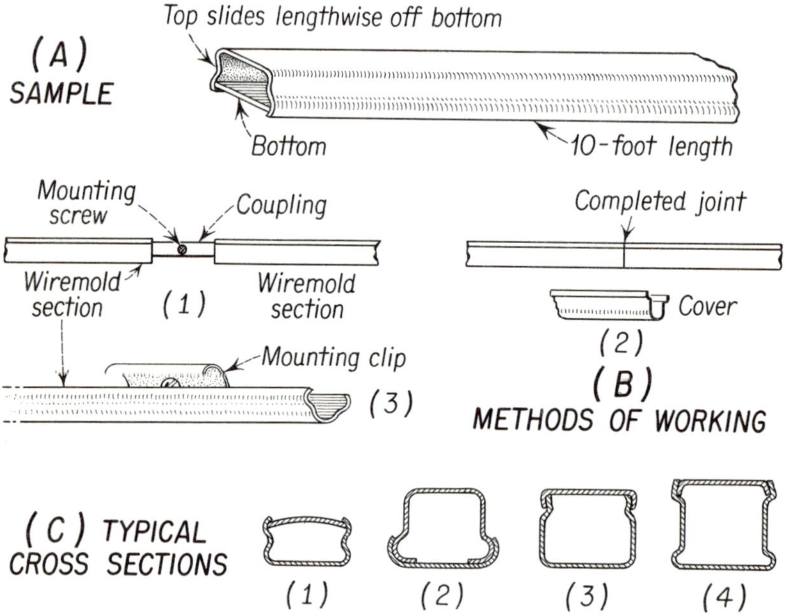

Fig. 14–21 Wiremold Metal Raceway

14–21(B)–2. A large number of fittings, elbows, tees, etc. are available and may be used between Wiremold lengths to form bends and junctions. Special boxes, switches, and outlets also are available. Wiremold metal raceway is attached to the supporting structure by means of metal mounting clips, as in Figure 14–21(B)–3. These clips are held by screws to walls, ceilings, etc.

Figure 14–21(C) shows endwise cross-sectional views of common shapes in which Wiremold is supplied. The height and width dimensions vary over a range of values to suit numerous installation requirements. The same familiar slide-on construction is employed with Wiremold of each shape.

14.10 ARMORED-CABLE WIRING

Figure 14–22 shows the familiar metal-armored cable which is used extensively for wiring and is sometimes called *BX*. Like Romex, this

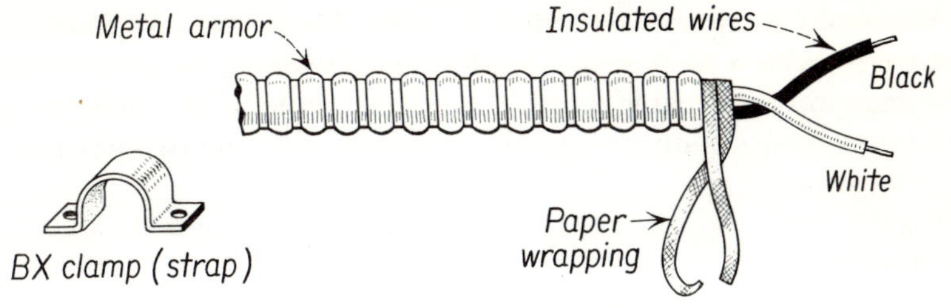

Fig. 14–22 Armored Cable

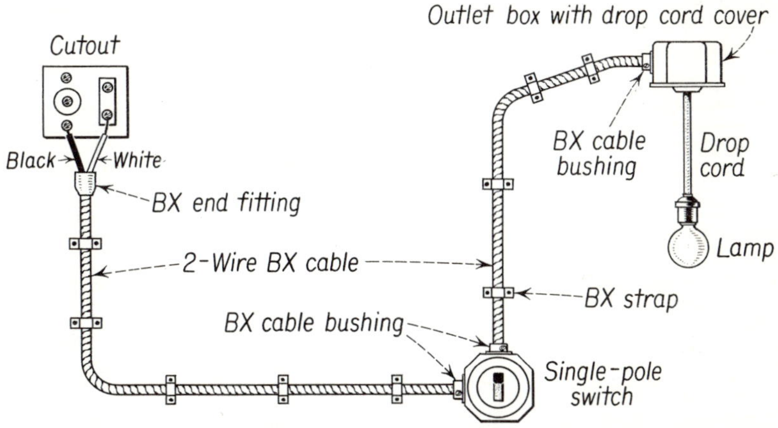

Fig. 14–23 Typical Open-Wiring Installation with Armored Cable

Fig. 14–24 Armored Cable Wiring to Outlet Box

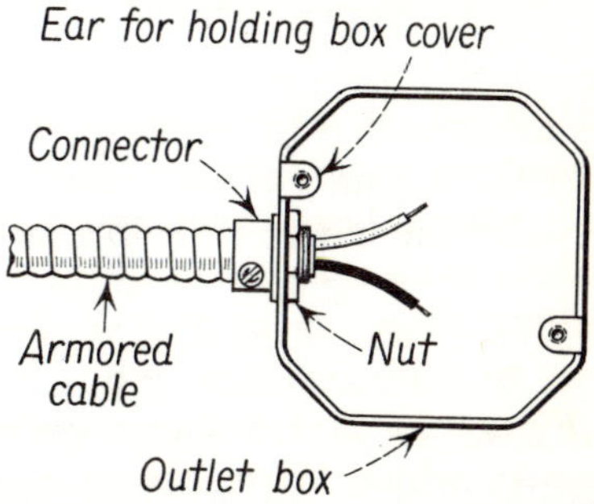

cable contains well-insulated conductors (two or three). The conductors are spirally wrapped with kraft paper and enclosed in the spiraled outer armor sheath of galvanized steel.

The steel jacket may be cut readily with a hack saw. After cutting, a split fiber bushing is inserted over the wires into the end of the jacket to prevent the sharp edges of the latter from cutting through the insulation of the wires.

Two-holed, semicircular clamps or straps are provided for fastening armored cable to supporting structures. Bushings also are provided for fastening the cable to junction, outlet, and switch boxes.

Unless armored cable is of the special type which has an additional inner lead sheath, its use is permitted only in permanently dry *indoor* locations.

Figure 14–23 shows an example of wiring with armored cable. Like the example given for Romex, this is a simple wiring installation including only a single switch and lamp, but it is illustrative of the manner in which this cable is run and supported. Figure 14–24 shows the use of a BX connector and nut to hold armored cable entering a box.

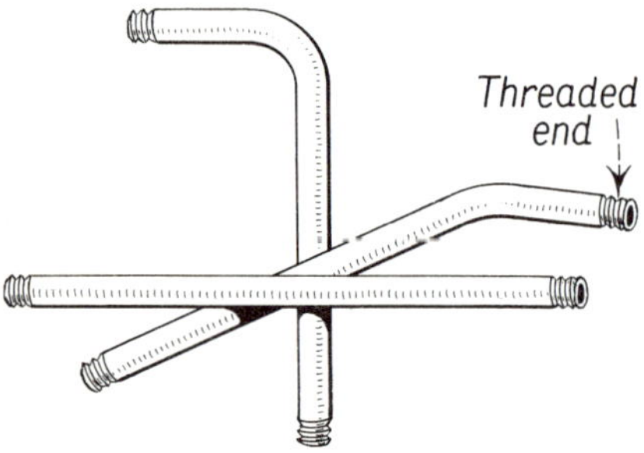

Fig. 14–25 Rigid Conduit

14.11 RIGID-CONDUIT WIRING

Conduit produced for electrical work is annealed steel piping. (See Figure 14–25.) The pipes must be cut, bent, and threaded as required to form the pathway for the wiring they will house. These pipes are attached to junction, outlet, and switch boxes to enclose completely the wiring system. Insulated conductors are pulled through the pipes with the aid of a flexible steel *fish tape*.

Two types of conduit are employed in electrical wiring: *rigid* and *thin-walled*. Rigid conduit is supplied in diameters of ½, ¾, 1, 1¼, 1½, 2, 2½, 3, 3½, and 4 in. Thin-walled conduit (called *electrical metallic tubing* and abbreviated *EMT*) is supplied in the same diameters but with thinner walls. This latter type of conduit is not threaded because the threads would break through the thin wall, but is secured to boxes by means of special connectors. Figure 14–27 shows the method of attaching both types of conduit to a box. Figure 14–28 gives a cross-sectional view of rigid conduit held to a box by means of threaded fittings.

Splices and joints in a conduit wiring system are made inside special boxes called *condulets.* Taps are made in a Type T condulet.

Figure 14–26 is an example of conduit-type wiring. While this is a simple application including only one lamp and one outlet operated from

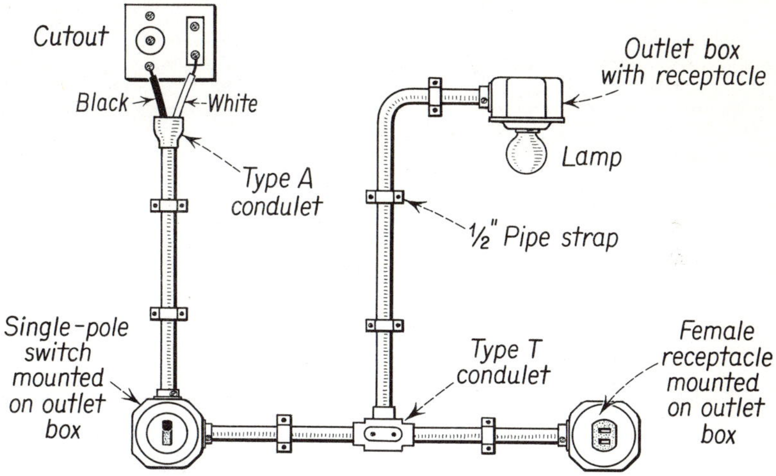

Fig. 14–26 Typical Open-Wiring Installation with Rigid Conduit

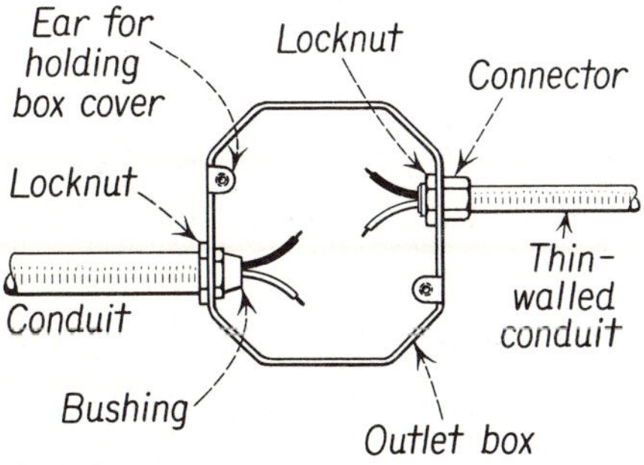

Fig. 14–27 Rigid Conduit Wiring to Outlet Box

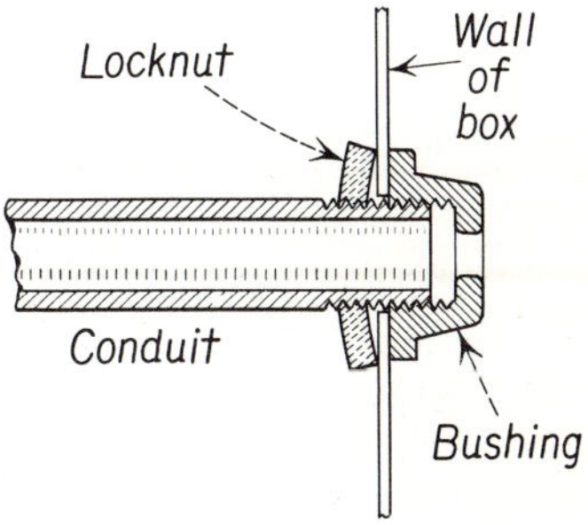

Fig. 14–28 Attachment of Conduit to Box

a single switch (all being surface-mounted units), it serves to illustrate this wiring method.

Rigid-conduit wiring is the most satisfactory of all the methods in use. It is encountered in both indoor and outdoor wiring, exposed and unexposed, especially in industry.

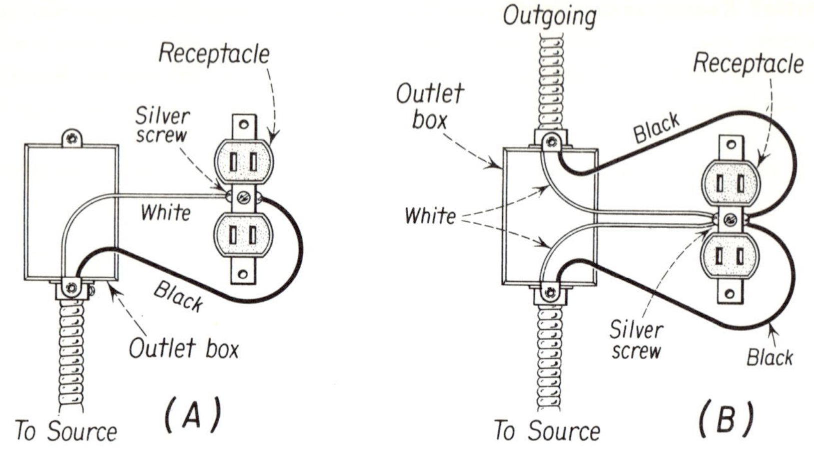

Fig. 14–29 Outlet Receptacle Wiring

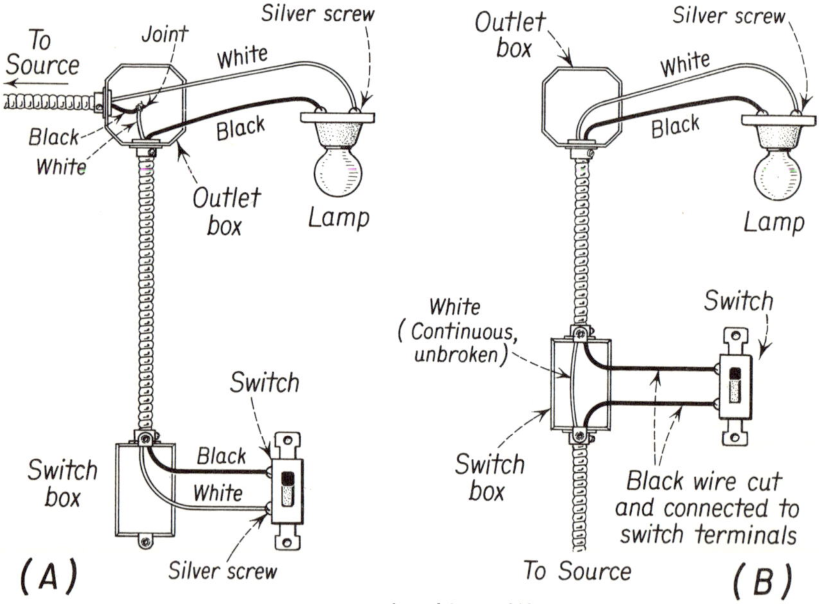

Fig. 14–30 Switch-and-Lamp Wiring

14.12 TYPICAL WIRING EXAMPLES

Figures 14–29 to 14–31 show selected examples of wiring in several typical cases. While armored cable (BX) is shown in Figures 14–29 and 14–30, the examples might apply also to Romex or conduit-type wiring, the circuits remaining the same in each case.

Outlet Receptacle. In Figure 14–29(A), the conductors are brought in from the power source and connected directly to the terminal screws of the receptacle. There is no further wiring beyond this point. A switch could be inserted into the circuit somewhere between the source and the outlet.

In Figure 14–29(B), the conductors are brought in from the power source and connected to the outlet. They then proceed from the outlet

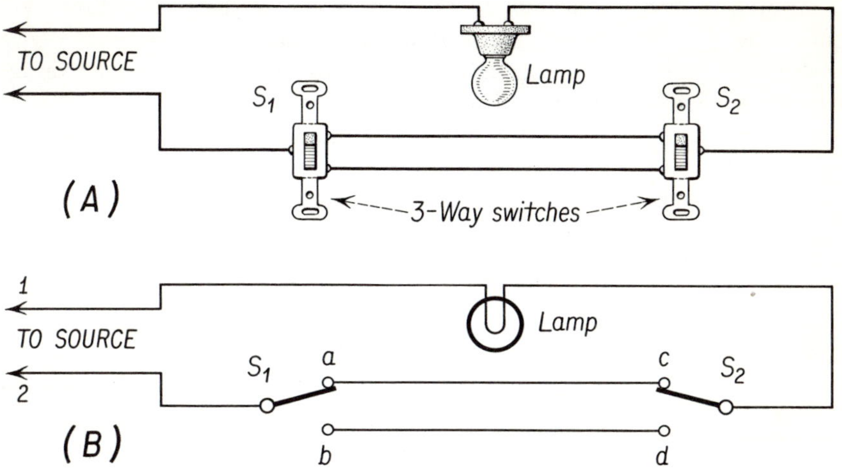

Fig. 14–31 Use of 3-Way Switches to Control Lamp from Two Different Points

receptacle (through the outgoing line) to some other point, such as a second outlet receptacle. The second unit then is in parallel with the first. The connections are made according to the method illustrated in Figure 14–6(B).

Switch-and-Lamp Wiring. Figure 14–30(A) shows the method of wiring a switch and lamp to the power source. Here, the switch is installed in the wall, the lamp in the ceiling, and the power line enters the outlet box to which the lamp fixture is attached. The circuit in Figure 14–30(B) accomplishes the same purpose as the first one, but in the latter the conductors from the power source enter the switch box, and connections then extend from the switch box to the lamp. One conductor [the white wire in Figure 14–30(B)] passes through the switch box to the lamp without interruption. But the other conductor (the black wire) is cut, and its ends are connected to the switch terminals.

Use of Three-Way Switches. It very often is handy to be able to turn a lamp or other appliance either on or off from either of two locations.

This is made possible by means of three-way switches. Figure 14–31(A) shows the arrangement pictorially while Figure 14–31(B) shows the circuit.

Referring to Figure 14–31(B), Switches S_1 and S_2 are separated by some distance. When the switches are in the positions shown, current can flow from Terminal 2 of the source, through Contact *a* of S_1, to *c* of S_2, through the lamp, and return to Terminal 1 of the source. Opening either S_1 or S_2 will extinguish the lamp.

Similarly, if S_1 is thrown to Position *b* and S_2 to *d*, the lamp will light. And if either switch then is opened, the lamp will go out.

From this, it may be seen that the lamp may be turned on or off with either switch S_1 or S_2. One switch often is installed at the top of a stairway and the other at the bottom, to operate a basement light.

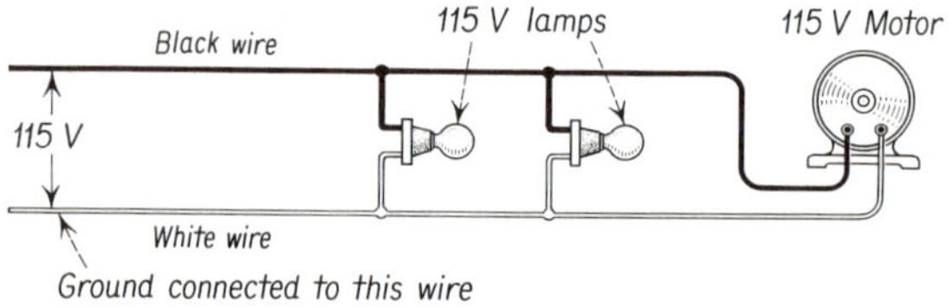

Fig. 14–32 Basic 2-Wire Supply

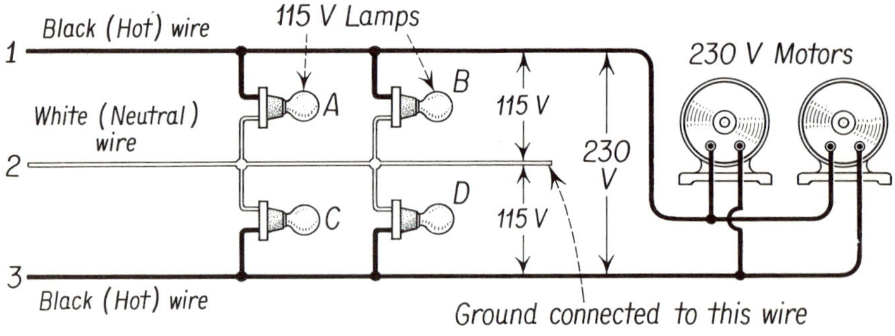

Fig. 14–33 Basic 3-Wire Supply

14.13 TYPES OF SERVICE

Electricity is brought into a building through either two-wire or three-wire service. Basic two-wire and three-wire circuits are shown in Figures 14–32 and 14–33, respectively.

In Figure 14–32, 115 v are available between the two wires. Thus, lamps, motors, and other devices rated to operate at this voltage may be connected directly across this line, as shown. In some localities, the line voltage may be 110 or 120.

One side of an a-c supply is grounded for safety purposes. That is, one wire (always the same one) is connected to earth, either by fastening it to a cold-water pipe or to a metallic rod or pipe driven deep into the earth. In order to identify the grounded wires in a circuit, these wires are provided with white (or light gray) insulation, while the "hot" wires

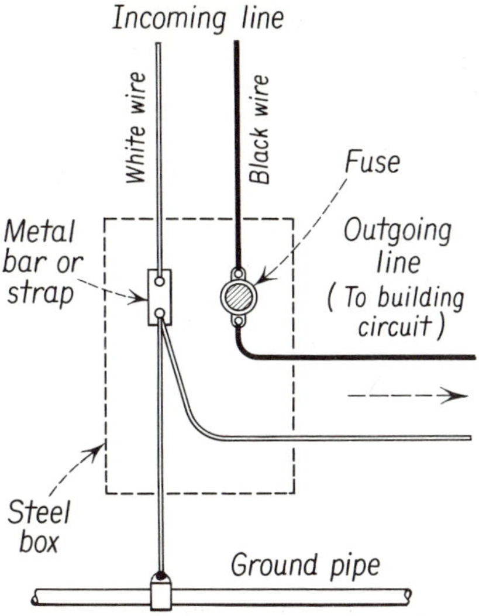

Fig. 14–34 Protection of 2-Wire System

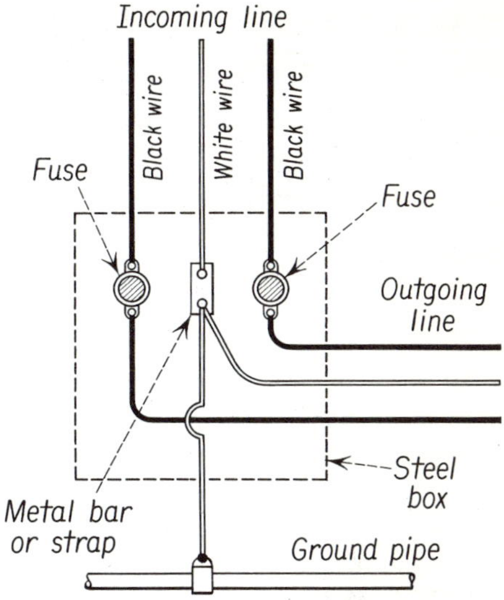

Fig. 14–35 Protection of 3-Wire System

have black insulation. These colors must never be swapped in wiring, the hot lead being kept black throughout the circuit. D-c service is not grounded.

In three-wire service (Figure 14–33), the center white wire (the *neutral* wire) is grounded, and the two outside black wires are hot. A voltage of 115 v is available between the neutral wire and *either* of the outside wires, while 230 v are available between the two hot wires. Thus, in Figure 14–33, the 115-v lamps, *A* and *B*, are connected to Wires 1 and 2, and the similar 115-v lamps, *C* and *D*, are connected to wires 2 and 3. In this way, two 115-v circuits are provided, so that neither circuit nor branch need be overloaded. The 230-v motors are connected to the two hot wires, 1 and 3. In this circuit, the center neutral wire is the one that is grounded. Not all three-wire service is 115 and 230 v. In some localities, 110 v and 220 v are supplied, and in others 120 v and 240 v.

14.14 CIRCUIT PROTECTION

Circuits are protected against burnout and fire by means of fuses or circuit breakers. Usually, there is one fuse (*service fuse*) in each hot line entering the service box and one additional (*branch-circuit fuse*) in the hot line of each circuit that branches out from the service lines.

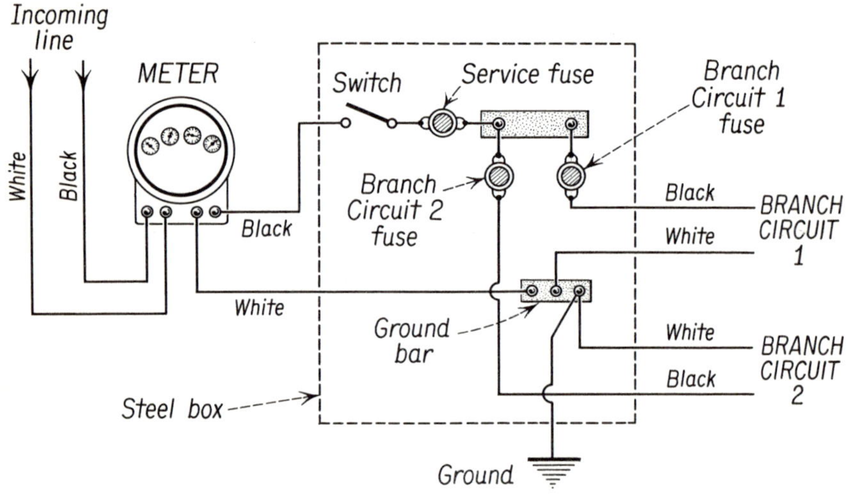

Fig. 14–36 Circuit Diagram of 2-Wire Service

For protection and safety, fuse holders or circuit breakers and any associated switches are installed inside a steel box with a closing door. Figure 14–34 shows the simple fusing and grounding of a two-wire line.

Figure 14–35 shows the fusing and grounding of a three-wire line. For simplicity, switches have been omitted from these simple examples.

Note carefully that fuses or circuit breakers always are connected in the hot line, *never in the ground line.* The reason for this is that the line is opened when a fuse blows or a circuit breaker opens. If the fuse were in the ground line, the ground would be removed from the system during blowout, and protection would be lost.

Branch-circuit fuses are rated to carry current levels that normally can flow safely in the branches concerned (for example, 10 or 15 amp in residential service). Service fuses are rated to carry the maximum total safe current of all the branches (for example, 20 to 30 amp in residential service).

In Figure 14–36, the positions of service and branch fuses are clearly shown. This is the basic circuit of a two-wire service in which two branches supply the building.

REFERENCES

1. NATIONAL ELECTRIC CODE, 1956 ed., National Fire Protection Association, 60 Batterymarch St., Boston 10.
2. Lincoln, E. S.: *Industrial Electric Wiring,* Duell, Sloan & Pearce, New York.
3. Perry, Edgar C., and Schafebrook, Harry V.: *Fundamental Jobs in Electricity,* McGraw-Hill Book Company, Inc., New York.
4. Richter, H. P.: *Practical Electrical Wiring,* 4th ed., McGraw-Hill Book Company, Inc., New York.
5. Smith, John J.: *House Wiring Guide,* Frederick J. Drake & Company, Inc., Wilmette, Ill.

Plate 15–1 Lighting is the best-known everyday service of electricity. (Courtesy, New York Life Insurance Co.)

Chapter 15

Electric Illumination

15.1 INCANDESCENT LAMPS

Lighting is one of the most familiar applications of electricity in everyday life. Our modern cities are plunged into darkness whenever disaster knocks out power facilities. The incandescent lamp is the most familiar lighting device. Not only is this type of lamp used for large-scale, general-purpose indoor and outdoor space lighting but it is found in flashlights, automobiles and other vehicles, instruments, therapeutic equipment, and toys.

In the incandescent lamp, a thin wire (*filament*) is mounted inside an evacuated glass bulb. Light energy is obtained from the glowing of this filament which is heated to incandescence by the flow of current. Either alternating current or direct current may be employed. (Except at very low frequencies, the filament is not extinguished when the a-c cycle passes through zero, because *thermal lag* will not permit the filament to heat and cool this fast. However, an incandescent lamp can be observed to blink on frequencies of 30 cps.)

Figure 15–1(A) shows the basic construction of the incandescent lamp. The resistance of the filament introduces a power loss (I^2R) which is evidenced in heating. At high heating temperatures, the filament glows brightly but does not burn up because very nearly all of the air and other gases have been removed from the bulb.

The filament metal is chosen principally for efficient conversion of electric power into visible light, mechanical workability, and resistance to mechanical breakage. Slender carbon filaments were used in Edison's early lamps, but these have been discarded in favor of metals. The principal materials that have been employed include tungsten, molybdenum, and platinum. After the air has been pumped out of the bulb, an inert gas, such as nitrogen, is introduced to forestall evaporation of the tungsten and the consequent darkening of the lamp by condensation of the vapor on the inside glass walls.

Figure 15–1(B) shows the practical construction of an incandescent lamp. The bulb stem is cemented into an end plug, usually brass, which is threaded for screwing into the power-outlet socket. One filament lead is connected to this threaded sleeve. The end of the plug contains an

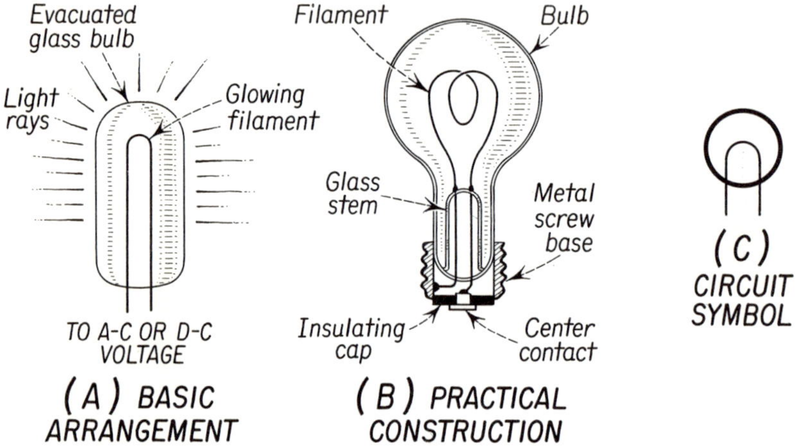

Fig. 15–1 Incandescent Lamp

insulating disc into the center of which is pressed a central contact button. The other filament lead is connected to this button.

Although the luminous intensity (in *candle power*) of incandescent lamps sometimes is specified, these lamps usually are rated in terms of the power they consume at their rated voltages. Common power ratings in watts are 5, 7½, 10, 15, 25, 40, 50, 60, 75, 100, 150, and 200, although this is not a complete listing.

15.2 ARC LAMP

The arc lamp once was widely used for street lighting but has become obsolete because of the greater convenience and economy of incandescent lamps. Furthermore, the arc lamp requires a d-c power supply for best operation, while the incandescent lamp operates on the alternating current available from modern power lines. The arc lamp is described here, however, because it is found today in large searchlights and in some motion-picture projectors (both applications favoring the high luminous intensity it provides) and in generators of ultraviolet rays.

The heart of the arc lamp consists of two pointed pencils of carbon. The points (Figure 15–2) first are brought together to complete the circuit, establishing current flow. Then they are pulled apart. When the points thus are separated by a short distance, ions travel across the gap (anions toward the positive carbon and cations toward the negative) and keep the current flowing. The ions strike the carbons with considerable velocity and their terrific impact causes the tips to glow white hot. The glow is especially dazzling at the positive carbon be-

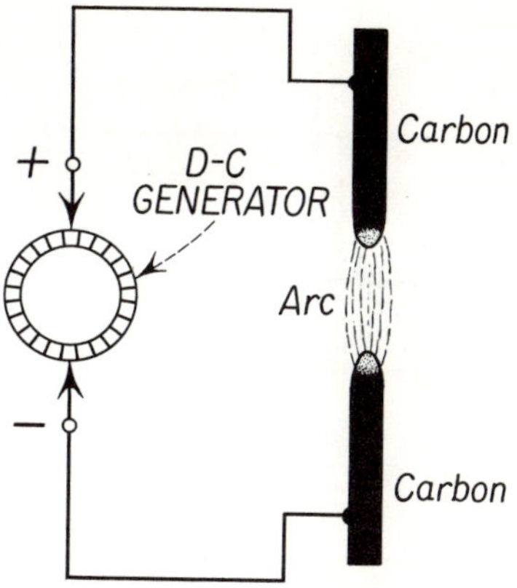

Fig. 15–2 Principle of Arc Lamp

cause of the higher energy of the negative ions striking this pencil. At this electrode, a white-hot *crater* is formed.

Figure 15–2 shows the basic arrangement of the arc-lamp circuit. The device takes its name from the brilliant, sustained, arc-shaped flame which is maintained between the tips of the carbon electrodes. Mirrors and/or lenses often are employed to concentrate the light of the arc.

The carbon rods are consumed during the operation of the arc lamp. Since the tips continually are wearing away, the positive about twice as fast as the negative, they must be moved progressively closer together to maintain a suitable gap, otherwise the arc will be extinguished. This is done by means of manual adjustment or through the action of an electromagnetic device which moves one or both carbons automatically.

15.3 GAS-DISCHARGE LAMPS AND TUBES

Luminous-tube lighting utilizes the generation of light by the passage of electricity through certain gases. (See Section 4.4, Chapter 4, for a discussion of current conduction in gases.) Light obtained in this manner is not as intense as that secured with incandescent lamps of the same size but is useful in many applications.

Figure 15–3 shows the basic arrangement of a gas-tube lighting circuit. After evacuation, the glass tube is injected with a controlled amount of suitable gas. Wire electrodes are sealed into each end of the tube. The tube may be kept straight, as shown in Figure 15–3, but it often is bent into the shape of letters of the alphabet, numbers, or patterns, since it frequently is the basis of illuminated signs.

The gas is ionized by a high voltage, usually between 10 and 25 kv, supplied by the secondary of a step-up transformer. The current drain is of the order of 10 to 50 ma rms.

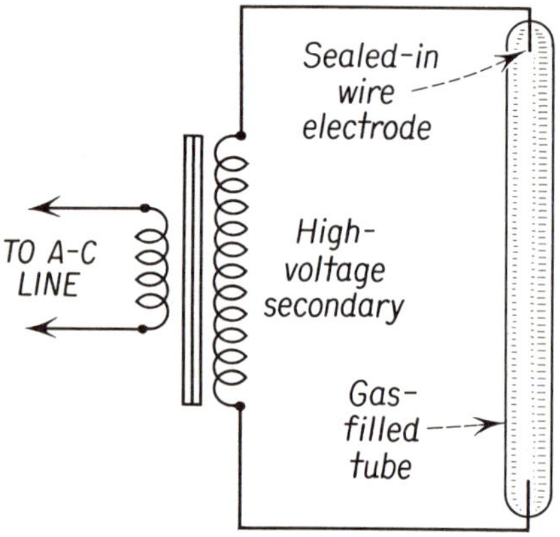

Fig. 15–3 Basic Gas-Tube Circuit

The color of light emitted by the tube depends upon which gas is used. Neon, for example, gives a pinkish-orange glow, hydrogen produces red, and argon gives blue.

15.4 MERCURY-VAPOR LAMP

The *mercury-vapor* lamp emits high-intensity blue light which is rich in both actinic and ultraviolet rays.

The mercury-vapor lamp consists of a glass or quartz tube filled with mercury vapor at low pressure. Electrodes are sealed in each end. A small amount of metallic mercury is present.

There are numerous starting mechanisms for these lamps, but the simplest scheme is to tilt the tube slightly. This causes the mercury to flow down the tube and momentarily to connect the two electrodes, initiating current flow. When the tube is tilted back to the horizontal position, the mercury column breaks apart, producing an arc. Mercury vapor resulting from this arc ionizes quickly and gives the sustained light which is characteristic of this type of lamp. The vapor is replenished continuously by the heat of the arc which vaporizes more of the liquid mercury.

Mercury-vapor lamps may be operated from either alternating current or direct current. Small mercury-vapor lamps used for therapeutic purposes and skin-tanning have quartz or special glass envelopes to pass the ultraviolet rays desired in these applications. A filament generates the heat required to vaporize a drop of mercury enclosed inside the bulb. Gaseous conduction then ensues between the filament terminals.

15.5 FLUORESCENT TUBE

The *fluorescent tube* is well known for its uniform, diffuse light and is widely used.

In this type of lamp, the inside walls of a glass tube are coated with a *phosphor;* that is, a fluorescent chemical. Ultraviolet rays are generated by mercury vapor resulting from the passage of current through the tube, and these rays cause the phosphor to glow, producing the glareless illumination which is characteristic of fluorescent lighting.

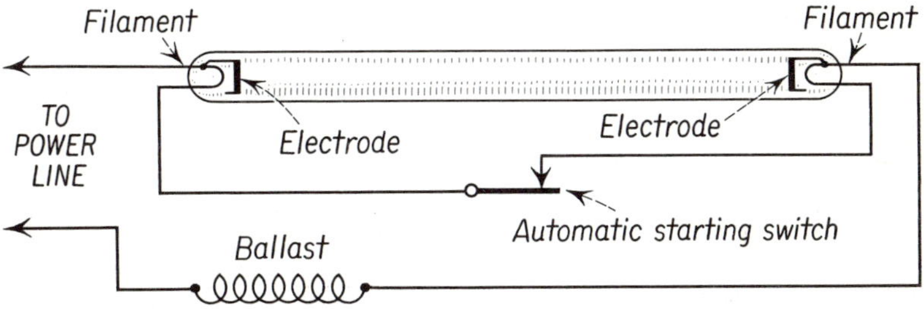

Fig. 15–4 Fluorescent Tube

High voltage supply is not needed for the operation of a fluorescent tube, since this tube takes advantage of the fact that gas tubes having heated filaments instead of cold electrodes will "fire" at ordinary power-line voltages (110 to 120 v).

Figure 15–4 shows a fluorescent-tube circuit. An automatic (thermally operated), normally closed spst switch (the *starter*) is wired in series with the two filaments. When the filaments reach normal operating temperature after the power has been switched on and the electrodes are heated sufficiently, the starter switch opens, allowing an increased potential to appear between the electrodes. This initiates the gaseous conduction in the tube, the inner walls of which fluoresce. The purpose of the ballast (which also might be a capacitor, but must be a resistor when the tube is used on direct current) is to stabilize the current.

15.6 ELECTROLUMINESCENT LIGHT SOURCES

Electroluminescence is a newly discovered phenomenon which permits light to be obtained directly from the action of an emf on certain phosphors.

Figure 15–5 illustrates the construction of a simple electroluminescent cell. This consists of a sandwich composed of an electroluminescent phosphor between two plates of glass. The inner surface of each glass plate is coated with a transparent metallic film which permits electrical contact to the phosphor while transmitting light rays.

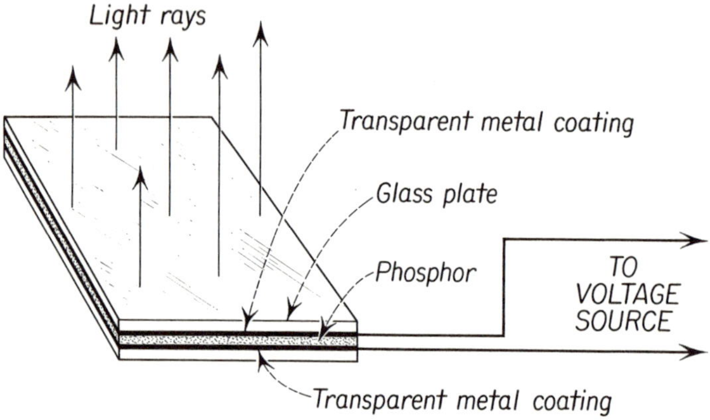

Fig. 15–5 Electroluminescent Cell

When an emf (between 600 and several thousand volts) is applied to the two faces of the phosphor via the transparent electrodes, the phosphor glows, and the light rays it emits pass through the glass plates.

At this writing, electroluminescence has not been applied on a commercial scale. However, it shows promise as a future source of electric illumination.

SUMMARY OF CHAPTER 15

The **Incandescent Lamp** utilizes heat produced by the I^2R loss in a thin, current-carrying filament *in vacuo* to produce visible light. The incandescent lamp is the most extensively used electric-lighting device.

The source of light rays in the **Arc Lamp** is the brilliant arc discharge produced by the flow of ions between the tips of carbon rods connected to a power generator. Carbon is the fuel in the arc lamp.

Gas Tubes use the ionization of such gases as argon, neon, hydrogen, etc., to produce colored light. High voltages are required to ionize the gas when cold electrodes are used. The use of hot filaments permits gas-tube operation at lower potentials. Characteristic gas-glow colors are: neon, *orange-pink;* hydrogen, *red;* and argon, *blue.*

The **Mercury-Vapor Lamp** employs ionization of mercury vapor to obtain an intense blue light. Small mercury-vapor lamps, used for therapeutic purposes, contain a hot filament which vaporizes a small drop of mercury.

A **Phosphor** coating the inner walls of a **Fluorescent Tube** glows when impinged upon by ultraviolet rays generated by the passage of current through the tube. A filament is situated in each end of the fluorescent tube, and the two are connected in series with each other and to the power line through an automatic *starting* switch and current-stabilizing *ballast*. When the filaments have heated the electrodes, the switch opens, and the increased emf initiates the gaseous conduction within the tube. The resulting ultraviolet rays cause the phosphor coating of the tube to glow.

Electroluminescence results from the direct action of an emf on a suitable phosphor. Electroluminescence has not been developed as a commercial source of electric light but has future importance.

QUESTIONS AND EXERCISES

1. (*a*) Describe the construction of an incandescent lamp. (*b*) What are two methods of rating these lamps?
2. (*a*) Explain the operation of an arc lamp. (*b*) Why is this lamp so named? (*c*) Which carbon pencil deteriorates faster?
3. (*a*) Describe the operation of cold-cathode gas-discharge tubes. (*b*) What color of light is generated by neon, argon, and hydrogen?
4. How can the discharge be initiated in a mercury-vapor tube?
5. Sketch a fluorescent-tube circuit and explain its operation.
6. Draw an electroluminescent cell and explain the phenomenon of electroluminescence.

Chapter 16

Telegraph and Telephone Fundamentals

16.1 KEY-AND-SOUNDER TELEGRAPH

Figure 16–1 shows the essentials of a conventional telegraph circuit. Station *A* is the transmitting, or sending, station. At this station are located a battery and *key*. The key is a special type of manually operated switch. It is normally open and closes the circuit when depressed.

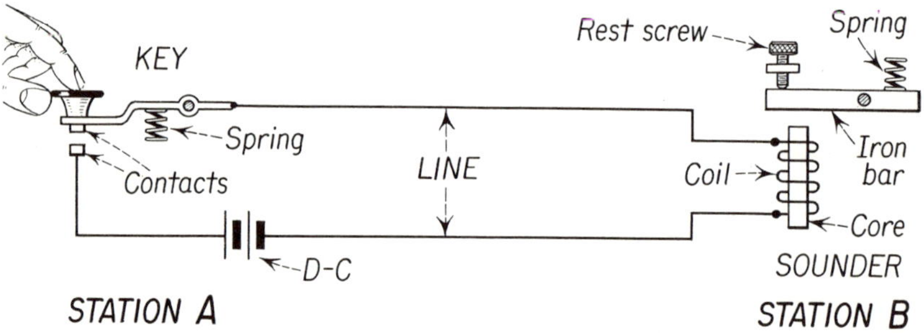

Fig. 16–1 Basic Arrangement of Key-and-Sounder Telegraph

Station *B*, the receiving station, is situated at a distance. At this station, there is located a *sounder*. The sounder consists of an electromagnet over the core of which is pivoted a thick iron bar provided with a spring to return it to this position of rest. The two stations are connected together by means of a two-wire line.

When the key is depressed, current flows over the line and through the sounder coil. This causes the core to become magnetized and to attract the bar. The bar strikes a metallic barrier, just above the core, sharply enough to emit a single click and remains in the downward (attracted) position as long as the key is held down. When the key is released, the line current is broken, and the sounder bar is released. On its return to its initial resting position, the bar strikes the rest screw and emits a second click.

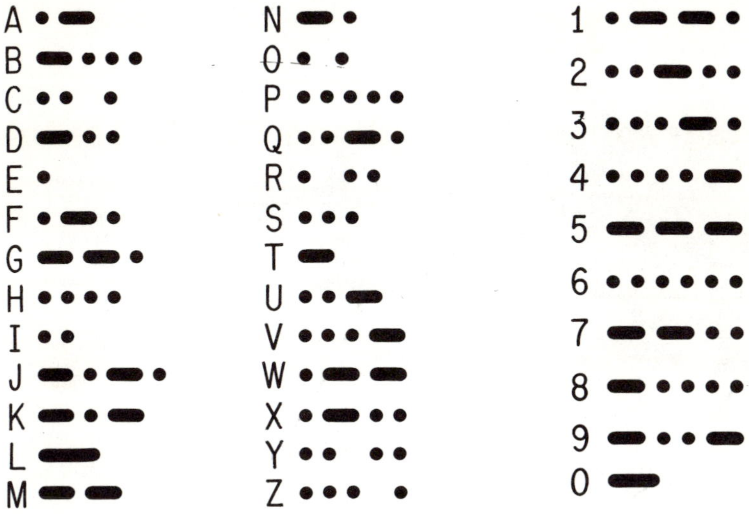

Fig. 16–2 (A) American Morse Code

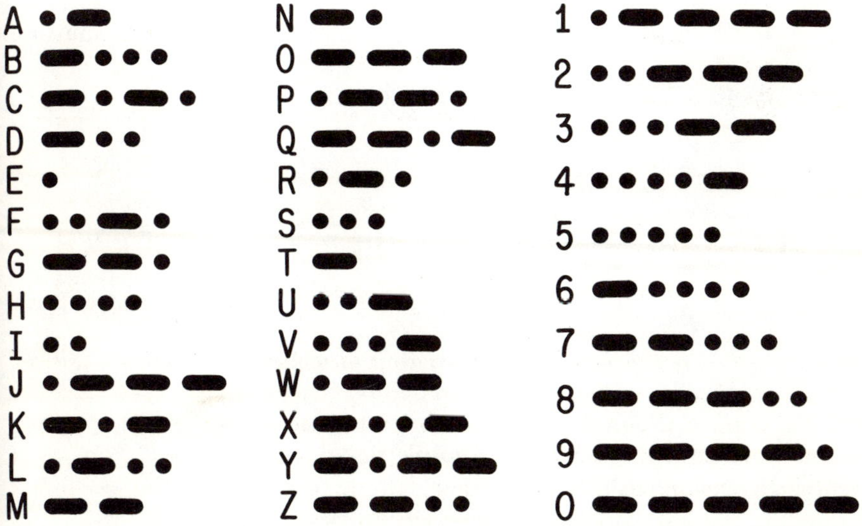

Fig. 16–2 (B) International Morse (Continental) Code

Intelligence is transmitted and received by means of a code whereby combinations of clicks are interpreted as letters of the alphabet, numerals, and punctuation marks. Figure 16–2(A) shows such a code (minus punctuation marks). This is the *American Morse Code.* Considerable auditory skill is necessary for the successful reading of this code because of the dot or dash spacing often employed in a single letter. Note that *L*, *T*, and zero each is one dash, but each is of different length.

Fig. 16–3 Commercial Telegraph

Figure 16–3 shows a commercial telegraph set consisting of a key, sounder, and switch on a single base. This combination provides a complete station (except for the battery), the key for sending and the sounder for receiving.

Telegraph stations often are so far apart that the line resistance reduces the currents to a level too low to operate the sounders. This is corrected by installing a type of booster station between transmitting

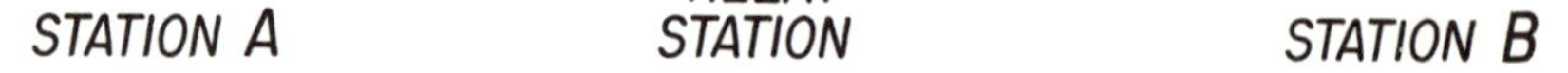

Fig. 16–4 Use of Relay in Telegraph Circuit

and receiving stations. A relay at this intermediate station is sensitive enough to be operated from the weak signal currents, and it closes a second battery circuit to send new, strong currents to the receiving station.

Figure 16–4 shows the introduction of a relay station into the circuit. The current that is made and broken into dots and dashes by the key at Station *A* operates the relay at the intermediate station. The relay reproduces the dot-and-dash current signals, sending similar current pulses from the second battery, B_2, to actuate the sounder at Station *B*. On a very long telegraph circuit, a number of relays may be installed at strategic points to boost the declining signals.

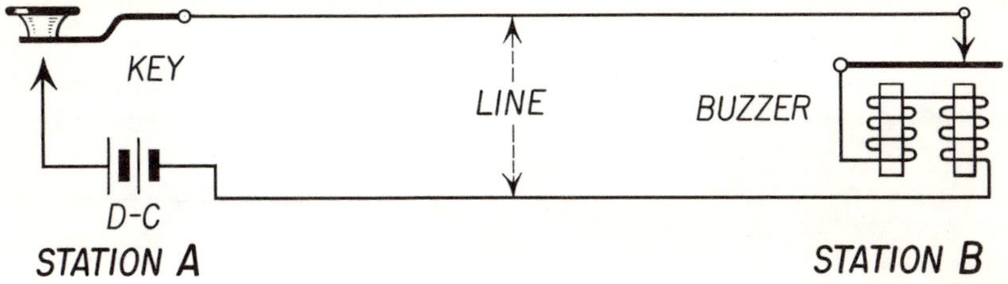

Fig. 16–5 Buzzer Telegraph

16.2 BUZZER TELEGRAPH

Considerable sound-reading skill is required to distinguish the dots and dashes of the key-and-sounder telegraph described in Section 16.1, preceding. Another type of telegraph, employing sustained tones and a different code somewhat less difficult to "read," is illustrated in Figure 16–5.

Here, the transmitting station is identical with the one described previously. At the receiving station, however, a buzzer is used instead of a sounder. (See Section 5.14, Chapter 5, for a description of buzzer operation.) As long as the key is held closed at Station *A*, the buzzer at Station *B* emits a continuous tone. A short buzz is read as a dot and a long buzz as a dash. The International Morse (or "Continental") code is employed and is shown in Figure 16–2(B).

Relays may be employed in long-distance buzzer telegraph circuits in the same manner as in key-and-sounder circuits (Figure 16–4).

16.3 SIMPLE TELEPHONE

Figure 16–6 gives the basic telephone circuit. Figure 16–6(A) shows the wiring, with the transmitter and receiver represented pictorially; Figure 16–6(B) shows the same circuit with standard symbols.

The main component at the transmitting station is the *transmitter* or *microphone*. This consists of a thin, metal disc or *diaphragm* to the back of which is fastened a small, enclosed button consisting of two carbon or metal blocks with carbon granules loosely packed between them. The main component at the receiving station is the *receiver* or *earphone*. This unit contains a thin, soft-iron disc (diaphragm), and very close behind it are situated the poles of a strong permanent magnet. A coil of a great many turns of fine wire is wound around each pole, and the two coils are connected in series-aiding.

A battery is inserted in the line, as shown. Under ordinary conditions, a small, steady current flows through the microphone button and the receiver coils. When sound waves strike the diaphragm of the microphone, as when speaking toward it, the diaphragm vibrates in unison with the waves. Its resulting front and back motion alternately compresses and loosens the carbon granules. When the granules are

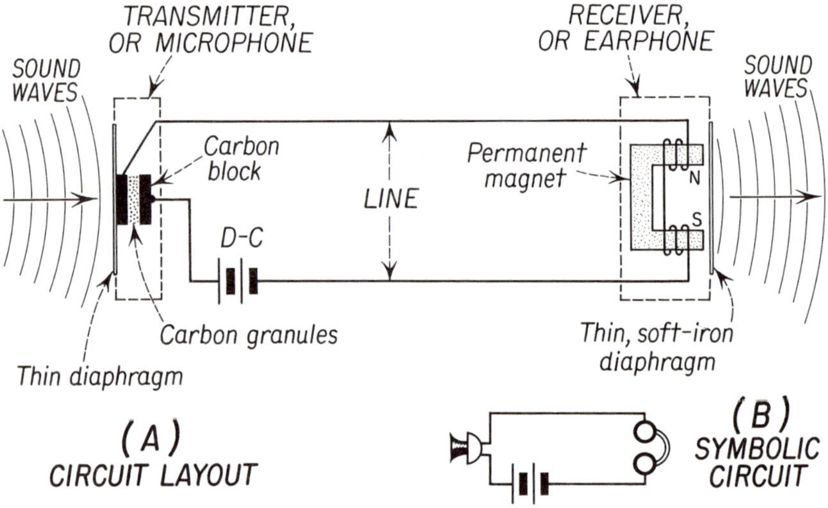

Fig. 16–6 Simple Telephone

compressed, the resistance of the button decreases; when they are loosened, it increases. This variation in resistance causes the current in the line to fluctuate in sympathy with the sound waves. This is called *modulating* the current. At the receiving end of the line, the fluctuating current passing through the magnet coils causes corresponding fluctuations in the magnetic field, and this causes the receiver diaphragm to vibrate back and forth, producing sound waves which reproduce the sound that originally actuated the microphone.

In some instances, particularly in long-distance communication, telephone stations must be isolated from the line, or direct current must be kept out of the line. This is accomplished by using transformer coupling, as shown in Figure 16–7. The current fluctuations, owing to the sound-controlled resistance of the microphone, pass through the primary winding of Transformer T_1. The secondary of this transformer presents to the line alternating currents having frequencies (*audio frequencies*) corresponding to the vibration frequencies of the initiating sound waves. At the receiving end of the line, these alternating currents actuate the receiver through Transformer T_2. The turns ratios of T_1 and T_2 may be chosen such that the best impedance match is obtained between transmitting and receiving circuits and the line.

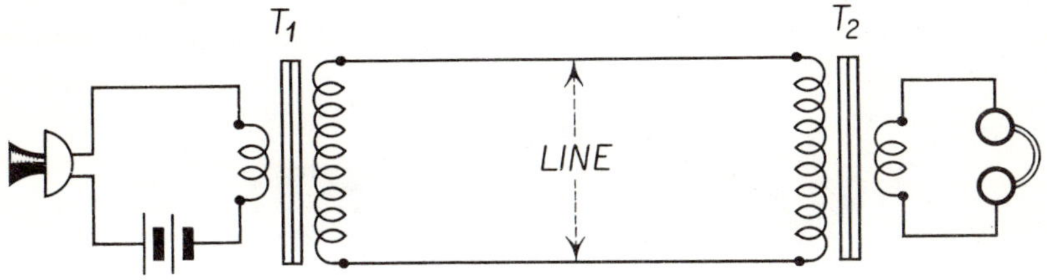

Fig. 16–7 Transformer-Coupled Telephone

16.4 RINGING CIRCUITRY

It is well known to the reader that talking, listening, ringing, and signaling the operator all are carried out over the same telephone lines. Some of the circuitry is complex and beyond the scope of this book.

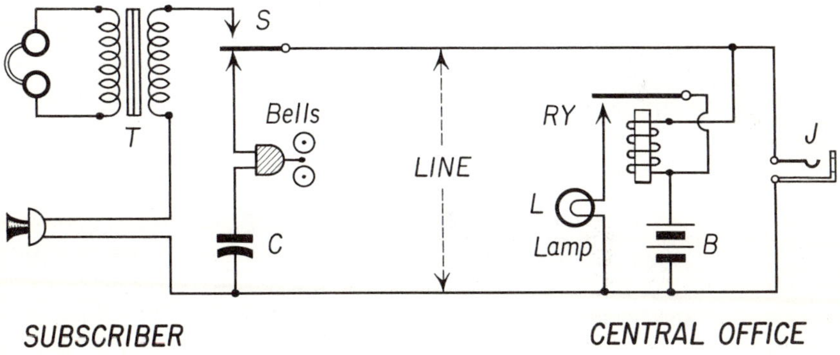

Fig. 16–8 Telephone System, Showing Ringing Circuitry

However, the simple circuit given in Figure 16–8 serves to show how a system might operate.

In this illustration, Switch S is actuated by the weight of the telephone receiver or handset. Its blade is against the lower contact, as shown, when the telephone is idle. Direct current from Battery B at the central station cannot pass through the subscriber's bell circuit because it is blocked by Capacitor C. But alternating current can operate the bells through the capacitor, and this is accomplished when the operator temporarily connects an a-c generator to the line.

When the subscriber lifts his receiver, Switch *S* moves to its upper contact. This closes the battery circuit, actuating Relay *RY* and lighting Lamp *L* to signal the operator. With the operator's earphone plugged into Jack *J*, the subscriber can talk to the operator, and can listen through Transformer *T*. To complete the call to another subscriber, the operator inserts a plug (connected to the line coming in from the other subscriber) into Jack *J*. (See Section 3.4, Chapter 3, for an explanation of jack-and-plug operation.)

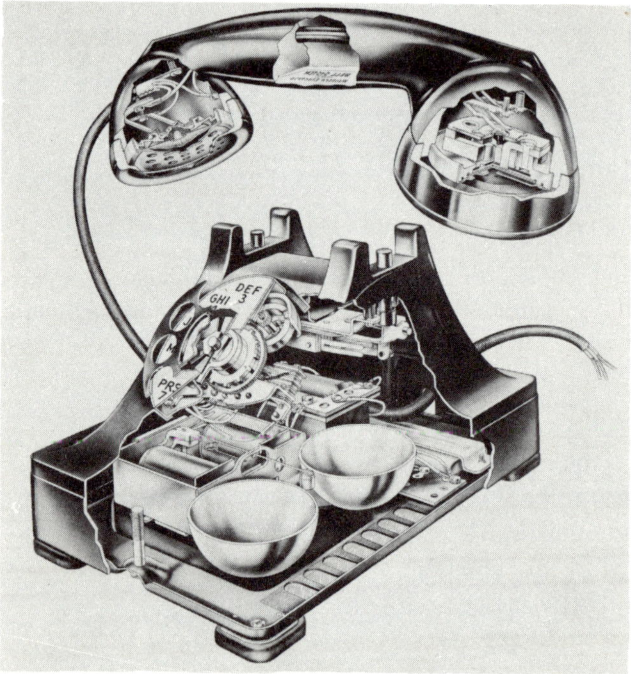

Plate 16–1 Modern Telephone Set. The handset is shown above the stand. The microphone is in its left end and the receiver in its right. In the stand, below, the ringer and bells are in the center foreground and slightly above them is the assembly of springs and contacts which are actuated when the dial turns. The conductors leaving the cord at the right go to the central office. (Courtesy, Bell Telephone Laboratories)

16.5 BASIC MULTIPLE-TELEPHONE CIRCUITRY

Multiples of the circuit in Figure 16–8, each terminating in a jack (*J*) on the central-station switchboard (but operated from a single battery, *B*, or d-c generator), form the basis of multiple-telephone circuitry.

The operator is able to interconnect any subscribers by "patching" a cord (insulated pair of wires) between their jacks on the switchboard. Similarly, the operator is able to ring any subscriber by sending the a-c ringing signal over that subscriber's line. When the subscriber's telephone already is in use, Switch S will be open because the receiver is "off the hook," and the ringing current therefore cannot pass through the bell circuit.

A multiple-telephone circuit need not necessarily include a central office, although it may be connected externally to that office. For example, all of the telephones within a large office operate into the office switchboard, but this switchboard also is connected, when desired, to the central office for outgoing calls. The orderly handling of intraoffice calls is taken care of by the office operator on the office switchboard.

SUMMARY OF CHAPTER 16

In **Telegraphy**, intermittent currents are transmitted over a wire line by making and breaking a d-c circuit by means of a **Key**, a special form of normally open switch. At the receiving station, these currents are converted into sounds (interpreted as dots and dashes) by a **Sounder** or **Buzzer**. The sounder produces clicks; the buzzer, sustained tones.

When telegraph signals must traverse a long-distance line, they may be built up at an intermediate station by the action of a **Relay** which closes a second battery circuit.

In a **Telephone Circuit**, sound waves are converted into a fluctuating current by a sound-controlled resistance device, the **Microphone** or **Transmitter**. At the receiving point, these fluctuations produce a varying magnetic field in the **Earphone** or **Receiver** and cause a diaphragm to vibrate in the field. The vibrating diaphragm reproduces the original sound.

The electromagnets of the telephone **Signal Bells** are connected to the line through a capacitor. The blocking action of the capacitor prevents the flow of direct current from the telephone battery or generator through the bell circuit. The operator rings the bells to signal the subscriber by sending over the line an a-c ringing current which operates the bell circuit through the capacitor.

QUESTIONS AND EXERCISES

1. Describe the difference between the sound signals emitted by: (*a*) sounder; (*b*) buzzer in a telegraph circuit.
2. Why are relays employed in long-distance telegraphy?
3. Describe the construction of a simple telephone microphone.
4. Describe the construction of a simple telephone receiver.

5. What prevents the d-c supply of a telephone circuit from ringing the bells at a subscriber's station?
6. Draw a simple key-and-sounder telegraph circuit.
7. Draw a telegraph circuit containing a relay.
8. Draw a buzzer telegraph circuit.
9. Draw a simple telephone circuit.
10. Draw a telephone circuit with ringing bells and operator's signal lamp.

Chapter 17

Introduction to Electronics

17.1 DIODE TUBE

The two-element tube, also called a two-electrode tube or *diode,* was described as a rectifying device in Sections 13.1 and 13.2, Chapter 13. All diode tubes have the same basic construction as the directly heated and indirectly heated types illustrated in Chapter 13, and they operate on the same principle. These tubes are available in many sizes and with a wide variety of electrical ratings, however, to suit a range of requirements.

The important electrical characteristics of the diode tube are *filament* or *heater voltage* (e_f, e_h), *filament* or *heater current* (i_f, i_h), *plate voltage* (e_p), *plate current* (i_p), *plate-to-cathode* or *plate-to-filament capacitance* (c_{pk}, c_{pf}), and *peak inverse plate voltage.*

The last two characteristics were not encountered in Chapter 13. Plate-to-cathode capacitance exists because the plate and cathode electrodes are metallic structures placed close to each other and therefore must evidence some capacitor effect. All types of tubes have this inter-electrode capacitance which, because of the small physical size of the electrodes, is low in value (usually not more than a few micromicro-farads). Peak inverse plate voltage is the maximum instantaneous voltage which may be applied safely between the plate and cathode when the tube is nonconducting (that is, plate negative and cathode positive). If this voltage is exceeded, flashover is apt to occur between the electrodes, and the latter may be damaged. When an a-c plate voltage is applied to the tube, the peak value of the negative half-cycle must not exceed the rated peak inverse value.

Plate current also may have a peak value which cannot safely be exceeded. It was explained in Chapter 13 that the output of an a-c-operated diode tube is a pulsating direct current. The pulsations consist of positive half-cycles of the applied alternating current and have peak, average, and rms values.

Figure 17–1 shows the characteristics of a typical twin diode of the indirectly heated cathode type (Type 6AL5).

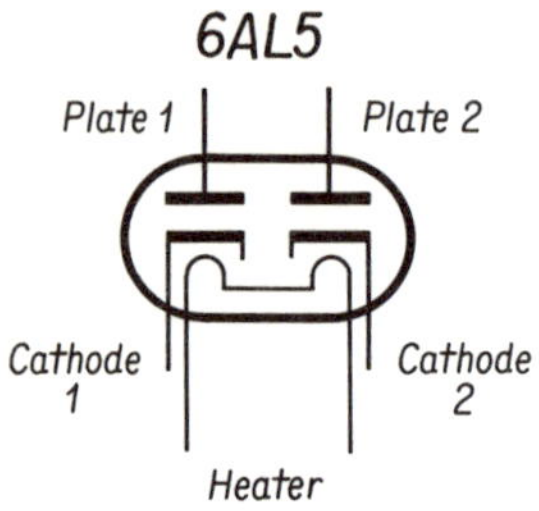

HEATER VOLTAGE, A-C or D-C	6.3 volts
HEATER CURRENT	0.3 ampere
CAPACITANCE (Plate 1 to Cathode 1)*	2.5 μμfd
PEAK INVERSE PLATE VOLTAGE	330 volts (max.)
PEAK PLATE CURRENT (Per plate)	54 ma (max.)
A-C PLATE VOLTAGE (Per plate)	117 volts rms
D-C OUTPUT CURRENT (Per plate)	9.0 ma

* Also Plate 2 to Cathode 2

Fig. 17–1 Twin Diode Tube

17.2 TRIODE TUBE

The flow of electrons from cathode to plate in a tube may be controlled by means of electric charges applied to a third electrode, called the *grid,* inserted between the cathode and plate. The grid must

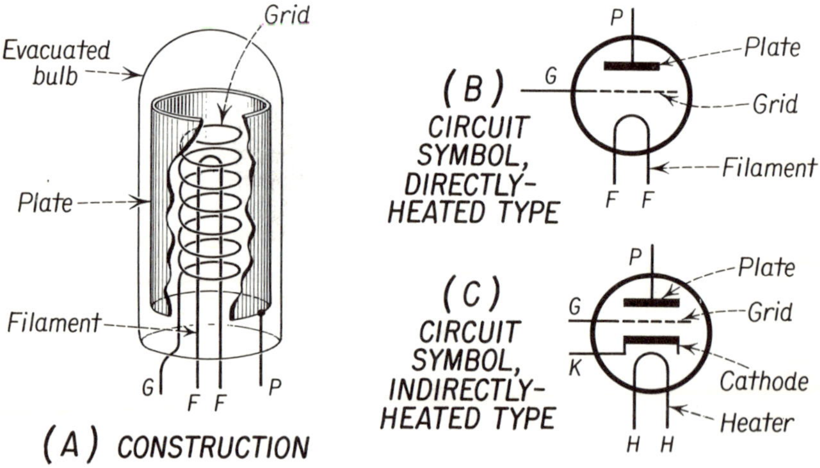

Fig. 17–2 Triode Tube

have an openwork construction which allows electrons to pass through it.

A workable grid may be constructed in any of a number of ways: cylinder of wire gauze, wire cage, wire coil, perforated metal cylinder, etc. Figure 17–2(A) shows one construction in which the grid is a coil of wire concentric to the plate and surrounding the filament.

A three-element tube, such as the filament-grid-plate type, is called a *triode* (that is, three electrodes). Like diodes, triodes may be of the directly heated type [Figure 17–2(B)] or of the indirectly heated type [Figure 17–2(C)].

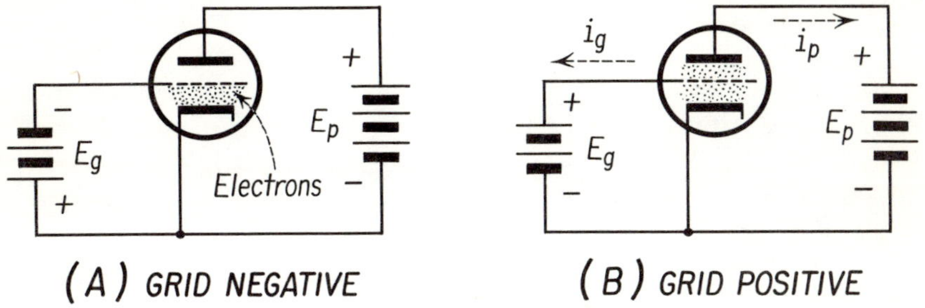

Fig. 17–3 Triode Action

Figure 17–3 illustrates basic triode action. A plate voltage, E_p, is applied between plate and cathode with the plate positive [Figure 17–3(A)]. If a voltage, E_g, of sufficient amplitude then is applied between the grid and cathode as shown, the negative charge of the grid will repel the electrons emitted by the cathode and will confine them to the cathode-grid space. Since no electrons then can reach the plate, there is no plate-current flow. If the grid voltage is reversed [Figure 17–3(B)], the positively charged grid attracts the electrons. A few of them will strike the grid and give rise to an electron current, *grid current* i_g, which flows between grid and cathode. But most of them are traveling fast enough to fly through the grid mesh and are attracted to the positively charged plate. On reaching the plate, they set up the plate current, i_p. A lower value of plate current flows when the grid voltage is zero than when it is positive.

When plate current is flowing, its value may be varied by changing E_p while keeping E_g fixed. Its value also may be varied by changing E_g while holding E_p constant. A small change in grid voltage gives a *large* change in plate current. This sensitivity of control is the advantage offered by all grid-type tubes and is the foundation of many electronic devices.

The family of curves in Figure 17–4 illustrates triode performance. These data show (1) variation of i_p with E_p (2) variation of i_p with E_g, and (3) variation of E_p with E_g. A study of the curves shows the control characteristic afforded by the three-electrode tube.

The curve in Figure 17–5 shows more clearly the effect of grid voltage upon plate current at one specific plate-voltage level. Note that a definite value of i_p flows when $E_g = 0$. Plate current is suppressed completely when E_g reaches a high negative value (*cutoff*) and rises to a limiting value (*saturation*) when E_g attains a critical, high positive value. The cutoff and saturation points vary with tube type. They are governed principally by the electrode size and spacing.

The upper and lower portions of the E_g–I_p curve (Figure 17–5) show considerable curvature (*knees*), while the mid-region (*a* to *b*)

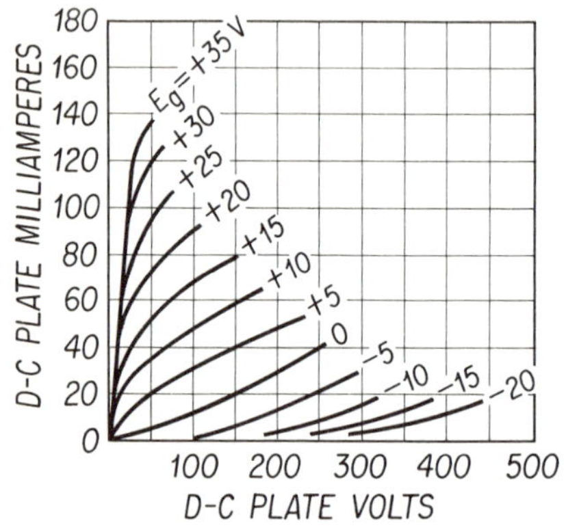

Fig. 17–4 Triode Response Curves

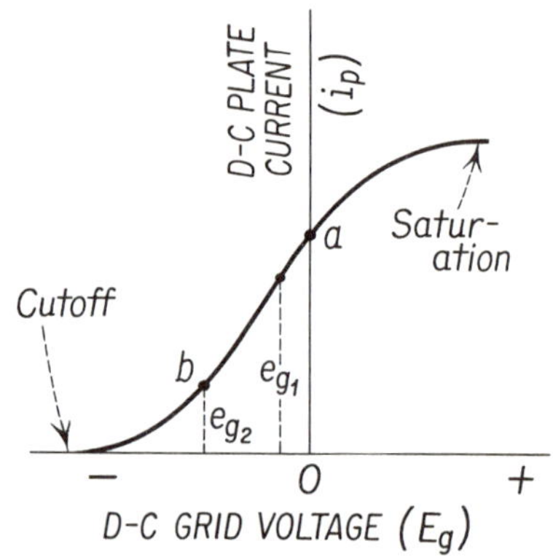

Fig. 17–5 Eg-Ip Characteristic of Triode

is linear. In order to utilize this linear response, in many applications a steady negative d-c voltage having a value between e_{g1} and e_{g2} is applied constantly to the grid. This is called a *bias voltage,* and the tube is controlled by varying this bias. When the bias voltage is chosen properly for a given tube, no grid current will flow. Plate current then is controlled entirely by varying the grid *voltage* above and below the bias value.

The extent to which plate current may be controlled by grid voltage changes is expressed by *transconductance.* Transconductance (g_m) in micromhos is the ratio of a *change* in plate current, at a fixed plate voltage, to the change in grid voltage (about the bias value) causing it. (These changes are represented by placing the letter *d* ahead of the changing quantity, thus dI_p signifies $I_{p1} - I_{p2}$, the difference between two plate current values.)

The numerical value of the transconductance may be calculated with the aid of the following equation.

(17–1) $$g_m = (1{,}000 dI_p)/dE_g$$

where dI_p = plate current change (in milliamperes)
dE_g = grid voltage change (in volts)
g_m = transconductance (in micromhos)

Example: A certain triode has a grid bias of -10 v. If this voltage is lowered to -9.5 v (that is, ½ v lower than the bias level), the corresponding plate current is 10 ma. If it is increased to 10.5 v (the same amount above the bias value), the plate current increases to 11 ma. What is the transconductance of this tube?

$$dE_g = 10.5 - 9.5 = 1.0 \text{ v}$$

$$dI_p = 11 - 10 = 1.0 \text{ ma}$$

$$g_m = (1{,}000 dI_p)/dE_g = (1{,}000 \times 1)/1 = 1{,}000 \text{ micromhos}$$

Another important characteristic of the grid-controlled tube is *amplification factor* (μ). Numerically, this is the ratio of a change in plate voltage (which will give a certain plate-current change) to a change in grid voltage which will give the same plate-current change when the plate voltage is held constant. Thus,

$$\mu = dE_p/dE_g \qquad (17\text{–}2)$$

Example: In a particular triode tube, a plate-voltage change from 150 v to 250 v produces a plate-current change from 10 ma to 20 ma. The same plate-current change is obtained at the operating plate voltage when the grid voltage is changed from -10 v to -5 v. Calculate the amplification factor of the tube.

$$dE_p = 250 - 150 = 100 \text{ v}$$

$$dE_g = 10 - 5 = 5 \text{ v}$$

$$dI_p = 20 - 10 = 10 \text{ ma}$$

$$\mu = dE_p/dE_g = 100/5 = 20$$

Since the triode plate circuit has current and voltage factors, it follows that this circuit must offer a definite amount of resistance to current flow. Figure 17–6 shows a triode plate-cathode circuit connected in series with a d-c source (E_B) and load resistor (R_L). With a current, i_p, flowing through the circuit, a voltage drop $e = i_pR_L$ is developed across the load resistor. The voltage between the plate and cathode of the tube accordingly is equal to the supply voltage minus this drop, or $e_p = E_B - i_pR_L$. The internal plate-to-cathode path of the tube appears as a resistance $r_p = e_p/i_p$.

While this *static* value of resistance is important in some applications of the tube, the *dynamic* value of plate resistance is of significance in most electron-tube circuits. This value is the ratio of a change in plate voltage to the corresponding change in plate current.

$$r_p = dE_p/dI_p \qquad (17\text{–}3)$$

where E_p is in volts, I_p is in amperes, and r_p is in ohms.

The electrical characteristics of the grid and plate circuits of the triode are interrelated in the following ways:

$$g_m = \mu / r_p \tag{17-4}$$

$$\mu = g_m r_p \tag{17-5}$$

$$r_p = \mu / g_m \tag{17-6}$$

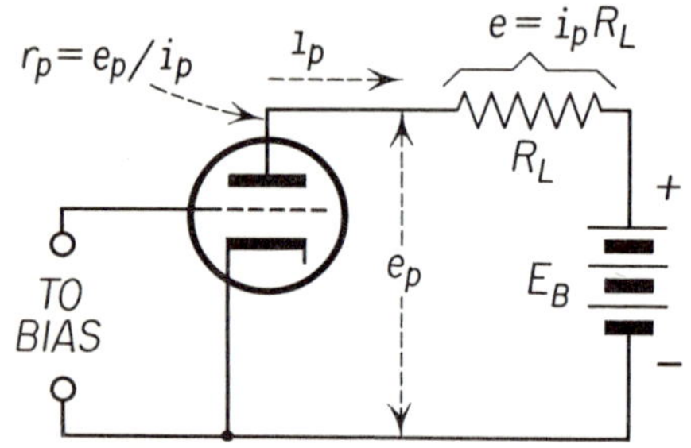

Fig. 17–6 Plate Voltage and Resistance Relationships

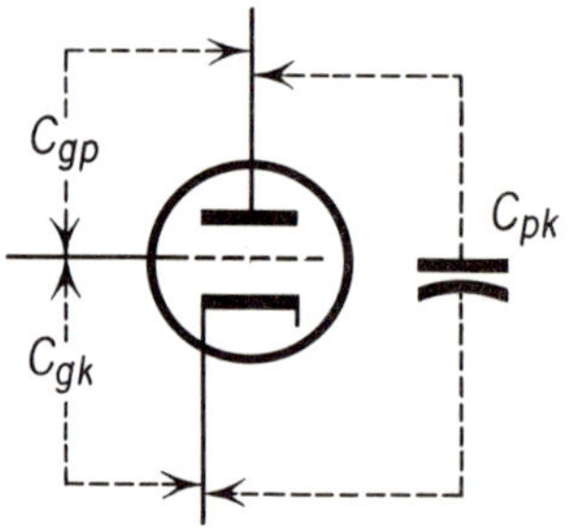

Fig. 17–7 Triode Interelectrode Capacitances

Like the diode tube, the triode exhibits capacitance between its electrodes. Figure 17–7 shows the positions of these capacitances. They are the grid-cathode or *input* capacitance (C_{gk}), grid-plate capacitance (C_{gp}), and plate-cathode or *output* capacitance (C_{pk}).

17.3 TETRODE TUBE

The grid-plate capacitance often causes malfunctioning of tube circuits. The chief effect of this mischief is the tendency of the capacitance to transmit an a-c signal *through* the tube. The grid-plate capacitance is reduced by inserting a second grid electrode between the first grid and plate. This auxiliary grid, termed a *screen grid* or simply *screen,* acts as an electrostatic shield between the original grid (now called the *control grid*) and plate. However, because it has a mesh structure, electrons can pass through it. The screen reduces the capacitance of the grid-plate path by breaking it up into two series-connected capacitances (control grid-to-screen and screen-to-plate).

The result of adding the screen is the four-element, or *tetrode,* tube. Figure 17–8 shows the arrangement of this type of tube, with the heater omitted from the diagram for simplicity. Tetrodes may be of the filament type as well as the cathode type.

In practical tetrode circuits, the screen is connected to a d-c voltage somewhat lower than the plate voltage. An external capacitor is connected from screen to the negative d-c terminal so that the screen is effectively connected to negative (ground) for alternating current but not for direct current.

In a screen-type tube, plate current is largely independent of plate voltage over a wide range. The reason for this is the shielding action

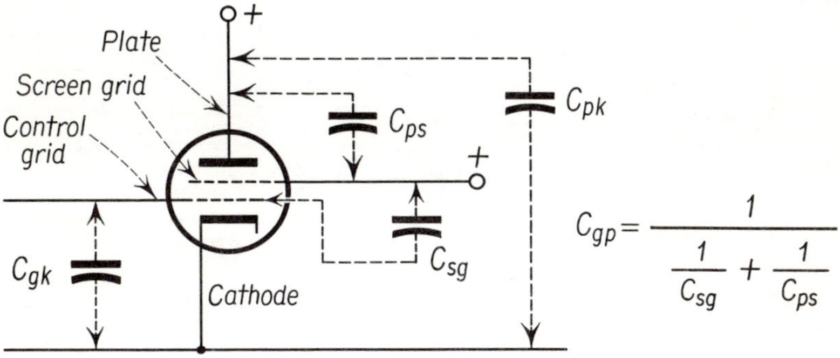

Fig. 17–8 Tetrode Tube, Showing Capacitances

of the screen which restricts the influence of the plate voltage in attracting electrons from the cathode. The screen, being positively charged, attracts these electrons which pass readily through its large openings and proceed to the plate. Screen voltage thus is more effective than plate voltage in determining the value of plate current.

17.4 PENTODE TUBE

Electrons strike the plate of a tube with sufficient impact to knock electrons from the outer orbits of the plate atoms. This action is termed *secondary emission.* In diodes and triodes, the secondary electrons are drawn back to the plate and cause no marked mischief. In the tetrode, however, they are attracted by the positively charged screen, and the plate current decreases as the result of this robbery.

Secondary emission is curtailed by placing a gridlike electrode between the screen and plate. (See Figure 17–9.) This electrode is called a *suppressor grid,* or simply *suppressor.* The suppressor usually is connected, internally or externally, to the cathode, although it sometimes is connected to ground (negative d-c terminal). The negative charge of the suppressor then repels the secondary electrons back to the plate. The suppressor adds a fifth electrode, and tubes containing this electrode accordingly are called *pentodes.*

Pentodes provide high values of g_m, μ, and r_p. As in tetrodes, pentode plate current is independent of plate voltage over a wide range.

This is shown by the family of plate-current curves in Figure 17–10. Such plots often are referred to as *constant-current curves.*

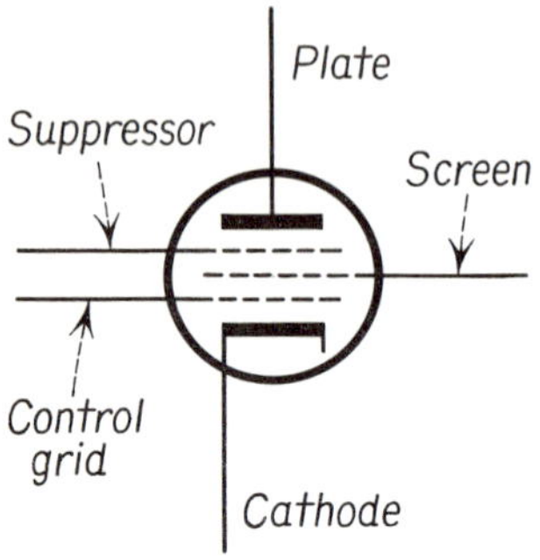

Fig. 17–9 Pentode Tube

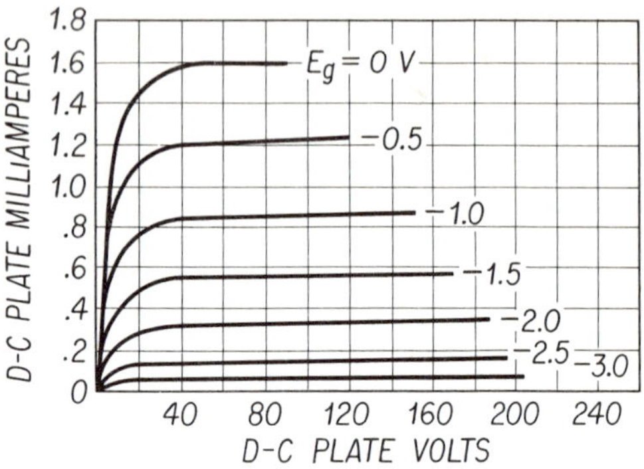

Fig. 17–10 Pentode Response Curves

17.5 BEAM POWER TUBE

Improved performance is obtained by concentrating the electrons in a directed beam from cathode to plate. This action is accomplished

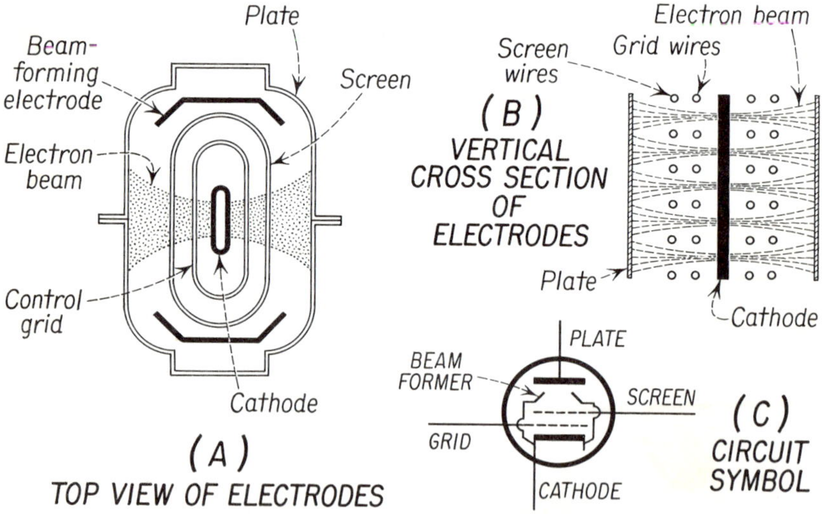

Fig. 17–11 Beam Power Tube

by means of beam-confining electrodes [Figure 17–11(A)] which are connected electrically to the cathode.

The *beam power tube* has the ability to handle higher *plate power* than other types and with higher efficiency. (D-c plate watts = $e_p i_p$.)

Beam power tubes have constant-current curves resembling those of pentodes (Figure 17–10).

Figure 17–11(C) gives one circuit symbol for the beam power tube. However, the beam former often is omitted, or a suppressor is shown instead.

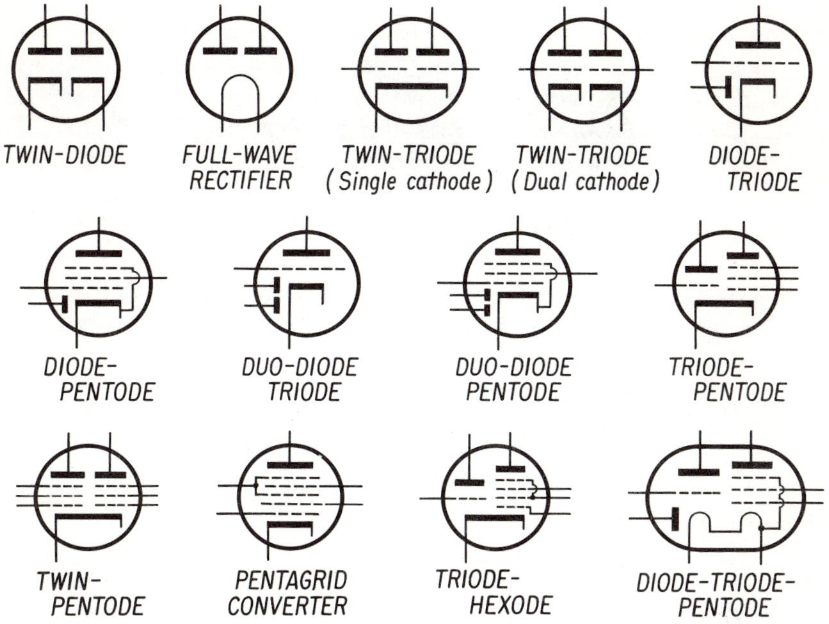

Fig. 17–12 Combination Tubes

17.6 COMBINATION TUBES

For compactness, two or more sets of tube elements often are enclosed in a single envelope. Combination tubes of this type are supplied in both directly heated filament and indirectly heated cathode types. Thus, a diode frequently is combined with a triode or pentode. Often also, two triodes are placed into a single bulb.

Figure 17–12 shows some of the combination tubes in current use. In most instances, the filament or heater electrodes have been omitted from these diagrams for simplicity. There are many more multipurpose tubes not necessarily within the scope of this book.

17.7 VACUUM-TUBE AMPLIFIERS

Vacuum-tube amplifiers fall into two categories—*voltage amplifiers* and *power amplifiers*. Some amplifiers are arranged to amplify a-c signals, while others handle d-c signals. It is the purpose of the voltage amplifier to deliver a high signal-voltage output for a low signal-voltage input. This type of amplifier provides *voltage gain.* The power amplifier delivers a signal-power output and may or may not provide voltage gain at the same time. This amplifier does provide *power gain.*

Voltage Amplifier. Figure 17–13 shows several voltage-amplifier circuits. Figure 17–13(A) is a single-stage, resistance-capacitance-coupled amplifier, so called because the a-c signal is coupled into the tube through the *R-C* network, $C_1 - R_g$ and is coupled out, after amplification, through the *R-C* network, C_2R_p. In this circuit, the d-c grid bias is developed as a voltage drop (i_pR_k) resulting from the flow of plate current through the cathode resistor, R_k. This voltage is applied to the grid through the high resistance, R_g.

The alternate half-cycles of a-c signal voltage, applied to the circuit, swing the grid voltage above and below its bias level. This causes the plate current to swing in sympathy above and below its resting (static or quiescent) value. [See Figure 17–13(E).] Figure 17–13(F) shows how amplification is obtained. Here, the $E_g - I_p$ curve is similar to the one given in Figure 17–5. Grid voltage E_{g1} is the bias, and i_{p1} is the corresponding static plate current. The negative half-cycle of the input signal swings the bias down to E_{g2}, and the plate current to i_{p2}. The positive half-cycle swings the grid voltage to E_{g3} and the plate current to i_{p3}. Because of the steepness of the $E_g - I_p$ curve, the output signal is larger than the input, therefore amplification has occurred.

The plate-current fluctuations pass through load resistor R_p in Figure 17–13(A). A fluctuating voltage drop (alternating current superimposed on direct current) is developed across this resistor and is coupled out of the circuit as an amplified a-c output signal by Capacitor C_2. The output signal is larger in voltage than the input signal, and voltage amplification has taken place. Pentodes provide higher voltage amplification than triodes and often are employed in *R-C*-coupled circuits. The value of voltage amplification, or gain (the ratio of output-signal voltage to input-signal voltage) may be calculated with the aid of the following equation.

(17–7) $$VG = (\mu R_p)/(r_p + R_p)$$

where VG = voltage gain
μ = tube amplification factor
r_p = plate resistance in ohms of Tube V
R_p = load resistance in ohms in the plate circuit

This equation allows one to find what gain to expect with a given tube

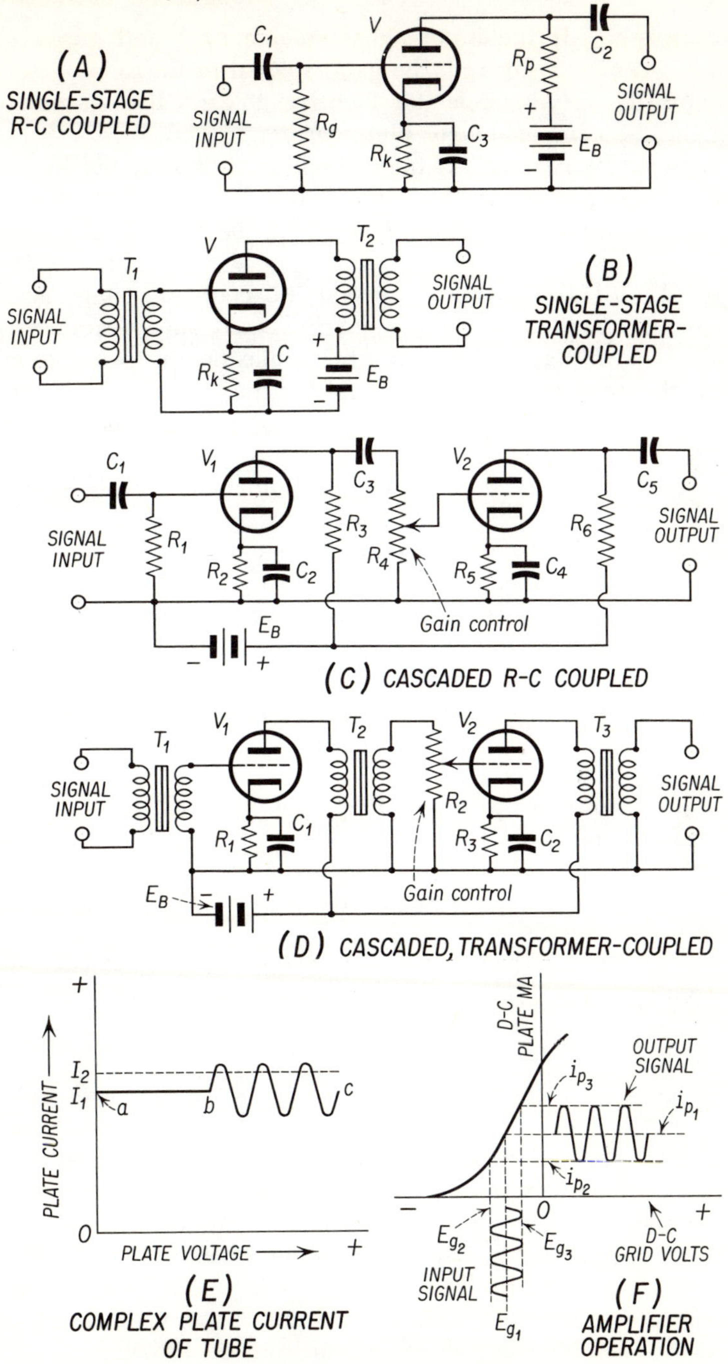

Fig. 17–13 Voltage Amplifiers

and load resistor. If the input-signal voltage (E_i) and output-signal voltage (E_o) can be measured, the gain is equal to E_o/E_i.

In Figure 17–13(A), Capacitor C_3 serves to carry the a-c component (but not the d-c component) of plate current around resistor R_k. This process is called *by-passing.* If the cathode resistor were not by-passed, the grid voltage (which is the drop across R_k) would fluctuate and change the operating point of the tube along the $E_g - I_p$ characteristic curve.

Figure 17–13(C) shows how several *R-C*-coupled voltage amplifier stages may be operated in cascade for increased amplification. In an arrangement of this sort, the over-all amplification is equal to $n(A)$, where n is the number of identical amplifier stages and A is the voltage amplification of one stage. If the stages are not identical, their individual gains simply are added to determine the over-all amplification. It is seen that grid resistor R_4 has been made a potentiometer. Since it may be set for various voltage values, it serves as a *gain control.* This is the principle of the volume control in amplifiers and radios.

Resistance-capacitance coupling offers simplicity and compactness. However, in some applications, it is necessary to couple the a-c signal into and out of an amplifier stage by means of transformers. Figure 17–13(B) shows a basic, transformer-coupled amplifier. One of the advantages of transformer coupling is the absence of high voltage drops in the plate circuits, such as those encountered with large resistances in *R-C*-coupled amplifiers.

In a transformer-coupled amplifier, the turns ratios of the transformers must be included in the calculation of over-all gain. Thus, if the gain of the tube circuit is 10 and Transformers T_1 and T_2 in Figure 17–13(B) each have a step-up turns ratio of $2\frac{1}{2}$:1, the over-all gain will be $10(2.5 \times 2.5) = 62.5$.

Figure 17–13(D) shows how transformer-coupled stages can be cascaded for increased voltage gain. This circuit is similar to the cascaded *R-C*-coupled amplifier of Figure 17–13(C) except for the transformer coupling.

In all amplifiers, the gain tends to fall off at very low and very high frequencies, owing to the effects of shunting capacitances and, in the transformer-coupled circuit, to frequency-sensitive parameters of the transformers. This is termed *frequency response.* The output wave form of a good amplifier is a replica of the input wave form. The output signal simply is larger than the input signal. The ability so to reproduce the signal faithfully is termed *fidelity.* The output signal is said to be *distorted* when it is not of the exact shape of the input signal. A high-fidelity amplifier introduces a minimum of distortion.

Power Amplifier. Figure 17–14 shows two circuits of power amplifiers. In Figure 17–14(A), the construction of Tube V is such that the static plate current is high, and a signal voltage applied to the grid-cathode input circuit through C_1 and R_1 causes heavy plate-current swings.

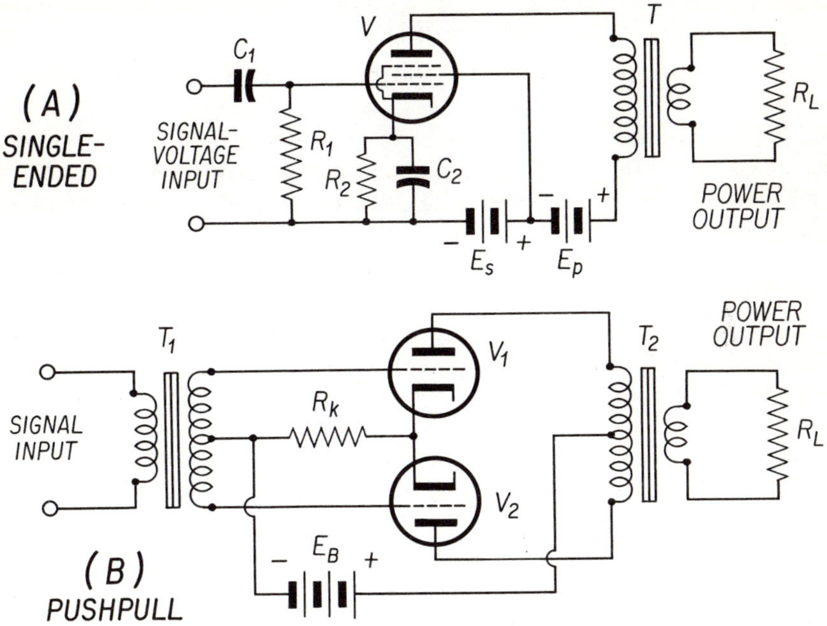

Fig. 17–14 Power Amplifiers

The d-c plate power is equal to E_pI_p (where P is in watts, E_p in volts, and I_p in amperes). Transformer T serves to match the plate impedance (E_p/I_p) to the load resistance, R_L, and to couple a-c signal power out of the amplifier to the load. (See Section 10.11, Chapter 10, for an explanation of impedance matching with transformers.) The output power equals E^2/R_L (where E is the signal voltage developed across load resistor R_L, P is the power in watts, and R_L is the load resistance in ohms).

The *plate efficiency* of a power amplifier is equal to the ratio of a-c output power to d-c input power. It is expressed as a percentage.

$$n = (100P_o)/(E_pI_p) \quad (17\text{–}8)$$

where P_o = a-c output power (in watts)
E_p = d-c plate voltage (in volts)
I_p = d-c plate current (in amperes)

Example: A power amplifier develops 10 v across a load resistance of 5 ohms. The tube has a plate voltage of 300 vdc and a plate current of 200 ma. What is the plate efficiency?

$$P_o = E^2/R_L = 10^2/5 = 100/5 = 20 \text{ w}$$

$$I_p = 200 \text{ ma} = 0.2 \text{ amp}$$

$$n = (100 \times 20)/(300 \times 0.2) = 2{,}000/60 = 33.3\%$$

Another important parameter in power amplifiers is *plate dissipation.* This is the difference between input and output power [$P_d = (E_pI_p) - P_o$] and represents the power expended as heating in the plate of the tube.

Figure 17–14(B) shows a "double-ended" or *push-pull* power-amplifier circuit. This arrangement consists of two identical halves with center-tapped input and output transformers for signal transfer. Due to the center-tapped secondary of T_1, when the input signal swings the grid of tube V_1 positive, it swings the grid of V_2 negative, and vice versa. D-c grid bias for both tubes is developed by the voltage drop due to the combined plate-current flow through cathode resistor R_k.

The push-pull amplifier has approximately twice the power output of a single-ended stage.

Each of the power amplifiers shown in Figure 17–14 may be preceded by a voltage amplifier to provide high sensitivity to weak signals. In many instances, the preceding amplifier need supply only voltage to the grids of the power amplifier. In some instances, discussed later, a small amount of signal *power* must be supplied to the power-amplifier grid. When this is done, still another parameter is considered in rating the amplifier. This is *power amplification,* or *power gain,* and is the ratio of power output (P_o) to grid power input (P_i): $PG = P_o/P_i$. Thus, a power gain of 100 is provided by an amplifier when it delivers 25 w output for a grid input power (*drive*) of 250 milliwatts.

Power amplifiers are employed in audio-frequency and radio-frequency applications. Air-core transformers must be used at radio frequencies, although powdered-iron, ferrite, and metallic-dust cores sometimes are employed. Iron cores are used at audio frequencies.

When the d-c grid bias of a power amplifier is set to a point midway on the linear portion of the $E_g - I_p$ curve and the signal-voltage peaks are limited to this straight region, the amplifier is designated *class-A.* A class-A amplifier has low distortion, low power output, and low efficiency (20 to 35 per cent). No grid current flows at any time in a true class-A amplifier.

In a *class-B* amplifier, the d-c grid bias is adjusted for plate-current cutoff. The plate current is approximately zero in the absence of an input signal. Plate current flows only during the positive half-cycles of input-signal voltage which need to be quite high in amplitude. Plate current does not flow through the entire half-cycle of signal voltage but nearly so. Because of the large input-signal voltage swings, the grid is driven positive, and grid current flows. In audio amplification, push-pull circuits always are used for class-B operation, in order to supply the missing half-cycle of output-signal voltage (each tube in the push-pull circuit supplying one half-cycle) and to reduce distortion.

Class-B amplifiers have a plate efficiency of approximately 50 per cent and are capable of higher power output with a given pair of tubes than in class-A amplification. However, because grid current flows, the input signal source must supply power (usually equal to approximately $0.1E_pI_p$ of the class-B tubes) to drive this amplifier.

Class-AB operation, also used with push-pull stages, is intermediate between class-A and class-B. In the class-AB circuit, plate current flows for more than one half-cycle (180 deg) of input-signal voltage but not for an entire cycle. The d-c grid bias is set at a point between cutoff and the class-A value. Class-AB_1 amplifiers draw no grid current, but class-AB_2 amplifiers draw a small amount. Class-AB amplifiers have plate efficiencies between those of class-A and class-B (AB_2 higher than AB_1), deliver less output power than class-B but more than class-A, and require considerably less driving power than class-B.

Class-C amplifiers are employed in communications only in r-f service (because of their inherent high distortion). However, they are also employed to operate relays in industrial circuits. In this type, the grid bias is adjusted to a value higher than cutoff (usually twice cutoff). A very high input-signal swing thus is required. Plate current flows only during a portion of the positive half-cycle of the input signal. This amplifier has the highest plate efficiency (70–75 per cent) and the highest power output.

17.8 OSCILLATORS

When the output of an amplifier is coupled back to the input in the proper phase, the tube will amplify any signal present at its input and will feed the amplified signal back around to the input to be amplified still more. This process continues, the signal passing around the loop to be reinforced on each trip.

This process is called *regeneration,* and the mechanism whereby it is accomplished is termed *positive feedback.* (This is called *positive* because the signal is reinforced and must drive the grid positive. If the phase of the feedback is reversed, the opposite effect is obtained, that is, the signal is reduced. The feedback then is *negative* and the mechanism is called *degeneration.*) Figure 17–15(A) shows how feedback is accomplished with a simple transformer connected between input and output of an amplifier.

An amplifier arranged for large amounts of positive feedback is called an *oscillator.* The oscillator is a nonmechanical generator of

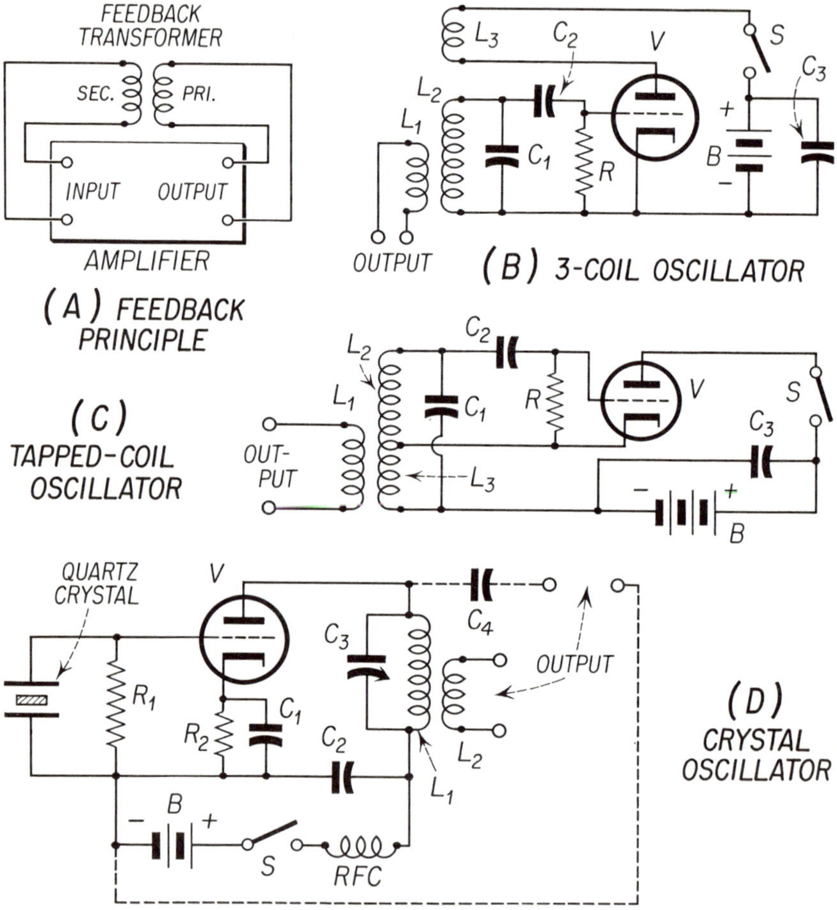

Fig. 17–15 Oscillators

alternating current and voltage. All high-frequency alternating current is generated by oscillators of some kind. There are many types of oscillator circuits. Not all can be described here. Figure 17–15 shows a few of these circuits.

Figure 17–15(B) shows a basic oscillator circuit. Feedback occurs, from the plate (output) to grid (input) circuit, as a result of the inductive coupling between Coil L_3 (plate coil) and Coil L_2 (grid coil). This circuit operates in the following manner: Assuming the polarity of L_3 to be correct for positive feedback, when Switch S is closed, a sudden rush (transient) of plate current flows through L_3. This causes magnetic lines of force to expand outward from L_3. These lines cut the turns of L_2 and induce a voltage across the latter coil such that the grid of Tube V becomes more positive and the cathode negative. This increase of the grid voltage in the positive direction increases the plate current through L_3 which causes the lines to expand still further and induce a higher voltage across L_2, making the grid still more positive and increasing the plate current further. This action would proceed without limit, except that the plate current saturation point (Figure 17–5) eventually is reached, and the plate current stops increasing.

Since induction depends upon a *change* of flux, the induction between L_3 and L_2 first decreases and then ceases as saturation is reached. This reduces the grid signal voltage and therefore the plate current. Just as increasing grid voltage caused lines of force to expand from L_3 and, through the induced voltage across L_2, to make the grid more positive, this decreasing grid voltage now causes the lines to collapse and make the grid more negative. This action continues to reduce the plate current and would continue indefinitely except that the plate-current cutoff point is reached, and current momentarily stops flowing. The cycle of events then is repeated.

The plate current (and the voltage across L_2) rises and falls with a definite frequency which depends upon the values of C_1 and L_2. These two components form a resonant circuit which *tunes* the oscillator. (See Section 9.7, Chapter 9, for instructions regarding calculation of resonant frequency.) Continuous variation of frequency (tuning) may be obtained by varying either C_1 or L_2. Power may be taken from the oscillator through Coil L_1 which is inductively coupled to L_2. Capacitor C_3 serves to by-pass the a-c component of plate current around the d-c source.

Oscillators may be designed and adjusted to produce almost perfect sine waves. Oscillators are employed in the generation of both audio and radio frequencies. Air-core coils are employed in r-f oscillators, and iron cores at audio frequencies.

Figure 17–15(C) shows a somewhat simplified version of the oscillator circuit just described. Here, the grid and plate coils have been combined into a single inductor which has been center-tapped (L_2 — L_3). The lower portion of this coil serves as the plate winding, and the upper portion as the grid winding. Operation of the circuit is the same as for the oscillator described previously.

Figure 17–15(D) is the circuit of a *crystal-controlled* oscillator.

The heart of this oscillator is an accurately ground and electroded quartz crystal connected in the grid-cathode (input) circuit of Tube V. The quartz crystal is *piezoelectric;* that is, it will vibrate in synchronism with an alternating voltage applied to its electrodes, and, conversely, when it is vibrated it will produce an alternating voltage. The frequency depends principally upon the dimensions of the crystal and is maintained very accurately by the vibrating crystal.

When the parallel-resonant circuit (L_1C_3) in the plate circuit of the tube is tuned to the crystal frequency, enough energy is fed back through the grid-plate capacitance of the tube to keep the crystal oscillating. The alternating voltage impressed upon the grid by the crystal then is amplified by the tube and stepped up by the parallel-resonant circuit. This oscillator will operate only at the natural frequency of the crystal, and that is the main advantage of the circuit. *Self-excited* oscillators, like those discussed previously, will operate on any frequency to which they are tuned and often will be set accidentally to an undesired frequency.

Crystal-controlled oscillators are employed principally at radio frequencies, since the crystals would be cumbersomely thick at audio frequencies. These oscillators are noted for their precision of frequency and stability.

In Figure 17–15(D), the *radio-frequency choke coil* (*RFC*), in conjunction with by-pass capacitor C_2, serves to keep the high-frequency component of plate current out of the d-c source, *B*. Output power is taken from the oscillator either through Coil L_2, which is inductively coupled to the tuned coil L_1, or by means of capacitance coupling through Capacitor C_4, as indicated by the dashed lines.

17.9 SOUND REPRODUCTION

The electrical reproduction of speech and music is a widespread popular application of amplifiers.

Most signal voltages available for this purpose (whether derived from phonograph records, magnetic tape, magnetic wire, motion-picture film, or microphones) are quite small. A sound system therefore must contain several voltage-amplifier stages, the purpose of which is to build these tiny voltages up to a level sufficient to drive a power amplifier. The power-amplifier output then is coupled to a loud-speaker. The latter is of the general nature of an enlarged telephone receiver.

Figure 17–16 shows in composite form the basic arrangement of the electronic sound system. In the pickup for record reproduction (*A*), the phonograph needle either conveys vibrations to a piezoelectric crystal to produce the signal emf, or it causes a coil to vibrate in a magnetic field for the same purpose. In *B*, sound vibrations have been recorded as magnetic flux variations along a wire, and this wire is passed between spools under a coil-and-pole device (called a *pickup head*) in which an alternating current is induced by the changing flux of the wire. In *C*, a similar effect is obtained with plastic tape coated with a magnetic oxide of iron. In *D*, sound has been recorded on motion-picture film as changes

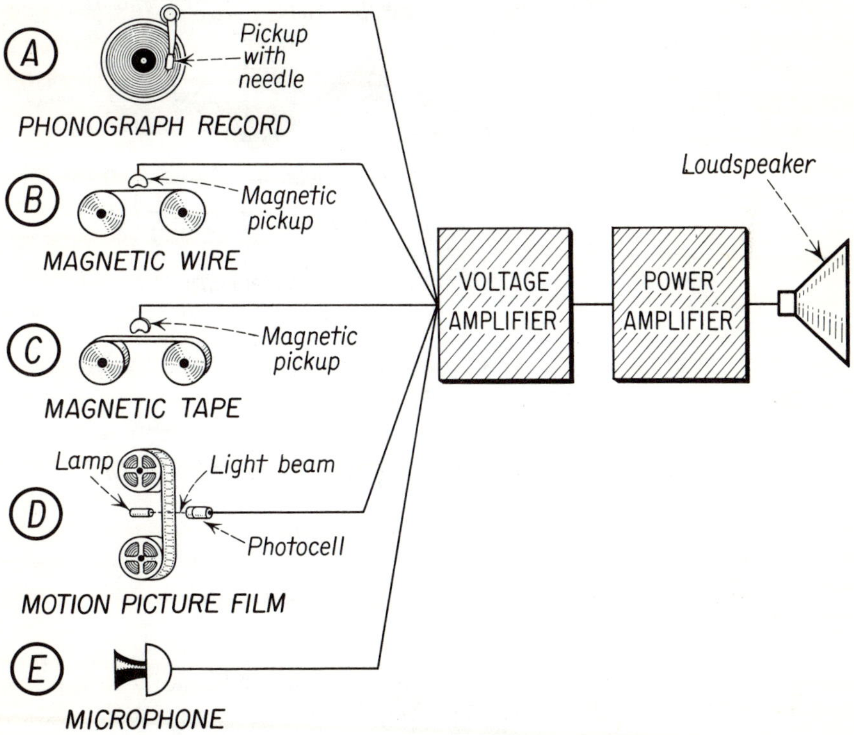

Fig. 17–16 Bases of Electronic Sound Systems

in the emulsion. As this film passes between reels, its *sound track* interrupts (modulates) a beam of light, and the resulting variations in light intensity are translated by a photocell or phototube into corresponding voltage variations. The principle of the carbon microphone (Pickup *E*) has been explained in Section 16.3, Chapter 16. Other microphones include the *dynamic* type in which sound waves cause a coil to vibrate in a magnetic field and thus to set up an alternating emf, the *velocity* type in which a thin ribbon vibrates in a magnetic field, and the *crystal* type in which the sound waves cause a piezoelectric crystal to vibrate and generate an emf.

Medium *A* is the basis of jukeboxes and home phonographs, as well as transcribing machines used in radio broadcasting. *B* and *C* underlie both the hobby and professional practice of long-period recording and playback, and are the foundation of modern dictating machines. *D*, of course, is the talking-picture scheme. The direct microphone input (*E*) is the basis of the public-address and paging systems.

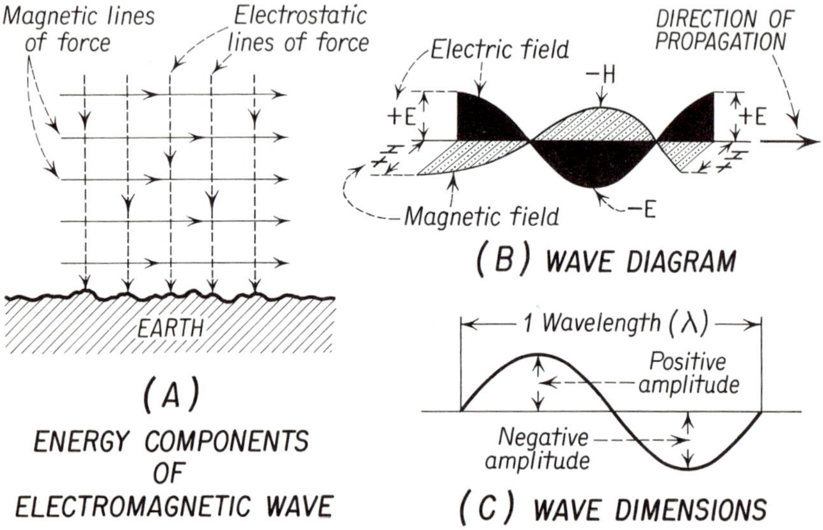

Fig. 17–17 Electromagnetic Waves

17.10 RADIATION AND ELECTROMAGNETIC WAVES

Low-frequency alternating currents are confined readily to the lines through which they flow. At high frequencies, however (that is, above 10 kc) not all of the energy returns to the conductor when the surrounding magnetic field collapses between cycles, but some of it escapes into space and may be detected at a considerable distance from the source.

This energy is considered to be radiated in the form of *waves* which travel away from the source at the speed of light, 186,000 miles per second. These *radio waves* are electromagnetic in nature, just as are light waves.

In a radio wave, the energy is divided into an electrostatic field and a magnetic field. The lines of force of these two fields are at right angles to each other and are perpendicular to the direction in which the wave travels (*direction of propagation*). Figure 17–17(A) illustrates the arrangement of the electrostatic and magnetic lines of force. The wave is considered to be propagated into the page. A wave is termed *vertically polarized* when its electrostatic lines are perpendicular to the surface of the earth and *horizontally polarized* when they are parallel to the surface. Figure 17–17(B) is another diagram showing the right-angle relationship of electrostatic and magnetic fields and the perpendicular direction of propagation. Here, the alternating electrostatic field, E (solid black), is vertical, and the magnetic field, H (crosshatched), is horizontal. Propagation is from left to right.

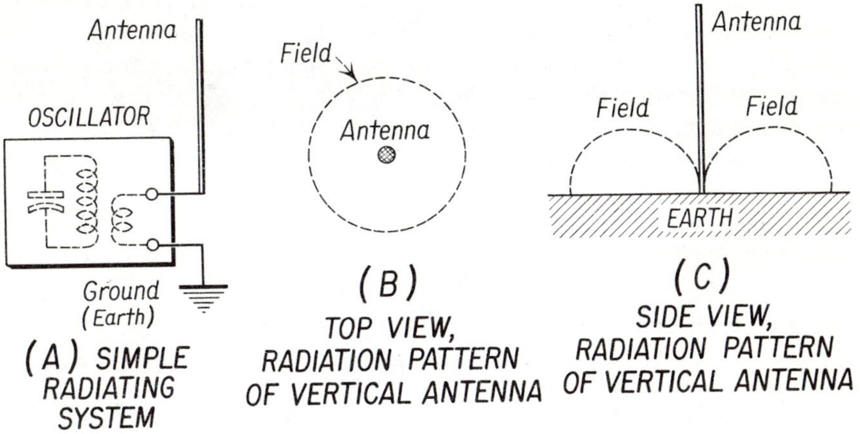

Fig. 17–18 Apparatus for Electromagnetic Radiation

Figure 17–17(C) shows the common dimensions of a wave—*amplitude* and *wave length.* Wave length is measured in meters (m). 1 m = 39.37 in. The symbol for wave length and for the meter is λ, the Greek letter *lambda.* Wave length and frequency are related in the following manner.

(17–9) $$\lambda = 300/f_{\text{mc}} = 300{,}000/f_{\text{kc}}$$

where λ = wave length (in meters)
f_{mc} = frequency (in megacycles)
f_{kc} = frequency (in kilocycles)

Also

(17–10) $$f_{\text{mc}} = 300/\lambda, \text{ and } f_{\text{kc}} = 300{,}000/\lambda$$

Example: The frequency 980 kc is assigned by the FCC to radio broadcast stations. What is the wave length of signals transmitted at this frequency?

$$\lambda = 300{,}000/f_{\text{kc}} = 300{,}000/980 = 306.12 \text{ m}$$

The apparatus for radiating waves can be complex. Figure

17–18(A), however, shows the simplest electronic system for this purpose. This consists of an oscillator, similar to the ones shown in Figure 17–15, with one terminal of its output coil connected to a rod or wire (*antenna*) supported above the surface of the earth, and the other terminal connected to the earth (*grounded*).

The field distribution about a vertical antenna of this type appears in top view as in Figure 17–18(B). This shows simply that if the strength of the field is measured at all points of the compass at a distance from the antenna, points of equal strength, when connected together, will form a circle. Actually, this pattern is not a flat disc but is rather like a doughnut, as the vertical cross section in Figure 17–18(C) shows. Various types of antennas possess various radiation patterns, depending upon antenna length and whether horizontal or vertical.

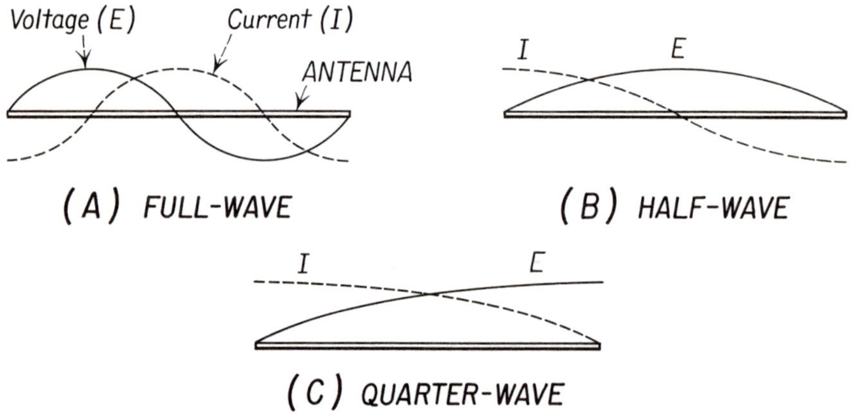

Fig. 17–19 Energy Distribution in Antennas

Currents oscillate back and forth from one end of an antenna to the other and give rise to a distribution of current and voltage maximum and minimum points along the wire. The number of such points depends upon the length of the wire with respect to the wave length of the signal. In Figure 17–19(A), for example, the antenna is a full wave length long. One wave length of current and one of voltage therefore are distributed as shown. Figure 17–19(B) shows a similar distribution for a half-wave antenna, and Figure 17–19(C) for a quarter-wave antenna. In each example, note that a current maximum (*loop*) and voltage minimum (*node*), and vice versa, are found at the same point.

When suitable apparatus is used, radio waves may be reflected, directed, refracted, and diffused in much the same manner as are light rays.

Not all waves leaving an antenna travel by the same route. The principal waves are the *ground wave* and *sky wave*. The ground wave

Figure 17–20(A) follows the curvature of the earth for a distance and finally becomes so reduced in amplitude (*attenuated*) that it cannot be detected with normal instruments. Sky waves travel upward at various angles, depending upon their wave length and the antenna dimensions, and strike the *ionosphere*—a shell of ionized particles surrounding the earth. This shell is a little under 100 miles above the surface of the earth

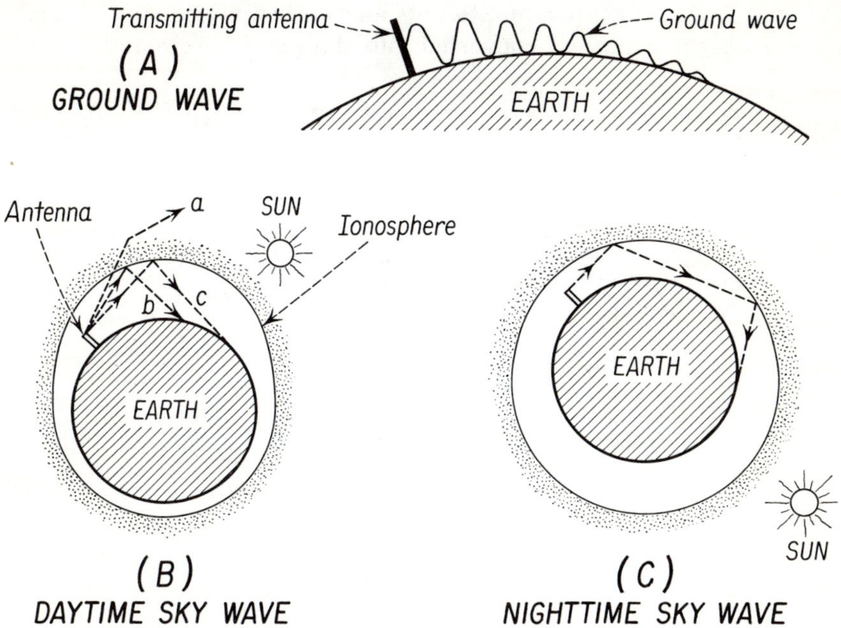

Fig. 17–20 Earth and Sky Components of Radiated Waves

and is higher on the heated, sunny side [Figure 17–20(B)] than on the cool, dark side [Figure 17–20(C)]. Waves striking the ionosphere (and even penetrating it for a short distance) are reflected back to earth. [See waves *b* and *c* in Figure 16–20(B).] If the angle is steep, however, as in the case of Wave *a*, the wave penetrates the ionosphere with some refraction and is lost in space. The sky waves may be detected at the point at which they return to earth and somewhat beyond but not in the region (the *skip zone*) between the point where the ground wave ends and the point of sky-wave return. The skip zone of Wave *c* in Figure 17–20(B) is seen to be longer than that of Wave *b*. In Figure 17–20(C), the ionosphere is shown closer to the antenna (nighttime condition), and one sky wave is shown traveling at such an angle as to be reflected twice by the ionosphere before returning to earth.

Sky-wave reflection accounts for the fact that a short-wave radio station might not be heard a few miles away but is picked up very clearly at a point several thousand miles distant.

17.11 RADIO TRANSMISSION

The principle of radiation is utilized in radio transmission. Radiotelegraph signals are transmitted by breaking the waves up into short and long trains, corresponding to dots and dashes, which may be picked up at a distant point and converted into short and long sounds. Radiotelephone signals are transmitted by modulating the radio waves through the action of sound waves, much as the direct current in the telephone circuit (Section 16.3, Chapter 16) is modulated by the sound-controlled resistance of the microphone.

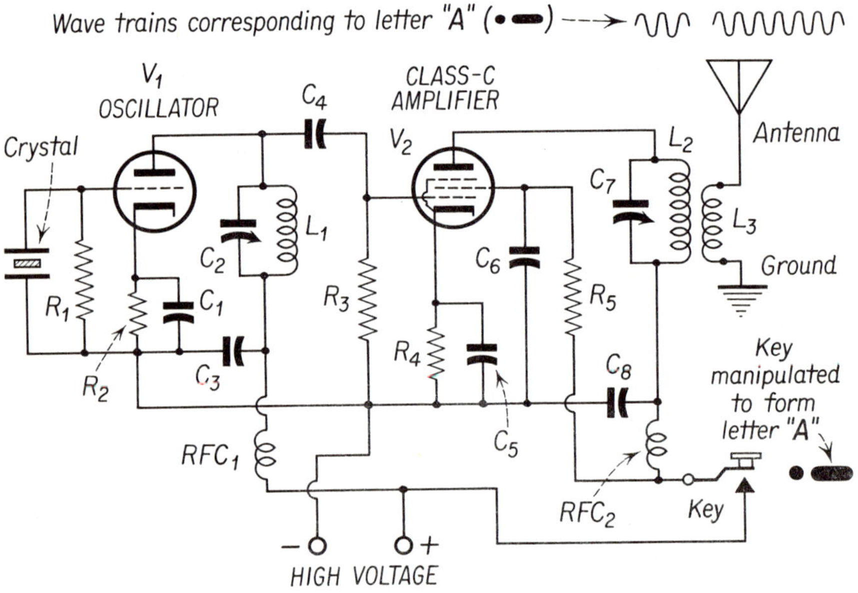

Fig. 17–21 Radiotelegraph Transmitter

An r-f oscillator may be coupled directly to an antenna and ground and will radiate radio waves, but a combination of oscillator and r-f power amplifier is more efficient. Figure 17–21 shows a simple radiotelegraph transmitter. Here, the oscillator stage is identical with the crystal oscillator shown in Figure 17–15(D). The r-f output of this oscillator is capacitance-coupled to a pentode r-f power amplifier through Capacitor C_4. The oscillator is adjusted by tuning its plate resonant circuit to the crystal frequency (adjustment of C_2). The amplifier is tuned to this frequency by adjusting C_7. The r-f output of the amplifier is coupled to an antenna and ground.

The oscillator is operated continuously during a transmission period because its stability is higher if it is not switched on and off. The amplifier does not operate, however, until the key is depressed, because Tube V_2 receives no plate and screen voltage when the key is open. As long as the key is closed, a continuous train of waves is emitted from the antenna. Code signals can be transmitted by manipulating the key to break the train into appropriate wave groups corresponding to dots and dashes. Thus, when the key is closed for one short interval and one long (the letter "A"), an "A" train of waves leaves the antenna, as shown.

Figure 17–22 shows how an audio amplifier section may be added to the radiotelegraph transmitter to produce a radiotelephone transmitter.

The crystal oscillator and class-C r-f amplifier generate an r-f wave of constant amplitude (*continuous wave,* or *cw*), as shown by pattern *e*. Sound waves impinging upon the microphone produce an a-f wave, such as *a*, at the secondary of transformer T_1. This wave is amplified by the pentode speech amplifier (V_3), a voltage amplifier, to have higher amplitude, *b*, but the same frequency and shape. It is amplified again by the triode power amplifier or *driver* (V_4), as *c*. The push-pull class-B amplifier (V_5–V_6) develops an a-f power wave, *d*, which is applied in series with the d-c power of the plate and screen of the r-f power amplifier, V_2. The plate and screen voltage of V_2 thus consist of an a-f voltage superimposed upon a d-c voltage, and both plate-screen voltage and current rise and fall in synchronism with the audio.

The result is that the continuous r-f wave (*e*) becomes modulated; that is, molded into the shape of the a-f wave, as in pattern *f*. This process is called *amplitude modulation* (am), since it alters the amplitudes (heights) of successive r-f half-cycles. The dotted lines enclosing the positive and negative r-f peaks in pattern *f* is called the *modulation envelope* and is seen to have the shape of the a-f wave. The r-f component is called the *carrier wave,* or simply *carrier*.

In another system of modulation, *frequency modulation,* the amplitude of the r-f carrier wave is held constant while the a-f modulating wave varies the carrier frequency in sympathy with the audio frequency.

17.12 RADIO RECEPTION

The simplest practical radio receiver consists of (1) a tuner for separating a desired signal from all others, (2) a detector for converting the received r-f signal into an a-f signal, and (3) a reproducer for converting the audio frequency into sound.

Figure 17–23 shows a circuit filling these requirements. This type

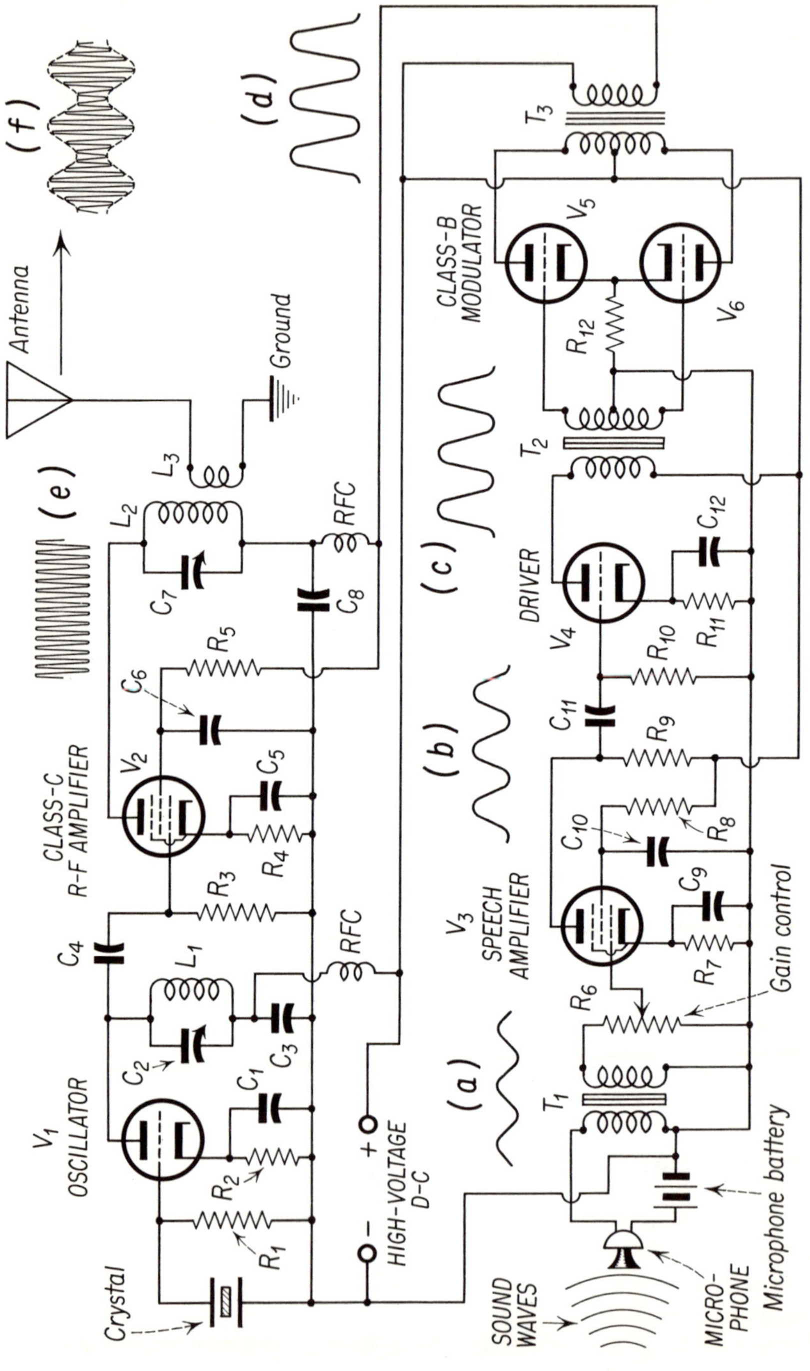

Fig. 17–22 Amplitude-Modulated (A-M) Radiotelephone Transmitter

is called a *crystal receiver* from the fact that it employs a simple crystal diode as the detector. The diode is a special form of semiconductor rectifier. (See Section 13.3, Chapter 13, for a description of semiconductor rectification.) The r-f diode employs germanium or silicon as the semiconductor. The tuner is composed of the r-f transformer (L_1–L_2), the secondary of which forms a resonant circuit with the variable capacitor, C_1.

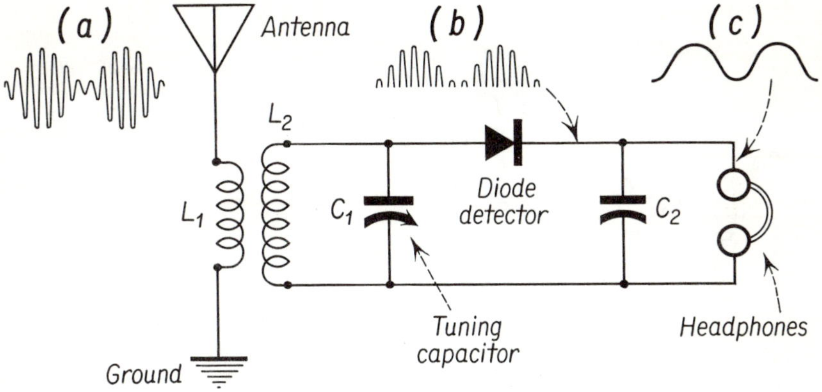

Fig. 17–23 Crystal Radio Receiver

An incoming a-m wave (pattern *a*) picked up by the antenna causes currents to flow through L_1 to ground. This induces a voltage of the same wave form across L_2 when the L_2C_1 circuit is tuned to the carrier frequency. The diode rectifies this wave, passing only positive half-cycles, as shown by wave form *b*. This current passes through the headphones. However, the headphones cannot respond to the extremely rapid alternations of the radio frequency, so the radio frequency is by-passed to ground through Capacitor C_2, and only the modulation envelope (pattern *c*) passes through the headphones. The latter then reproduces the sound which originated at the transmitting station.

The crystal receiver is simple but lacks sensitivity. It gives best results only when it is connected to a good outside antenna and when located fairly close to the transmitting station. It cannot be used for the reception of c-w radiotelegraph signals (Figure 17–21).

Figure 17–24 shows a sensitive, tube receiver of the *regenerative* type. Notice the similarity between this detector circuit and that of the oscillator [Figure 17–15(B)]. Radio-frequency amplification is afforded by the positive feedback between coils L_3 and L_2. The strength of the feedback is controlled by adjustment of the variable capacitor C_3. The output of the detector is coupled into an audio amplifier, through Transformer T, to obtain a strong a-f signal. The tuning range of the receiver is governed by the sizes of L_2 and C_1.

The regenerative detector will make c-w radiotelegraph signals audible in the following manner: When C_3 is set for strong regeneration, the circuit oscillates at radio frequencies. When the oscillation frequency is slightly higher or lower than that of the received carrier wave, the two interact to produce a *beat note* equal to the difference between the two frequencies. This difference frequency will be audible if it lies within

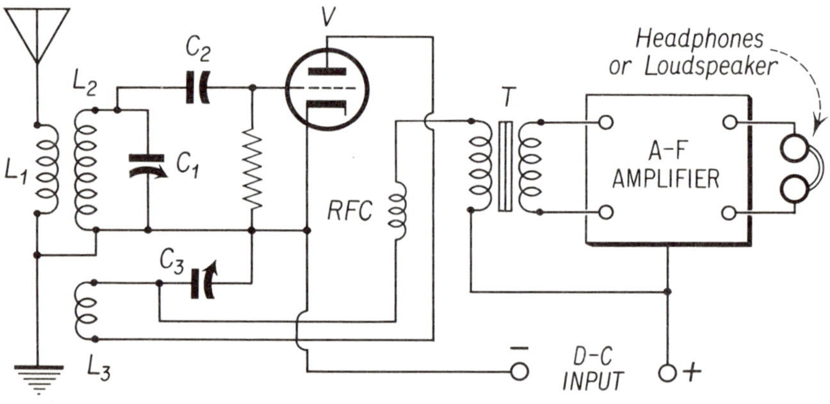

Fig. 17–24 Regenerative Receiver

the a-f range (that is, between 60 and 10,000 cps). For example, the received signal is 2 megacycles (2,000,000 cps) and the regeneration frequency is 2.001 megacycles. The difference frequency = 2,001,000 − 2,000,000 = 1,000 cps, an audible beat note.

The sensitivity of a receiver to weak signals may be increased by operating a tuned amplifier *ahead* of the detector. Such an amplifier might, for example, be used ahead of the simple crystal receiver. Because this amplifier must operate at high frequencies in order to handle the incoming signals, it is termed an *r-f amplifier.* Figure 17–25(A) shows the arrangement of a tuned r-f receiver. Figure 17–25(B) shows the circuit of a typical a-c-operated power supply used to operate receivers or transmitters from the power line. A full-wave rectifier (V) is employed. A capacitor-input filter (C_1-CH-C_2) smooths the d-c output of the tube. Various d-c voltages (E_1, E_2, E_3) are provided by taps on the voltage-divider resistor, R. Transformer secondary L_1 supplies low-voltage alternating current to light the tube heaters, L_2 lights the rectifier filament only, and the center-tapped winding, L_3, supplies high voltage to be rectified by the tube.

The most sensitive and selective of all receivers is the *superhetero-*

dyne. This circuit employs the beat-note principle to convert all incoming signals to a single low frequency at which they may be amplified with highest efficiency. The block diagram in Figure 17–26 shows the arrangement of the various channels in the superheterodyne circuit.

The *r-f amplifier* and *first detector* are continuously tuneable over the carrier-frequency range (such as the standard broadcast band, 545 to 1,600 kc) to be covered by the receiver. A signal frequency f is

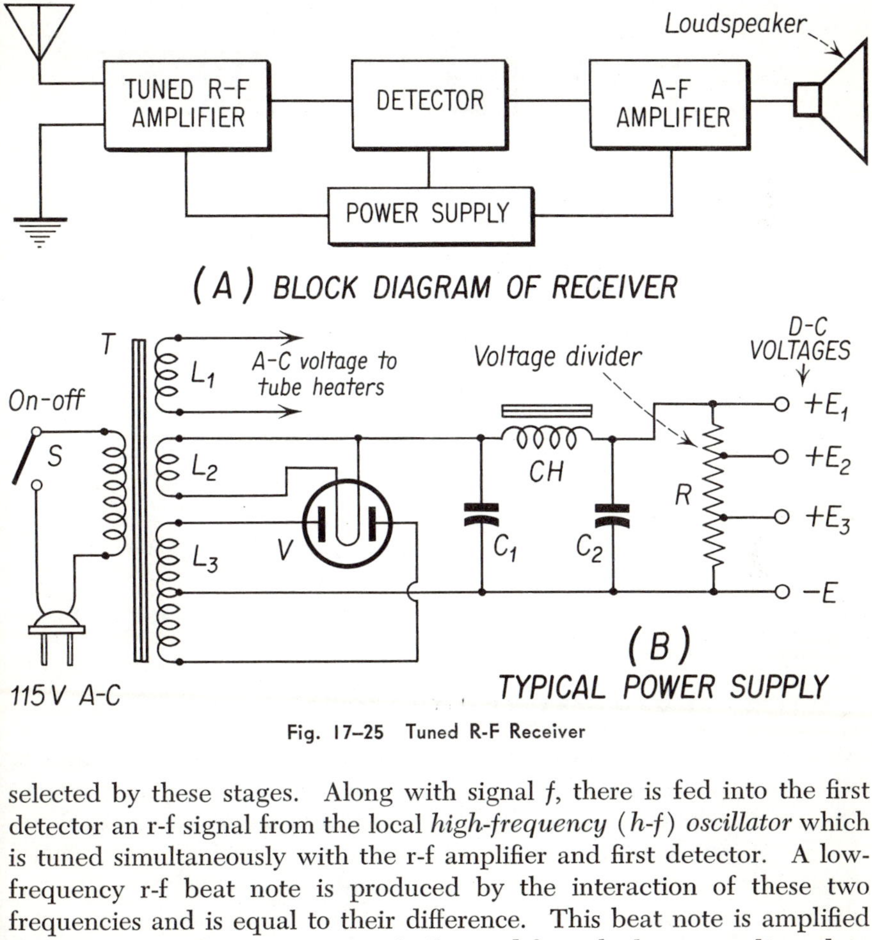

Fig. 17–25 Tuned R-F Receiver

selected by these stages. Along with signal f, there is fed into the first detector an r-f signal from the local *high-frequency* (*h-f*) *oscillator* which is tuned simultaneously with the r-f amplifier and first detector. A low-frequency r-f beat note is produced by the interaction of these two frequencies and is equal to their difference. This beat note is amplified by the *intermediate-frequency* (*i-f*) *amplifier* which is preadjusted to this single beat-note frequency, often 455 kc. The reason for this arrangement is that the i-f amplifier, being fixed in frequency, can be designed and adjusted for the most efficient performance. The h-f oscillator here may be set to $f + 455$ kc or to $f - 455$ kc. In compact superheterodyne receivers, the first detector and h-f oscillator are combined in a single combination tube, and the dual-purpose stage becomes known as a *converter.*

The amplified i-f signal is detected by the *second detector,* which is operated at the intermediate frequency, and the a-f output of this detector is presented to the *a-f amplifier.*

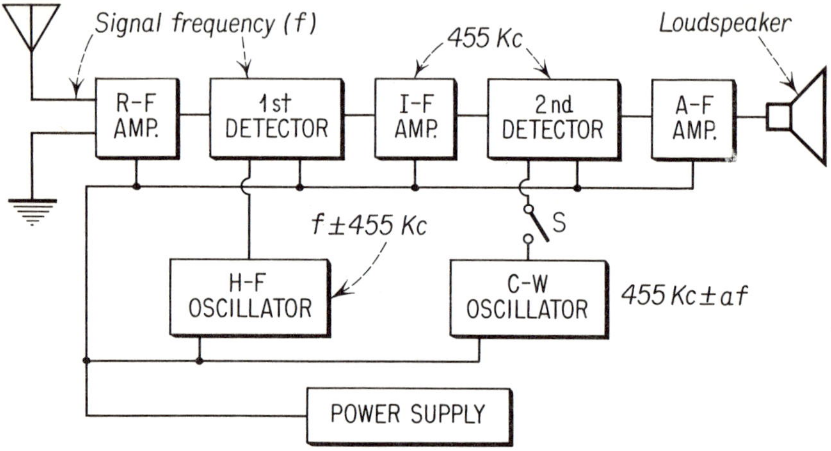

Fig. 17–26 Superheterodyne Receiver

A *c-w oscillator,* coupled into the second detector, is provided to make c-w telegraph signals audible. This is accomplished by setting the c-w oscillator frequency slightly higher or lower than the intermediate frequency so as to produce an audible beat note. The Switch, S, is added for cutting this oscillator out of the circuit when receiving radiotelephone signals.

17.13 PHOTOTUBE AND PHOTOCELL

A large part of the subject of electronics is devoted to the use of light to control electrical processes and the conversion of light energy into electrical energy. Some practical, everyday applications of *photoelectricity* are burglar alarms, object counters, safety devices, photographic exposure meters, door openers, sun batteries, lighting controls, television, talking pictures, picture transmission and reception (facsimile), and industrial smoke control.

Two classes of devices are available for supplying the linking medium between light and electricity; these are *phototubes* and *photocells.* Two phenomena are utilized in the application of both devices: *photoconductivity,* which refers to the ability of the tube or cell to change its resistance in accordance with changes in illumination and thereby to alter a current flow; and the *photovoltaic effect,* which is the ability of the device to generate an emf directly as the result of illumination. Photoconductivity usually is the property that is utilized in phototube circuits, while either photoconductivity or the photovoltaic effect may be used in photocell circuits.

Figure 17–27(A) shows the basic construction and simple circuit of a phototube. The cathode of this tube is a half-cylinder, usually of silver, coated on its inner surface with a thin layer of cesium oxide. The latter material is a light-sensitive compound, called *photoemissive* since

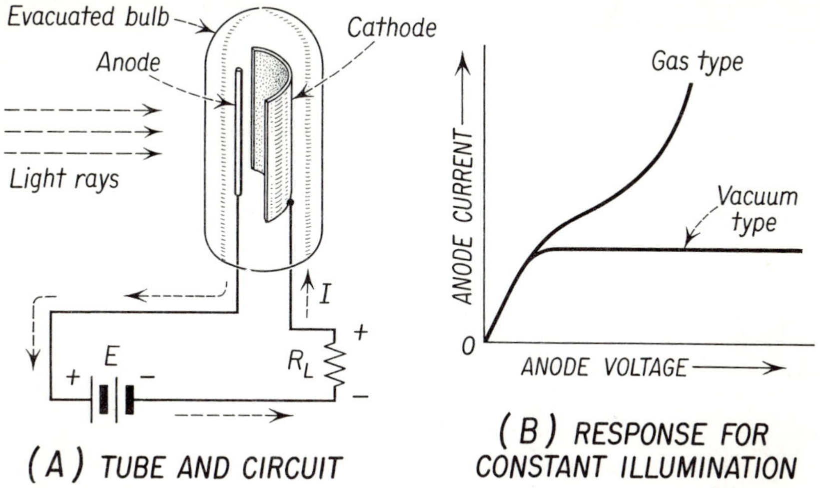

Fig. 17–27 Phototube

it emits electrons (*photoelectrons*) when light falls on it. The anode is a straight, vertical wire placed at the axis of the cathode cylinder. Depending upon the type of phototube and its intended application, the bulb is either evacuated highly or is gas-filled at low pressure.

When the phototube is connected into a circuit with a d-c source and load and its anode is made positive with respect to the cathode, as shown in Figure 17–27(A), an electron current (I) will flow through the circuit when the inner surface of the cathode is illuminated but will cease when the tube is darkened. At a given d-c voltage value, E, this current will be proportional approximately to the light intensity and is supported by the passage of electrons between the cathode and anode. The photoconductive device thus may be regarded as a light-controlled resistance.

When the tube is gas-filled, gas ions are available, in addition to photoelectrons, to increase the current strength. Figure 17–27(B) shows the comparative shapes of the anode current-vs-voltage curves for vacuum-type and gas-filled phototubes.

There are many models of photocells. All employ semiconductor materials. A familiar type is the selenium cell, which resembles a single-

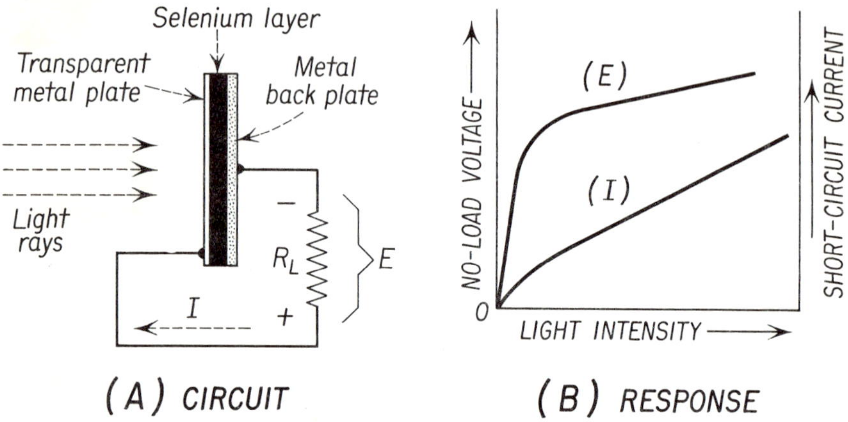

Fig. 17–28 Photocell as Photovoltaic Device

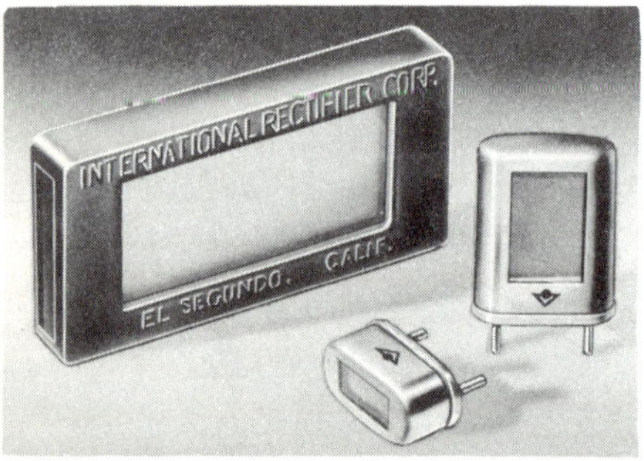

Fig. 17–29 Selenium Photo Cells (Courtesy, International Rectifier Corp.)

plate selenium rectifier (such as is described in Section 13.3, Chapter 13) except that the front electrode is transparent for the admission of light. Such a transparent metallic coating is obtained by vacuum-evaporating a thin film of metal, such as silver, on the surface of the selenium.

Figure 17–28(A) shows the construction of a selenium-type photocell. Here, a thin layer of processed selenium has been applied to a metallic backplate and crystallized. Then, a transparent front electrode, in the form of a thin metal plate, has been deposited on the surface of the selenium. Figure 17–29 shows a group of commercial selenium-type photocells.

When light rays strike the selenium through the front electrode, photoelectrons enter the backplate from the selenium, and an electron current (I) flows through the circuit in the direction of the arrow. A voltage drop accordingly is produced across the load resistor, R_L. Figure 17–28(B) shows typical response of a photovoltaic cell. Curve E shows how the open-circuit (no-load) voltage output of the cell varies with light intensity. The voltages shown here become lower as the output current is increased (R_L decreased). Curve I shows how the output current varies with light intensity. This is a representation of the "short-circuit" current (that is, current that flows when R_L is extremely low—theoretically zero ohms). The current values drop as R_L is increased.

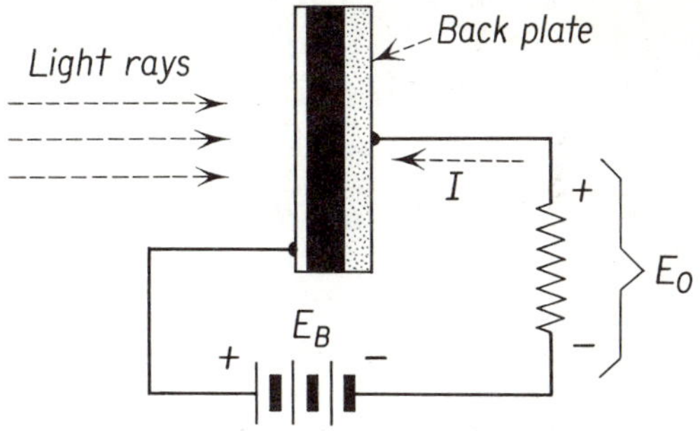

Fig. 17–30 Photocell as Photoconductive Device

The photovoltaic cell (also called a *self-generating* photocell) is the basis of the modern *sun battery* or *solar battery*. In this application, specially processed high-output cells are connected in series-parallel to supply useful voltage and power from sunlight.

Figure 17–30 illustrates the use of a photocell as a photoconductive device. The advantage of using a photocell in this manner is the higher output it affords over that of the same cell operated as a photovoltaic device. The d-c source, E_B, is connected so that the cell is reverse-biased. When the cell is darkened, the current flowing through it therefore is the very low back current of the semiconductor. When the cell is illuminated, however, its back resistance is lowered in proportion to the light intensity, and the current (I) and output voltage (E_o) increase accordingly.

Photocells are made commercially in several types, sizes, and shapes. Aside from selenium, other semiconductor materials employed in these cells include cadmium sulphide, copper oxide, germanium, lead sulphide, and silicon.

17.14 PHOTOELECTRIC CIRCUITS

Figure 17–31 shows several typical circuits employing light-sensitive devices.

Figure 17–31(A) is a light-operated relay employing a phototube (V_1) and a vacuum triode (V_2). A single d-c source, represented by Battery B, supplies anode voltage for the phototube and plate voltage for the triode.

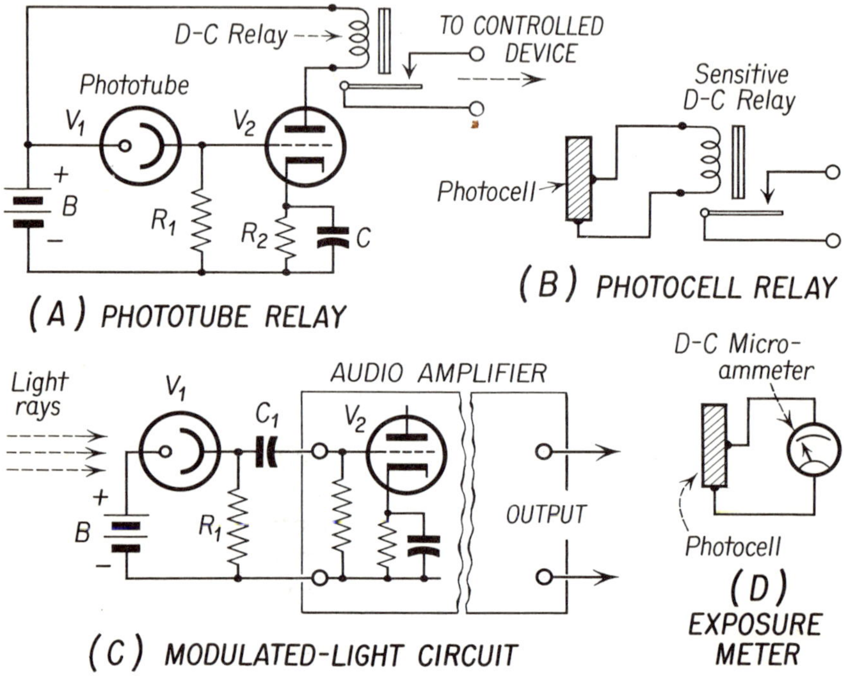

Fig. 17–31 Photoelectric Circuits

When the phototube is darkened, any current flowing through the circuit $B - V_1 - R_1$ is negligible. Because of the d-c grid bias developed across cathode resistor R_2, the triode plate current is quite low, and the d-c relay therefore cannot operate. When light falls on the phototube cathode, an electron current is passed by this tube and a voltage drop appears across resistor R_1, the top of this resistor becoming positive and the bottom negative. This increases the grid voltage in the positive direction and raises the plate current sufficiently to actuate the relay. Thus, the relay is held closed as long as the phototube is illuminated. If a contact is placed on the opposite side of the relay pole, the relay will be closed when the light is interrupted.

Figure 17–31(B) shows a similar circuit employing a self-generating photocell. The sensitive d-c relay in this circuit must be capable of operating on a few milliamperes and must have low resistance so as to utilize as nearly as practicable the short-circuit current of the photocell [Figure 17–28(B)]. The advantage of this circuit over the one described previously is its ability to operate without a power supply or tubes. However, it does require a more sensitive relay, and the latter usually is more expensive and delicate than one suitable for the phototube circuit.

Photoelectric relays are the basis of light-operated object counters, burglar alarms, door openers, automobile headlight dimmers, automatic lighting controls, and safety devices.

Figure 17–31(C) shows a circuit which may be used with chopped or modulated light. Such light is obtained from the sound track of sound motion-picture film passing in front of a lamp. Modulated light is used also in light-beam communication systems and is obtained by alternately brightening and darkening the filament of a lamp at the transmitter by means of currents from a microphone and amplifier.

The alternate presence and absence of light (in chopped light) or brightening and dimming (in modulated light) causes a modulated current to flow in the $B - V_1 - R_1$ circuit (alternating current superimposed upon direct current). The a-c component of the resulting voltage drop across resistor R_1 is coupled to the input of an audio amplifier by Capacitor C_1. The first stage of the amplifier, built around Triode V_2, is shown inside the block. This audio unit amplifies the a-c component in order to reproduce the sound originally modulating the light.

Figure 17–31(D) shows a simple light meter or photographic exposure meter. In this circuit, the d-c output of the illuminated photocell operates a d-c microammeter. The deflection of this meter is proportional to the light intensity, and the scale therefore may be graduated to read directly in foot-candles or other light units.

17.15 THYRATRON TUBE AND APPLICATIONS

The *thyratron* is a gas-filled triode or tetrode. Like the gas-filled diode (as compared to the vacuum diode), a thyratron is capable of higher anode current than a vacuum triode or tetrode of the same size. Thyratrons are important in industrial electronic devices which control heavy currents.

Thyratron action differs from vacuum-tube action chiefly in the performance of the grid as a control electrode. In the vacuum tube, the grid *always* is in control of plate current. As the grid voltage is made more positive, the plate current increases; and as the grid is made more negative (less positive), plate current decreases. If the grid voltage fluctuates, so does the plate current. In the thyratron, however, once the grid voltage has been set to a certain value (the *ignition potential*), the tube "fires"; that is, the gas discharge is initiated, and anode current flows. But then the grid has no further control in either changing or stopping anode current flow. Its voltage may be increased, decreased, or reduced to zero, but anode current will continue to flow in proportion to anode voltage and the circuit resistance until such time as the anode circuit momentarily is interrupted.

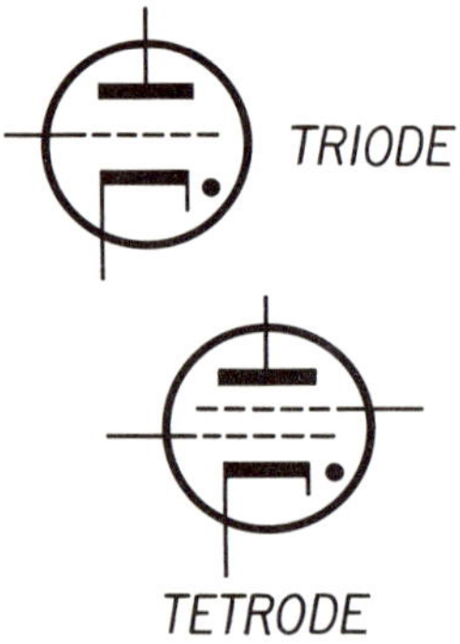

Fig. 17–32 Thyratron Tube Symbols

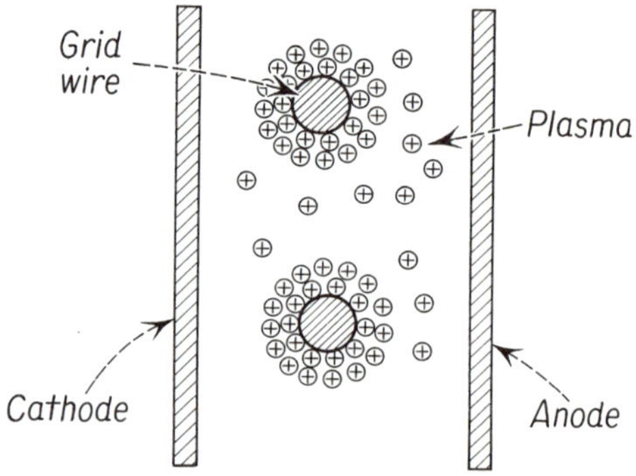

Fig. 17–33 Sheath Formation in Thyratron

The following mechanism underlies this action: Suppose a positive voltage is applied to the anode and a high negative voltage to the grid. Electrons leaving the cathode are repelled with high velocity through the grid mesh, and some of them ionize the gas molecules by collision. This gives rise to a quantity of positive gas ions in the space between the anode and cathode. These ions form the *plasma.* (See Figure 17–33.) The electrons travel away from the grid and many of the positive ions *to* the negatively charged grid. These ions cluster around the grid wires forming a *sheath,* as shown in Figure 17–33. The sheath causes the grid to appear as a positive element in the tube structure and prevents the attraction of electrons from the cathode-grid space by the anode charge.

If the grid voltage is made less negative (more positive), a point eventually is reached (the ignition potential) at which enough of the sheath has been "dissolved" to allow an arc discharge to take place quickly between anode and cathode, and anode current to flow. This current then is proportional to the anode voltage and the load resistance in the anode circuit. With the arc already started, a further positive increase in grid voltage can have no effect on anode current. Making the grid voltage more negative will thicken the sheath but without effect on the arc and anode current. However, breaking the anode circuit momentarily will extinguish the arc, interrupt the anode current, and restore control to the grid.

The grid ignition potential depends upon the anode voltage. In a gas tetrode, it depends also upon the shield-grid (fourth-electrode) voltage.

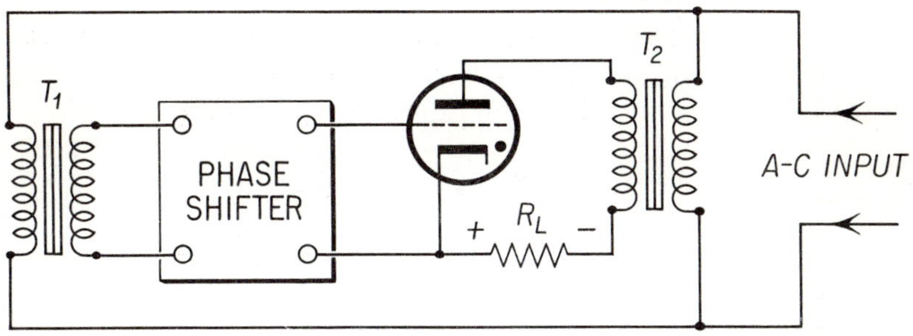

Fig. 17–34 Phase-Controlled Thyratron Rectifier

When grid and anode voltages both are alternating, there will be one point at which grid voltage and anode voltage have the correct values to ignite the tube. Anode current will flow from that instant to the end of the anode-voltage half-cycle, since the anode voltage falls to zero at the end of the half-cycle, and the arc discharge extinguishes.

This action is utilized to control the d-c output of a thyratron rectifier. Figure 17–34 shows the rectifier circuit. Here, the anode is supplied with a-c voltage through Transformer T_2, and the grid through T_1. In addition, there is included in the grid circuit a continuously variable *phase shifter,* to be described later in this section. The purpose of this device is to adjust the phase of the grid voltage so that the latter reaches its ignition potential at different parts of the positive half-cycle of anode voltage. Figure 17–35 shows some of the wave forms resulting from this phase variation. It is assumed that the thyratron in this instance fires at zero grid voltage.

In Figure 17–35(A), the grid voltage is zero when the positive half-cycle of anode voltage is zero. Anode current accordingly flows throughout the entire positive half-cycle, as indicated by the shaded portion. In Figure 17–35(B), the grid-voltage phase has been adjusted to 90 deg lagging the anode voltage. The grid voltage therefore is at its ignition potential when the anode voltage is maximum, and anode current can flow only during the latter half of the positive half-cycle. In Figure 17–35(C), the grid voltage is nearly 180 deg lagging anode voltage, with the result that anode current flows during only a brief portion of the half-cycle. If the grid voltage were exactly 180 deg out of phase with the anode voltage, the anode current would be zero.

The shaded areas represent the average value of direct current flowing through the load resistor (R_L in Figure 17–34). Since the adjustment of the grid-voltage phase allows the anode current to be varied between zero and maximum, this scheme yields a smooth control of the d-c output of the thyratron rectifier. Phase control of high-current thyratron rectifiers is widely used in industry.

Figure 17–36 shows several thyratron circuits. Figures 17–36(A) and (B) show circuits for phase shifters which may be used in the phase-controlled thyratron rectifier. In the circuit in Figure 17–36(A), the

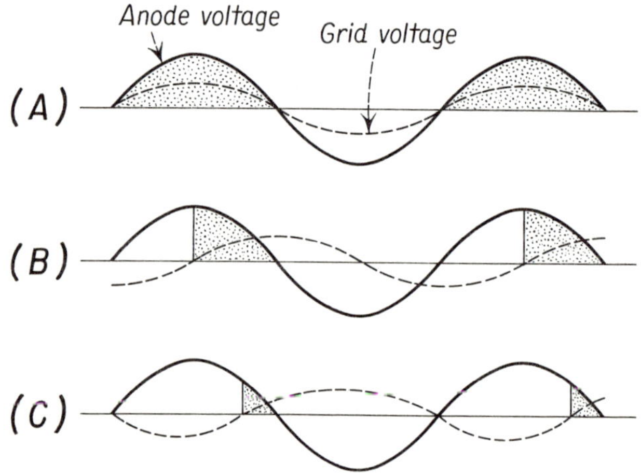

Fig. 17–35 Phase-Controlled Thyratron Waveforms

input alternating current divides into two parts, one (i_1) flowing through R_1 and R_2 in series, and the other (i_2) flowing through Capacitor C and Resistor R_3 in series. Because the left-hand branch of the circuit contains resistance only, the voltage drop at Point 1 is in phase with the applied a-c voltage. The voltage at Point 2, however, is not in phase with the applied voltage, since the right-hand branch of the circuit contains capacitance (C) and resistance (R_3) in series. Because of the capacitance, the voltage at 2 will lead the applied voltage by some angle between zero and 90 deg. The phase angle is large when R_3 is high; it is small when R_3 is low or C is high. The angle may be varied at will if a variable resistor (rheostat) is substituted for the fixed resistor, R_3.

By including variable capacitors and variable resistors in both branches of the circuit, as shown in Figure 17–36(B), a phase shifter is obtained which will provide a continuously variable shift between zero and 180 deg.

Figure 17–36(C) shows a *full-wave*, phase-controlled thyratron rectifier. Except for the full-wave feature, this circuit is similar to the half-wave rectifier shown in Figure 17–34.

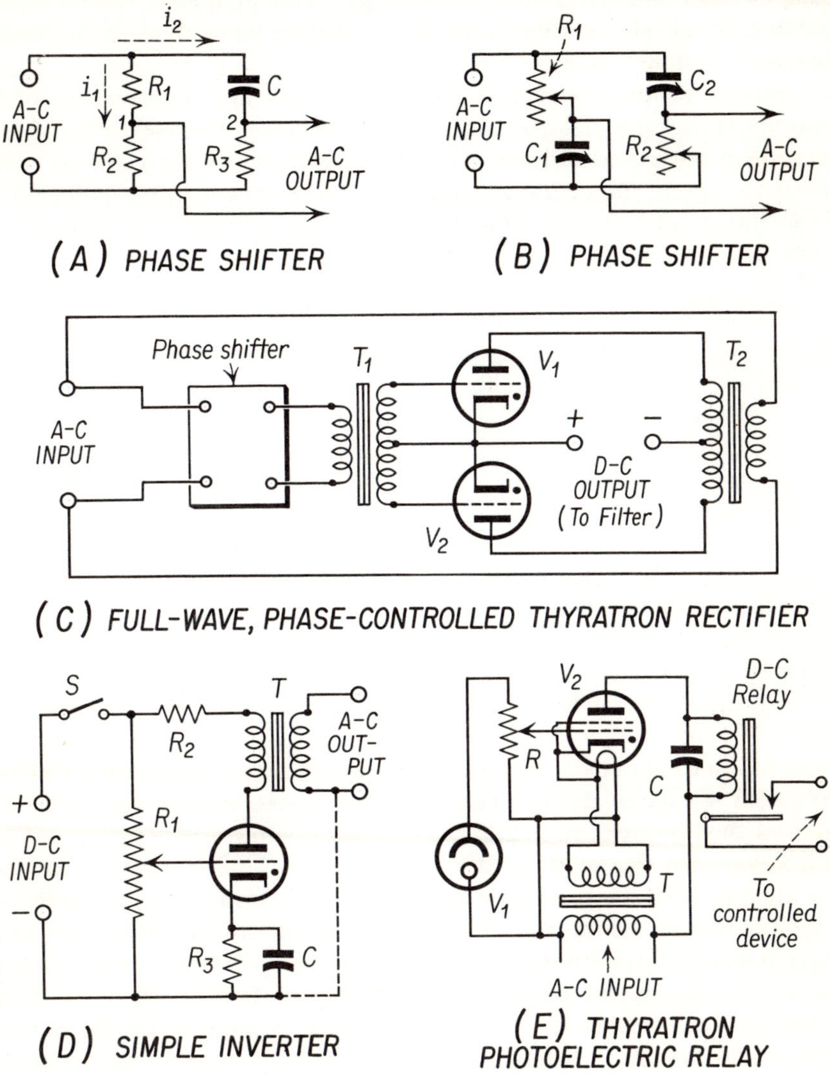

Fig. 17–36 Thyratron Circuits

Figure 17–36(D) is the circuit of a d-c–a-c *inverter.* An inverter is a nonmechanical device for changing direct current into alternating current at useful power levels.

Operation of this circuit, which is only one type of thyratron inverter, takes place in the following manner: The d-c input voltage is higher than the ignition potential of the thyratron. The tap is set along Resistor R_1 in such a position as to select the correct grid voltage for this tube. When Switch S is closed, the tube fires. Current then flows through the high-resistance cathode resistor (R_3) and into Capacitor C, through the tube, the primary winding of Transformer T, and the low-resistance current-limiting resistor (R_2). As Capacitor C charges, the cathode voltage increases rapidly and reduces the cathode-to-anode voltage of the tube almost to a point under the ignition potential. However, the inductance of the transformer primary maintains the current flow, and Capacitor C finally becomes fully charged. The voltage across this capacitor eventually becomes higher than the d-c input voltage to the circuit. When the lines of force about the primary have collapsed, the anode-cathode potential of the tube drops, but the cathode potential (provided by the charged capacitor, C) is higher than the grid or anode voltage. This cuts off the anode current (tube deionizes).

Capacitor C now discharges through resistor R_3. When the discharge progresses far enough to reduce the capacitor voltage, and accordingly the cathode voltage, to a value at which the anode-cathode potential of the tube once again is equal to the ionization potential, the tube re-fires. The cycle of events then is repeated. The resulting rise and fall of anode current in the primary of the transformer induces an alternating voltage across the secondary, and this is the voltage that is delivered by the inverter circuit.

Figure 17–36(E) is the circuit of a simple thyratron photoelectric relay. The advantage of a thyratron in this circuit is its higher anode current which allows the use of a more rugged d-c relay than could be used with a vacuum tube. This circuit operates directly from the a-c line and thus requires no rectifier and filter. Transformer T is poled in such a way that the anode of the phototube (V_1) swings positive at the same instant that the anode of the thyratron tetrode (V_2) is positive. If the phototube is darkened, it passes no current, and the thyratron grid receives no voltage from this source. But if the phototube is illuminated, it passes current which produces a voltage drop across Resistor R. This voltage drives the grid of V_2 positive, and the tube fires, passing anode current and actuating the relay. Since the anode current consists only of positive half-cycles with gaps in place of the missing negative half-cycles, the relay would chatter from trying to follow this intermittent current flow, except for Capacitor C which remains charged during the current gaps and holds the relay closed. The tap is preset on Resistor R so that the thyratron grid receives the correct d-c voltage when the cell is illuminated.

17.16 IGNITRON TUBE

The *ignitron* is a special tube capable of handling very large currents. The evacuated chamber of this tube contains an anode (often a disc or cylinder of graphite) in its top, and a cathode consisting of a pool of mercury in its bottom. (See Figure 17–37.) A third electrode, called an *ignitor* and responsible for the name of this type of tube, dips into the mercury.

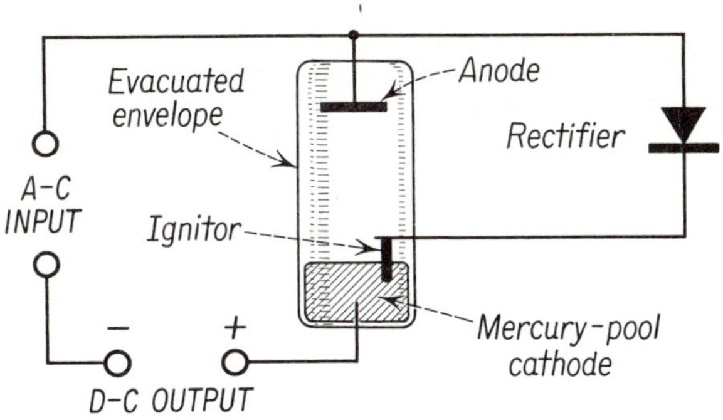

Fig. 17–37 Ignitron Rectifier

The ignitron tube may be connected into a rectifier circuit in the manner shown in Figure 17–37. Ordinarily, no current could flow across the cathode-anode gap. However, if a current pulse is passed momentarily between the ignitor and cathode, a spark will jump at the point of contact. This spark will liberate some electrons which will ionize the surrounding mercury vapor. With ionization initiated, a discharge occurs between cathode and anode, and anode current then flows.

Since the anode current ceases at the end of each positive half-cycle of anode voltage, the ignitor must be fired at the start of each new positive half-cycle. This is accomplished automatically by including the rectifier in the ignitor circuit. Because of the unilateral (one-way) conduction of this rectifier, the ignitor receives firing current only during positive half-cycles.

17.17 TRANSISTORS AND APPLICATIONS

Semiconductor diodes were described briefly in Section 13.3, Chapter 13. The *transistor* is a semiconductor *triode* (and, in a few instances, a tetrode).

Transistors are small devices, light in weight and rugged. They have no filaments and require no vacuum, yet they perform many of the functions, such as amplification and oscillation, which formerly were possible only with electron tubes. Although the transistor is a young component in the electronic field, it is commercially available already in a wide variety of models, sizes, and ratings, and its literature has become extensive. We can consider here only the rudiments of transistor construction and operation. Plate 17–1 shows a transistor, compared with a miniature vacuum tube.

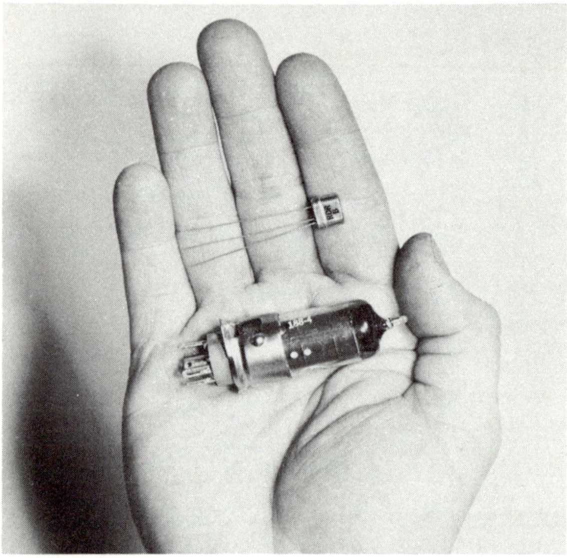

Plate 17–1 A transistor (top) shown with a miniature tube in socket (below) for size comparison. (Courtesy, Bogue Electric Mfg. Co.)

The *junction* transistor consists essentially of a thin wafer of germanium or silicon which has been processed to have three distinct, adjacent conduction regions or layers. A lead is attached to each region. Figures 17–38(A) and (C) illustrate this construction.

The processed region of the semiconductor is called *N-type* when it conducts electric currents by means of electrons, and *P-type* when conduction is by means of *holes.* (Holes are slower-moving, empty spaces left in the semiconductor crystal by vacated electrons. They exhibit positive charges.)

The central region is called the *base,* one of the outer regions the *emitter,* and the other outer region the *collector.* The base region is made as thin as practicable. When the transistor is in operation, the emitter acts like the cathode of a vacuum tube, the base like the grid, and the collector like the plate. Junction transistors are classed as *PNP*

[Figure 17–38(A)] and *NPN* [Figure 17–38(C)], depending upon the arrangement of the conduction regions.

The junction between a *P* region and *N* region acts as a semiconductor rectifier. Thus, when the *P* region is made positive with

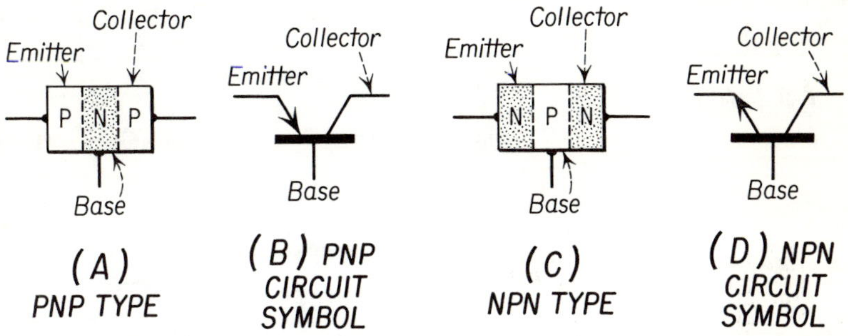

Fig. 17–38 Junction Transistors and Symbols

respect to the adjacent *N* region, high forward current flows; when the *P* region is biased negatively, low reverse current flows. Similarly, when the *N* region is made negative with respect to the *P* region, high forward current flows; when the *N* region is made positive, low reverse current flows.

In practice, d-c voltages are connected to the transistor with polarities such that the emitter is forward-biased by V_{ee} and the collector reverse-biased by V_{cc}. [See Figures 17–39(A) and (B).] Under these conditions, the emitter injects current carriers (electrons in the *NPN* transistor; holes in the *PNP* transistor) into the semiconductor body. These carriers are attracted by the oppositely charged collector, and most of them pass through the thin base layer. Thus, most of the emitter current (i_e) also flows in the collector circuit as collector current (i_c). In practical transistors, i_c is 80 to 90 per cent of i_e. Since the current flowing in the base lead is composed of emitter and collector currents of opposite polarity, it is less than either i_c or i_e.

The ratio of a collector-current change (di_c) to the corresponding emitter-current change (di_e), for a constant collector voltage (v_c), is termed the *current amplification factor* (α, alpha) of the *common-base* circuit, Figures 17–39(A) and (B).

(17–11) $$\alpha = di_c/di_e$$

where α = common-base current amplification factor
di_c = collector current change (in milliamperes)
di_e = emitter current change (in milliamperes)

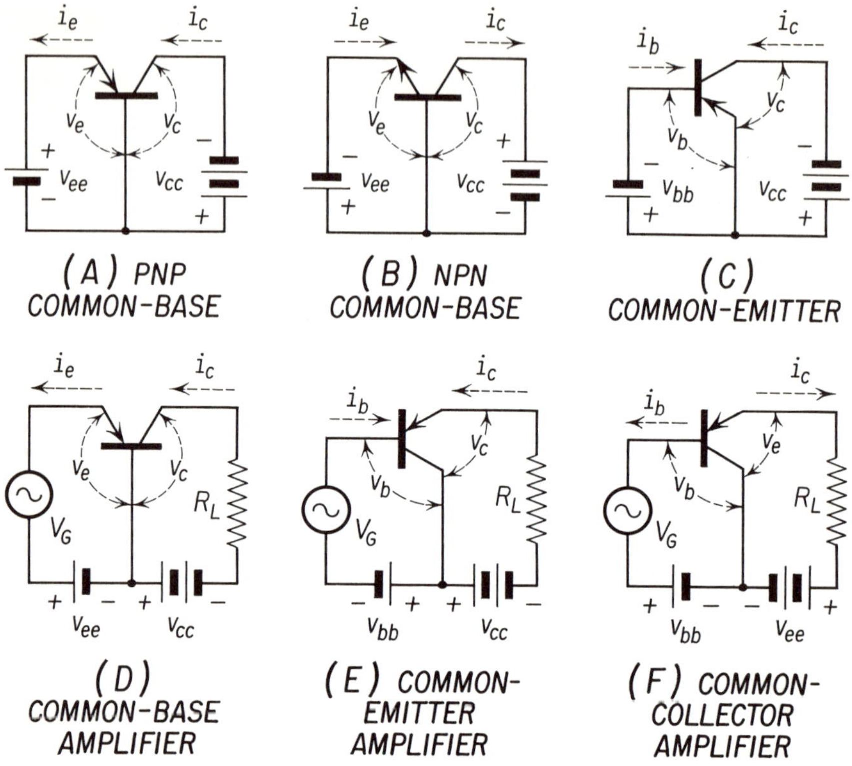

Fig. 17–39 Basic Transistor Circuits

Example: In a certain common-base circuit, an emitter-current change from 1 to 2 ma produces a collector-current change from 0.9 to 1.8 ma. Calculate the alpha of the transistor.

$$(a) \quad di_e = 2 - 1 = 1.0 \text{ ma}$$

$$(b) \quad di_c = 1.8 - 0.9 = 0.9 \text{ ma}$$

$$(c) \quad \alpha = di_c/di_e = 0.9/1 = 0.9$$

In the *common-emitter* circuit [Figure 17–39(C)], the base current (i_b) is varied. It has already been mentioned that i_b is smaller than i_c. Consequently, a small change in i_b might be expected to produce a large change in i_c. This is entirely correct. The current amplification (β, beta) for the common-emitter circuit is equal to $\alpha/(1 - \alpha)$ and is the ratio of a collector-current change to the corresponding base-current change, for a constant collector voltage (v_c):

(17–12) $$\beta = di_c/di_b$$

Example: A given transistor in the common-emitter circuit gives a collector-current shift from 5 to 7½ ma for a base-current change from 100 to 200 microamperes (μa). Calculate beta.

$$(a) \quad di_c = 7.5 - 5 = 2.5 \text{ ma}$$

$$(b) \quad di_b = 0.2 - 0.1 = 0.1 \text{ ma}$$

$$(c) \quad \beta = di_c/di_b = 2.5/0.1 = 25$$

Figure 17–40 shows typical d-c response curves for transistors. These plots are seen to have the same general shape as the constant-current curves of the pentode tube (Figure 17–10), with the common-

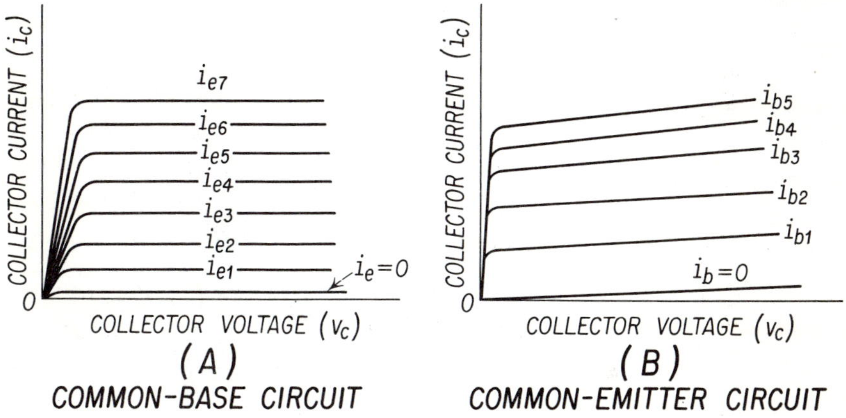

Fig. 17–40 Transistor Response Curves

base curves [Figure 17–40(A)] being more evenly spaced and flat than the common-emitter family [Figure 17–40(B)].

Since, as was pointed out earlier, the input circuit of the transistor (emitter) carries high forward current, and the output (collector) circuit carries low reverse current, the collector supply voltage, V_{cc}, may be made much higher than the emitter supply (V_{ee}). The transistor output impedance thus is higher than the input impedance. When approximately the same current flows in both input and output circuits [as in the common-base circuit, Figure 16–39(A)], voltage amplification and power amplification take place. The reason for this is that the same current value is flowing through a low-impedance input circuit but a high-impedance output circuit. This accounts for the ability of the transistor to amplify. The transistor is a more efficient amplifier than the vacuum tube, largely because it requires no filament-heating power.

Figure 17–39(D) shows a common-base amplifier circuit. Here, the a-c generator voltage (V_G) is introduced in series with the emitter d-c bias voltage V_{ee}. The resulting a-c component of emitter current is transmitted through the transistor (the amount flowing in the collector circuit being determined by α) and produces a voltage drop across the load resistance, R_L. This voltage is higher than the input voltage, V_G. Increasing V_{cc} and R_L results in a higher output-signal voltage and increased voltage gain.

Considerably higher voltage gain and power gain are afforded by the common-emitter circuit [Figure 17–39(E)] because of the larger current amplification factor (β) of this circuit.

The common-collector circuit gives less than unity voltage gain and the lowest power gain of all three circuits. The common-collector power gain is less than half of that of the common-emitter circuit.

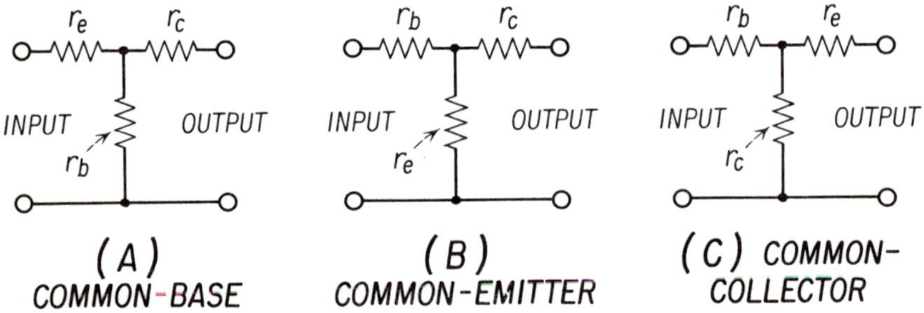

Fig. 17–41 Internal Resistances of Transistor

The common-collector circuit has the highest input impedance (often 0.5 to 1 megohm), the common-emitter medium input resistance (1,000 ohms average), and the common-base the lowest input resistance (often 50 to 60 ohms).

The current flowing through each of the transistor electrodes has a different value. This indicates that the transistor has a different internal resistance for each electrode. These resistances resemble the internal plate-cathode resistance (r_p) of the vacuum tube. The internal resistances of the transistor appear as a T-network, as shown in Figure 17–41. This simple picture holds only for direct current and low-frequency alternating current, however. The network becomes more complex at high frequencies because of reactive effects and other internal phenomena. Internal resistance values for a representative junction transistor (Type 2N34) are r_e 26 ohms, r_b 800 ohms, and r_c 2 megohms.

Figure 17–42 shows a two-stage, *R-C*-coupled audio amplifier employing transistors. Note the over-all resemblance to the *R-C*-coupled vacuum-tube amplifier. A single battery, *B*, supplies d-c voltages for both transistors (TR_1 and TR_2). Negative bias voltage for the transistor base electrodes is obtained from voltage-divider networks ($R_1 - R_2$ for transistor TR_1, and $R_4 - R_5$ for Transistor TR_2).

The collector resistance (r_c) of a transistor is high, while the base resistance (r_b) is low. This necessitates a step-down impedance-match-

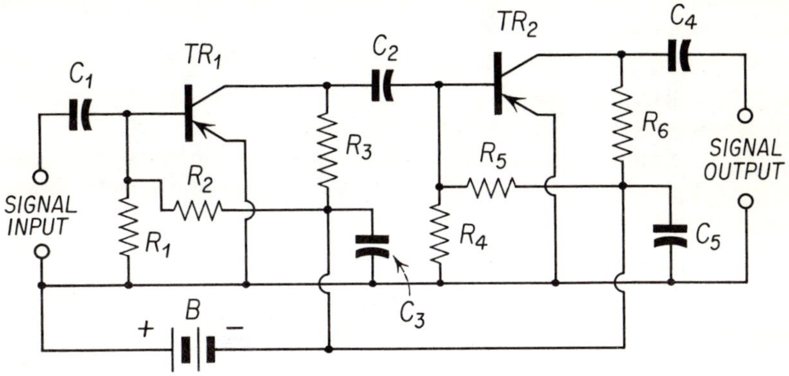

Fig. 17–42 Transistor RC-Coupled Amplifier

ing device for coupling between two transistor stages.* Obviously, then, the collector impedance of Transistor TR_1 in Figure 17–42 is not matched to the base impedance of TR_2, and maximum power therefore cannot

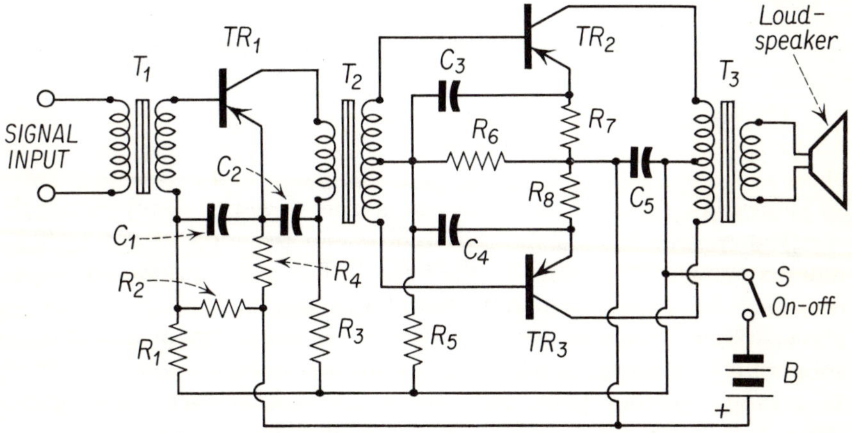

Fig. 17–43 Transistor Transformer-Coupled Audio Amplifier

be transferred between stages. For this reason, *R-C* coupling is not widely used in transistorized amplifiers. To compensate for the loss of power in this simple circuit, an additional one or two stages usually will be required for a desired over-all gain.

Power loss is prevented by employing impedance-matching step-down transformers between transistor stages. Such a transformer-coupled amplifier circuit is shown in Figure 17–43.

Figure 17–44 shows a simple, transistorized audio oscillator. This circuit is similar to the vacuum-tube oscillator described earlier in this

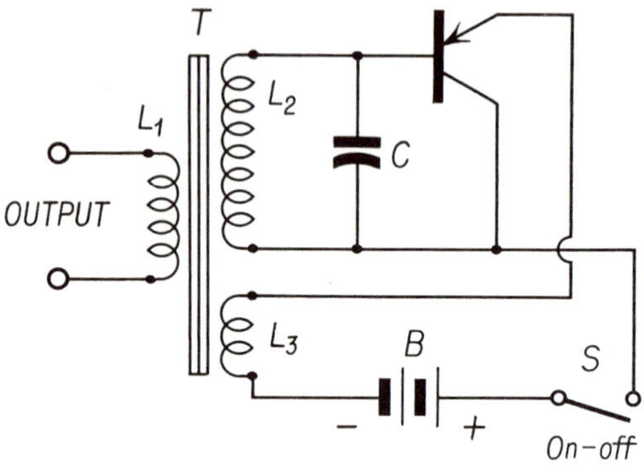

Fig. 17–44 Transistor Oscillator

chapter. Positive feedback occurs between the output (collector) winding (L_3) of Transformer T and the input (base) Winding (L_2). The frequency of oscillation is governed by the inductance of L_2 and the capacitance of C. Output is obtained from the coupling winding, L_1.

At this writing, transistors are being employed in amplifiers, radio receivers, hearing aids, test instruments, and certain industrial electronic equipment. Developmental work is in progress on the application of transistors to television.

17.18 X RAY

When fast-moving electrons strike a solid target of heavy metal, the energy they give up on impact is released in the form of a very-high-frequency radiation emitted by the target. This radiation, called *X rays,* has a frequency of 10 trillion megacycles or higher.

X rays have the property of penetrating solid, opaque substances and permit the examination of interiors not accessible to light rays. They will expose a photographic film, ionize a gas, and fluoresce certain chemicals. X rays are used extensively in science, industry, medicine, and dentistry. In medicine, they have therapeutic, as well as diagnostic value.

Figure 17–45 shows the construction of an X-ray tube. Inside the evacuated bulb, the source of electrons is a hot filament surrounded by a tubular, molybdenum focusing electrode. Mounted near by in the path of the emitted electrons is a solid tungsten anode or *target* having a slanting face. A high voltage (often 100 kv or more) is applied between the filament and anode. This high positive potential on the anode attracts electrons with extremely high velocity. On striking the target, they give rise to the X rays which are emitted by the slanting face of the target, as shown.

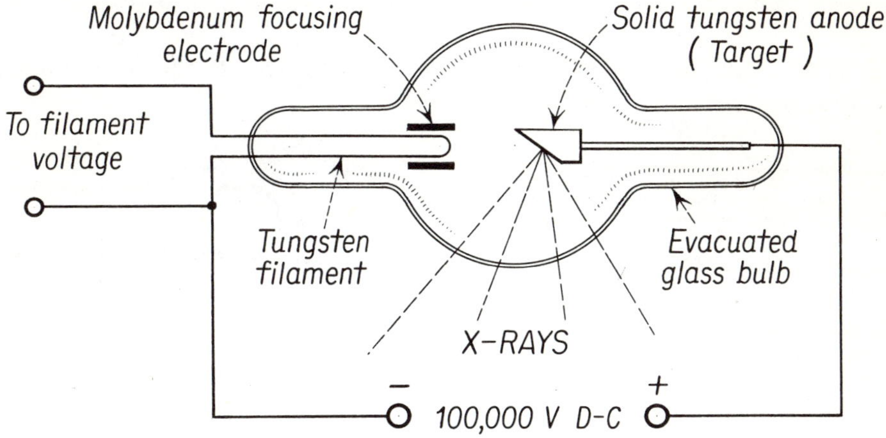

Fig. 17–45 X-ray Tube

17.19 INDUSTRIAL ELECTRONIC APPLICATIONS

Although electronic circuits and devices were used first in communications systems and were long associated with radio, they eventually proved themselves in industrial equipment. Some of the devices described earlier in this chapter, for example, which have wider application in industry than in communication, include the photocell, phototube, gaseous tube, thyratron tube, ignitron tube, and X ray.

Electronics is applied in industry in a very diversified manner. Some of its functions include automatic control of machines (automation), remote control of devices and processes, circuit and machine protection, temperature control, humidity control, and similar applications. Some of its functions in industrial instrumentation are explained in Section 18.11, Chapter 18. The subject has grown so wide in scope that a survey of it would fill a good-sized book. In this section, it is our aim only to describe a few circuits and systems which typify some of the accomplishments of industrial electronics.

Electron-Tube Relays. The sensitive control action of electron tubes is utilized industrially to boost the amplification of heavy-duty relays. This is useful in circuits in which the control voltage is too low to actuate a relay directly. Figure 17–46 shows several simple electron-tube relay circuits.

In Figure 17–46(A), a d-c relay is inserted in the plate circuit of a vacuum tube. The static plate current supplied by Battery E_B is insufficient to close the relay. But a suitable d-c voltage applied to the D-C INPUT terminals in the polarity shown will bias the grid positively, the plate current will increase, and the relay will be actuated. Because of the transconductance of the tube, a small d-c input signal produces a large plate-current increase. The operating voltage thus can be much lower than that which would be required to operate the relay without the tube.

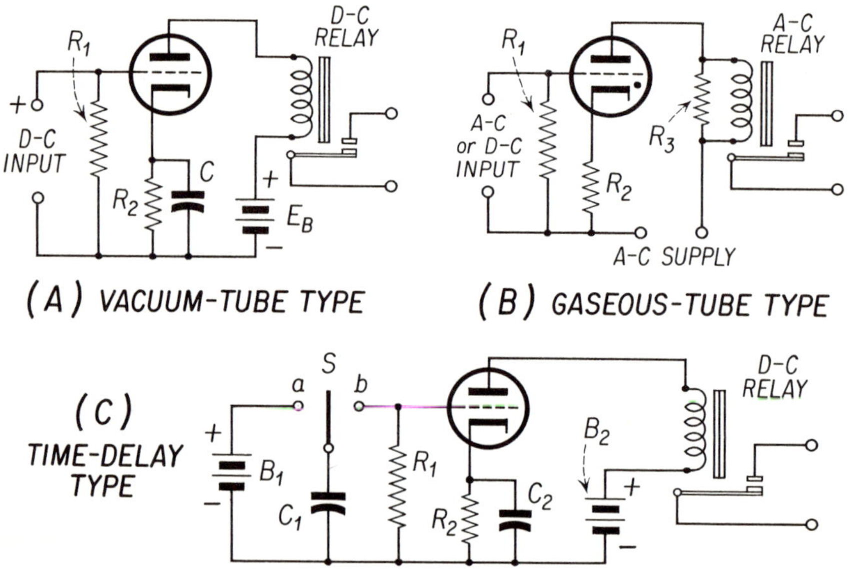

Fig. 17–46 Electron-Tube Relays

The circuit shown in Figure 17–46(B) is similar to the previous one, except that a gaseous tube is used here. Because the gaseous tube plate (anode) current is much higher than that of the vacuum tube (up to many amperes in thyratron tubes), a heavier-current relay may be operated. To prevent the circuit from "locking-in" in the manner characteristic of thyratrons, an alternating supply voltage is used. This causes the anode current to be interrupted each time the a-c cycle passes through zero, and control to be returned to the grid. The input signal may be either alternating or direct current.

Figure 17–46(C) shows one type of electronic time-delay relay circuit. The basic circuit is the same as Figure 17–46(A), the only additions being Battery B_1, Capacitor C_1, and Switch S. When S is in Position *a*, Capacitor C_1 is charged from Battery B_1 with its top electrode positive. When S subsequently is thrown to Position *b*, the grid is driven positive by the charged voltage of the capacitor, the plate current increases, and the relay closes. The charge in C_1 now leaks off through Resistor R_1, and when the voltage on C_1 has fallen to a value lower than the critical grid voltage required to maintain the high plate current, the relay drops out. The relay hold-in time interval thus depends upon the time constant R_1C_1.

Temperature Control. The sensitivity of the electron-tube amplifier and control circuits makes it possible to control the temperature of ovens, furnaces, and similar equipment with considerable precision.

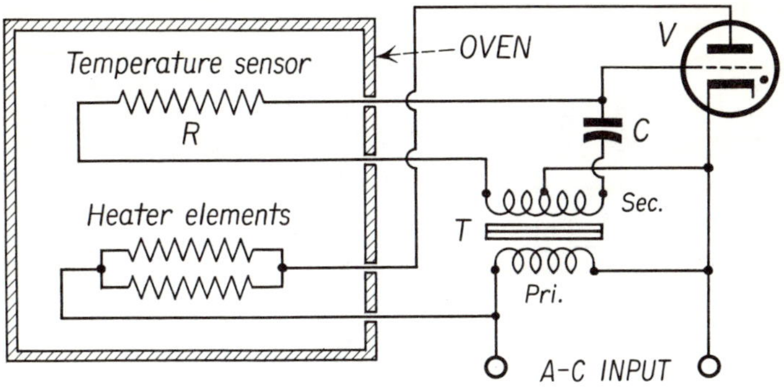

Fig. 17–47 Thyratron-Type Temperature Control

Figure 17–47 shows one type of thyratron circuit for temperature control. The grid circuit of Thyratron *V* contains a bridge-type phase-shift network composed of the two halves of the secondary winding of Transformer *T*, a temperature-sensitive resistor (*R*) inside the oven, and Capacitor *C*. As the temperature of *R* changes as a result of the oven heat, the proportionate change in the resistance of *R* unbalances the bridge network and causes more or less current to flow in the anode circuit of the thyratron, depending upon the direction of the bridge unbalance. The oven heater elements are connected in series with the thyratron anode, and their temperature is raised or lowered in accordance with these current changes.

The control characteristics of the circuit are determined by the performance of the phase-shift network. Proper selection of Capacitance *C* enables the operating point to be set wherever desired. Capacitance *C* often is set for a phase shift of 90 deg between grid and anode currents at the desired oven temperature. Any temperature deviation then will unbalance the bridge and the accompanying phase shift will cause more anode current to flow if the temperature dropped, or less current to flow if the temperature rose.

Flame-Failure Protection. The failure of flames or jets fed by oil or gas is a pronounced hazard in many industrial installations. Means must be provided for shutting off the oil or gas flow quickly if the flame becomes extinguished.

Figure 17–48 shows a typical electronic system for flame-failure protection. Light rays from the flame are focused by the lens upon the cathode of the phototube. The phototube is connected to a thyratron relay circuit similar to the one shown earlier in Figure 17–36(D). The relay output terminals switch the a-c power to a solenoid-type valve in the gas or oil line.

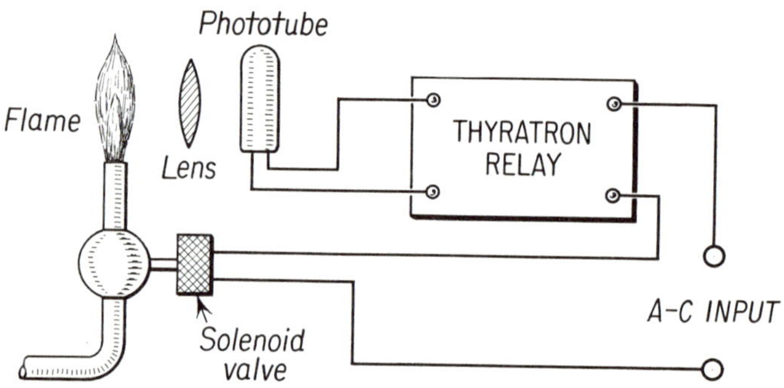

Fig. 17–48 Flame-Failure Protection

As long as the flame is present, the phototube is illuminated and the operation of the circuit permits the valve to remain open. As soon as the flame dies, however, the phototube is darkened and the relay drops out, switching the alternating current through the solenoid valve and shutting it off.

Welding Control. In resistance welding systems, such as spot, line, and seam welders, it is imperative that electric current be passed through the workpieces only for short, predetermined time intervals. The duty cycle (ratio of on-time to off-time) governs the quality and appearance of the weld. In modern welders, this timing operation is performed electronically, since the fractional-second *on-time* required is not readily controlled by mechanical timers.

Figure 17–49 shows the basic arrangement of a circuit for welding control. Here, the secondary of Transformer T_2 is connected in series with the primary of the welding transformer, T_1. If the primary of T_2 were opened, current flowing into T_1 would be reduced by the impedance of T_2 below the level required for welding. If, instead, the T_2 primary were short-circuited, the secondary impedance would be lowered, and the T_1 current would increase to the level required for welding.

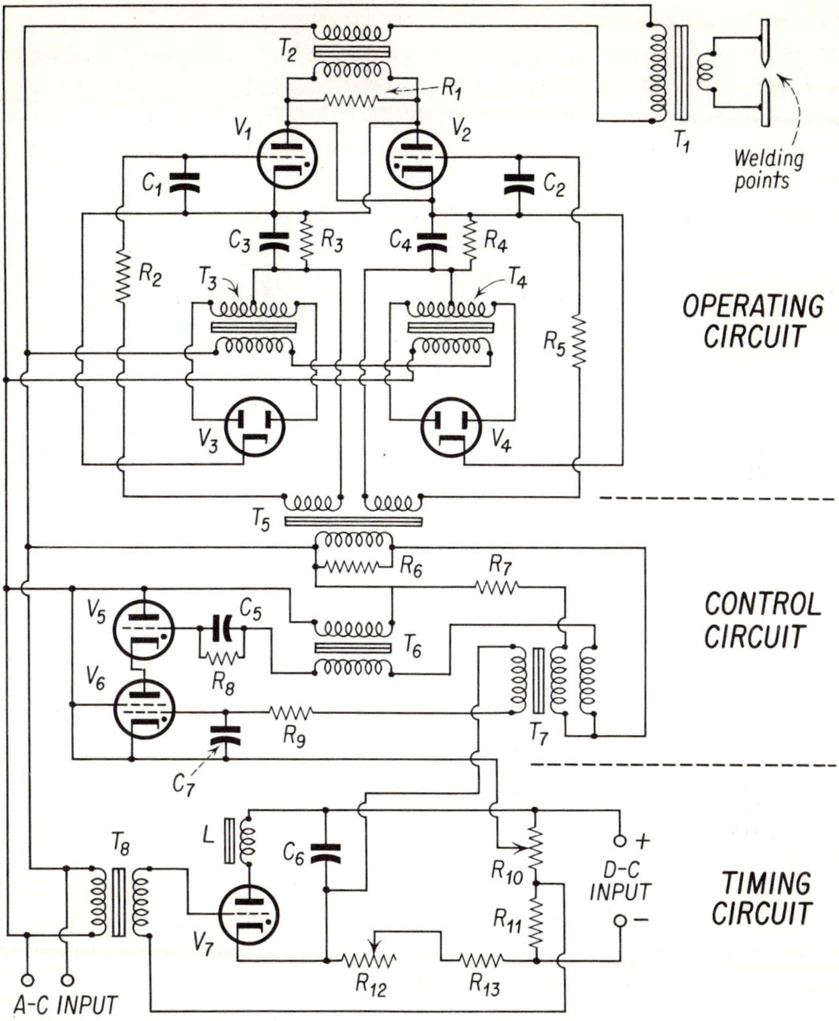

Fig. 17–49 Electronic Welding Control

In the electronic circuit, this opening and short-circuiting action is accomplished with the thyratron tubes, V_1 and V_2. The rest of this circuit controls the time interval during which the thyratrons are conducting or nonconducting. Thus, the complete system may be divided into *operating circuit, control circuit,* and *timing circuit,* as shown in Figure 17–49.

The thyratron tubes in the operating circuit receive their d-c grid-bias voltages from rectifier tubes V_3 and V_4 which, in turn, are powered by Transformers T_3 and T_4. These separate voltages bias V_1 conducting V_2 nonconducting when no signal is received from the rest of the circuit.

The a-c exciting signal (in phase with the thyratron anode voltage) is delivered by Transformer T_5. The primary winding of this transformer receives a full-wave voltage as the result of Thyratrons V_5 and V_6 conducting alternately. Thyratron V_6 is fired by a pulse from the timing circuit and causes current to flow through the T_5 and T_7 primary windings. Transformer T_7 is a peaking transformer with one secondary exciting the grid of V_5, and the other secondary the grid of V_6.

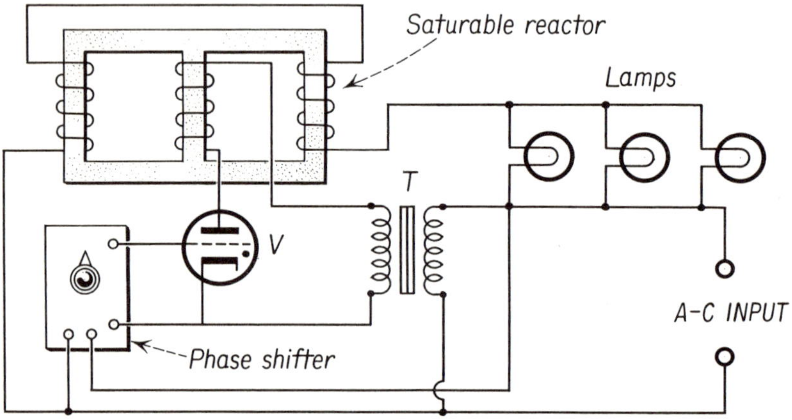

Fig. 17–50 Thyratron Control of Lighting

The anode of Thyratron V_7 in the timing circuit is d-c-operated, but its grid is supplied with peaked a-c voltage from the peaking transformer, T_8. Because the voltage drop across C_6 is zero when the direct current is applied and the anode-to-cathode voltage is zero, V_7 does not fire as soon as the direct current is applied. At this point, the V_7 negative grid voltage is equal to the d-c supply voltage minus the drop across R_{11}. As C_6 charges, the R_{12}-R_{13} voltage drop decreases and the capacitor voltage increases, soon reaching a value sufficient to ignite V_7 and discharge the capacitor through Inductor L. The reactive effect of the inductor afterward causes C_6 to charge a small amount in the opposite direction, making the V_7 anode negative and extinguishing this tube. The sequence then may be repeated the next time the power is switched on by the welder control switch. The time-on interval depends upon the time constant governed by the setting of Rheostats R_{10} and R_{12} and the capacitance of C_6.

Lighting Control. Figure 17–50 shows a thyratron circuit for the continuously variable control of lighting. This circuit allows high-current lamp banks, such as theater lighting systems, floodlights, etc., to be dimmed, brightened, or turned off by manipulating a small, light-duty control such as a rheostat or potentiometer.

The lamps are connected in series with a saturable reactor and the a-c power line. (See Section 10.14, Chapter 10, for an explanation of saturable-reactor action.) The control winding of the saturable reactor is connected in the anode circuit of a thyratron tube. A-C anode voltage is supplied by the secondary of Transformer *T*. Current passing through the control winding is direct because of the rectifying action of the tube.

A-C control voltage is applied to the thyratron grid through a continuously variable phase shifter. This enables adjustment of the anode current of the thyratron to any desired value between zero and maximum. As the anode current is increased, the impedance of the saturable reactor is decreased, and the lamps glow more brightly. As the anode current is decreased, the reactor impedance increases, and the lamps are dimmed. At a sufficiently low anode current, the reactor impedance is high enough to extinguish the lamps.

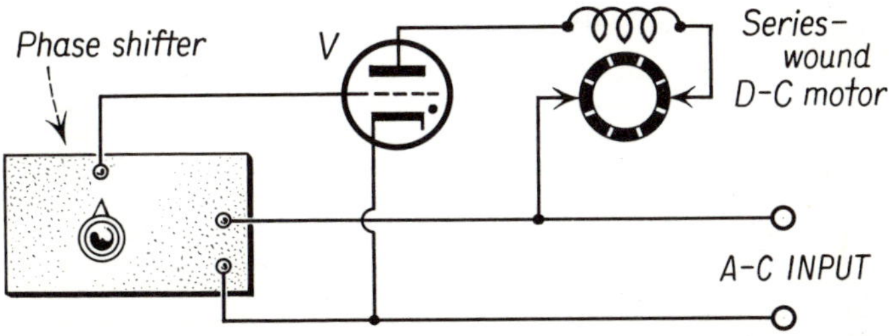

Fig. 17–51 Thyratron Control of Motor Speed

Motor Speed Control. If a d-c motor is connected in the anode circuit of a thyratron tube supplied with alternating current, rectifier action of the tube will supply direct current for running to the motor. The ability to control the anode current of a thyratron by varying the phase of the grid voltage provides a convenient method for controlling the speed of the motor. The variable control in the phase-shift network can be a light-duty component, since the current it must carry often is negligible while the motor current may be quite high.

Figure 17–51 illustrates this principle. A series-wound d-c motor runs on the rectified anode current of Thyratron *V*. The anode current consists of positive half-cycles. An a-c voltage is applied to the grid through the phase shifter. When the positive half-cycles of grid voltage are in phase with the anode-current pulses, maximum anode current flows, and the motor rotates at its fastest speed. When the grid voltage is out of phase with the anode pulses, the anode current is zero, and the motor stops. Between these two limits, the anode current and motor speed are continuously variable.

Selsyn Motor. While the *selsyn motor* is not specifically an electronic device, it is to be found in so many electronic systems that it appropriately should be described here. The name of this device is due to the self-synchronizing nature of its operation.

Briefly, in a simple selsyn system, two selsyn motors (which may be separated by a few feet or several miles) are connected together and to a common a-c supply. The rotors do not spin as they would in an ordinary motor. But if the rotor of one motor (the "transmitter") is turned to a given position, the rotor of the other motor (the "receiver") will follow it around and assume the same position. Thus, the shaft of the transmitter may be connected to some machine or device the motion of which is to be transmitted to a distant point, and the rotor of the receiver

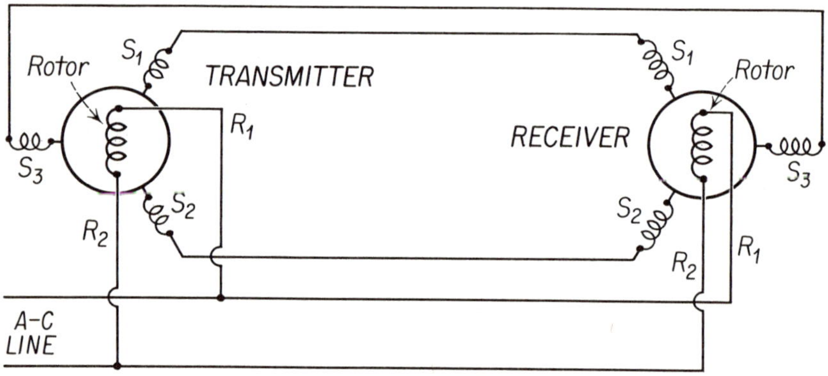

Fig. 17–52 Selsyn System

will follow automatically and indicate all positions of the machine at the originating station. The receiver rotor shaft also might be connected to a distant machine so that the latter will assume all positions of the transmitting machine. The system is two-way; that is, either motor may be used as the transmitter, and vice versa. The selsyn motor thus is a convenient remote-control device. This motor also is known by the names *autosyn* and *synchro*.

A two-pole rotor with slip rings is employed in the selsyn. The stator is a three-circuit, distributed, Y-connected winding. Corresponding stator windings of the transmitter and receiver motors are connected together, as shown in Figure 17–52. Corresponding rotor windings are connected together automatically through the common power line.

Operation of the selsyn system is explained as follows: The transmitter rotor (R_1–R_2) sets up voltages of different value but of the same phase in the three stator windings (S_1, S_2, and S_3) as a result of transformer action (R_1–R_2 acting as the primary of this transformer). Because of this, currents of correspondingly different values are set up in the motor stator windings, and a field is produced in the same direction as that in the transmitter. The tendency of the field produced by alternating current in the rotor of the receiver is to parallel itself with the field produced by the receiver stators. The torque which results pulls the rotor around to the same position as that of the rotor in the transmitter. When this "balance" is obtained, the current in the stator windings of both motors drops to zero, since the receiver rotor induces voltages in its stator windings which are equal in value but opposite in phase to those of the transmitter stator windings.

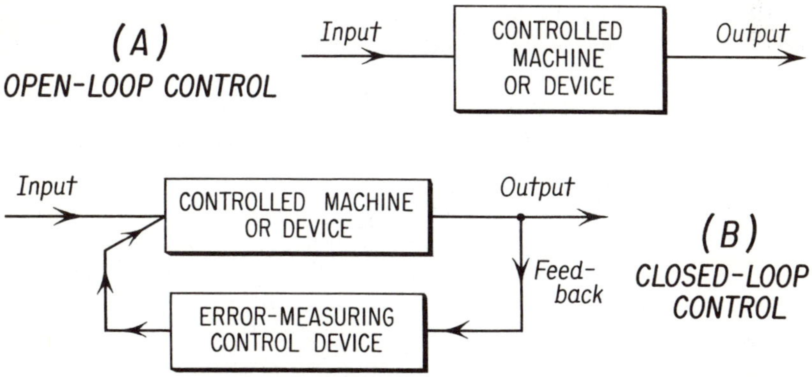

Fig. 17–53 Principle of Servomechanism

Servomechanism. Simple, ordinary control systems are unable to correct their own operation. An example of this is a manually operated speed control for a motor. The control may be set for a certain rate of speed, but if the motor should deviate from this rate, the human operator must reset the control. Another example is the control of lamps by means of a voltage control. This control may be set for a desired lamp brightness, but if the illumination should change as the result of a voltage shift, the operator must reset the control. Still another everyday example is the steering of an automobile. Such systems are described by the term *open-loop control* and are symbolized by the block diagram in Figure 17–53(A).

A *servomechanism,* on the contrary, is a self-correcting control system. It corrects its performance by taking a portion of the output (a process called *sampling*), measuring the amount of error by means of a suitable error-measuring control device, and feeding a proportional correction signal back to the input of the machine or device being controlled. The servomechanism thus is a *closed-loop* control system. Figure 17–53(B) shows its block diagram. An everyday example of a simple servomechanism is the thermostatic control of temperature. Power is fed into a heater. A thermostat samples the room temperature and compares it with the desired temperature (determined by the setting of the thermostat). If the temperature is correct or is lower than desired, the heater is allowed to continue in operation. If, instead, the desired temperature is exceeded, the thermostat opens the circuit, disables the heater, and allows the room to cool down to the desired temperature. The servomechanism is defined as a stable, feedback system of control.

The liquid-level gauge illustrated in Figure 18–41, Chapter 18, is an open-loop system, since movement of the float causes a meter to be deflected but there is nothing in the circuit to correct the level of the liquid. If, instead of the meter indicator, an electrically operated valve is inserted in the pipeline and arranged to open, whenever the current has dropped to zero or to some other predetermined low level, admitting more liquid to the tank, or to close when the liquid reaches the desired level, a compensating feedback loop will have been introduced, and we have a type of servomechanism.

Hundreds of kinds of servomechanisms have been developed. In electronic equipment of various types, they perform many tasks with precision and reliability, ranging from the automatic control of machines to the aiming of guns, automatic piloting of airplanes, and the keeping of a ship on course.

Automation. Automation is a new art. It is concerned with the automatic supervision and control of machines and devices and of industrial processes. Automation calls into play all that is available in industrial electronics, instrumentation, and servomechanisms.

By means of automation, a production machine, for example, may be run, without a human operator, to turn out piece after piece of finished work in accordance with instructions transmitted to it (often in the form of suitable electrical pulses) from a magnetic recording or electronic computer. Servomechanisms compare the quality and progress of the work continuously with a standard (which also may be in the form of recorded pulses) and feed correcting signals back into the machine control circuits to readjust tools and positioning. One human operator is able to supervise an entire group of automatized machines.

It should be noted that although automatized machines perform their tasks with astonishing cleverness, they have not been endowed with the ability to think. What they really do is to follow the instructions that have been laid down for them (by recording, or "programming") with great precision and without fatigue.

SUMMARY OF CHAPTER 17

According to the number of electrodes (elements), electron tubes are classified as **Diodes** (2 electrodes), **Triodes** (3), **Tetrodes** (4), and **Pentodes** (5). Special types include **Hexodes** (6), **Heptodes** (7), and **Octodes** (8).

Combination Tubes contain two or more separate and complete sets of elements in one envelope. Examples: **Twin Triode, Diode-Triode, Triode-Pentode.**

The basic tube elements are *filament* (or *heater*), *cathode, control grid, screen grid, suppressor grid,* and *plate.* The grid provides basic control action, the screen provides electrostatic shielding to reduce cathode-plate capacitance, and the suppressor minimizes the effects of secondary emission of electrons from the plate. Electrons are emitted primarily from either a filament or from a cathode which is heated by a filament (*heater*). The first is the *directly heated* or *filamentary type;* the second, the *indirectly heated* or *cathode type.*

Transconductance (g_m) is the ratio of plate-current change to grid-voltage change: $g_m = (1{,}000 dI_p)/dE_g$. **Amplification Factor** (μ) is the ratio of plate-voltage change to grid-voltage change which will produce the same plate-current change: $\mu = dE_p/dE_g$. **Plate Resistance** (r_p) is the ratio of plate-voltage change to corresponding plate-current change: $r_p = dE_p/dI_p$. $g_m = \mu/r_p$, $\mu = g_m r_p$, and $r_p = \mu/g_m$.

Voltage Amplifiers increase signal voltage, or exhibit *voltage gain.* **Power Amplifiers** exhibit *power gain* with or without attendant voltage gain. Amplifier circuits may be single-stage, multistage (cascaded), single-ended, push-pull, *R-C*-coupled, or transformer-coupled. Voltage gain, $VG = (\mu R_L)/(r_p + R_L)$, where r_p is the tube plate resistance and R_L the load resistance. In a power amplifier, **Plate Efficiency** (n) is the ratio of a-c output power (P_o) to d-c plate-input power (E_pI_p): $n = (100P_o)/(E_pI_p)$ %. **Plate Dissipation** is the difference between d-c plate input power (E_pI_p) and a-c output power (P_o). $P_d = (E_pI_p) - P_o$. **Power Amplification,** or **Power Gain,** is the ratio of a-c output power (P_o) to a-c grid input power (P_i): $PG = P_o/P_i$.

A **Class-A** amplifier is biased along the linear portion of its $E_g - I_p$ characteristic curve and does not draw grid current. Plate current flows during the entire 360 deg of the signal-voltage cycle. A **Class-B** amplifier is biased to cutoff, and its plate current flows over almost the entire positive half-cycle of signal (excitation) voltage. A **Class-C** amplifier is biased to twice cutoff or higher, its plate current flowing during only a small portion of the positive half-cycle of signal voltage. **Class-AB** amplifiers operate at bias voltages between those of class-A and class-B systems.

An **Oscillator** essentially is an amplifier with *positive feedback* provided between its output and input circuits for *regeneration.* Oscillators may be *self-excited* (in which the frequency is determined by L and C circuit values) or *crystal-controlled* (in which the frequency is determined by the dimensions of a quartz crystal).

Radio Waves are radiated at a speed of 186,000 miles per second from an excited wire called an *antenna.* Wave length (λ) is inversely proportional to frequency (f): $\lambda = 300/f_{mc} = 300{,}000/f_{kc}$. Two types of radiation take place from an antenna—*ground wave* and *sky wave.* The sky wave is propagated to great distances by the "skip effect." Radio waves may be broken into groups for telegraph code transmission, or modulated for radiotelephone transmission. A **Radio Receiver** must have a means for *detecting* the radio waves; that is, making them audible for communication purposes. The signal is intensified before detection by means of an *r-f amplifier* and after detection by means of an *a-f amplifier.* Tuning may be provided in the r-f amplifier as well as in the detector.

Devices for converting light energy into electricity or for controlling electric currents by means of light are classed as **Phototubes** or **Photocells.** The phototube primarily is a *photoconductive* device; that is, its resistance varies with the intensity of light falling on its cathode. The **Photocell** may be used either as a photoconductive or *photovoltaic* device (self-generator of electricity when illuminated).

The **Thyratron Tube** is a gas-filled triode or tetrode able to pass high values of anode current. Thyratron grid action differs from that in a vacuum tube in that the grid loses its ability to control the anode current, once it has initiated the arc discharge. Anode current flow may be stopped only by reducing the anode voltage momentarily to zero. In an a-c-operated thyratron, anode current is controlled by means of a *phase shifter* circuit operated into the grid.

In an **Ignitron Tube,** the cathode is a pool of mercury. Current flow through the tube is initiated by passing a spark between the *ignitor* electrode and the mercury pool into which it dips. This ionizes the mercury vapor in the vicinity and starts the discharge between cathode and anode. Ignitrons handle extremely high currents.

The **Transistor** is a semiconductor triode. A *junction* transistor is classed as *NPN* or *PNP* according to the arrangement of its separate conduction layers, and is so-named because of the rectifying junctions between adjacent N and P layers. One layer (*emitter*) acts like the cathode of a tube, injecting current carriers (*holes* or *electrons*) into the semiconductor; the second layer (*base*), which always is the central one, acts like the grid; the third (*collector*) acts like the plate. Input-current values influence the value of output current in the transistor. This device therefore has current gain which is expressed by the *current amplification factor, alpha* (α), or *beta* (β). $\alpha = di_c/di_e$, $\beta = di_c/di_b$. Basic **Transistor Amplifier Circuits** are the *common-base, common-emitter,* and *common-collector.*

In an **X-ray Tube,** high-velocity electrons strike a solid, positively charged metal target. The energy given up on impact is released in the form of X rays which are emitted by the target. The wave length of X rays is shorter than that of light, heat, or radio waves.

QUESTIONS AND EXERCISES

1. Describe the structure and operation of the following types of vacuum tubes: (*a*) diode; (*b*) triode; (*c*) tetrode; (*d*) pentode; (*e*) beam power tube.
2. In a certain triode, a grid-voltage change from 3 to 4 v produces a plate-current change from 5 to 6 ma. Calculate the transconductance.
3. In a certain tube, a grid-voltage change from 5 to 6½ v produces a certain observed plate-current change. A plate-voltage change from 10 to 40 v produces the same plate-current change. Calculate the amplification factor.
4. The transconductance of a given tube is 10,000 micromhos and the amplification factor is 48. Calculate the plate resistance.
5. In an *R-C*-coupled voltage amplifier circuit, the load resistance is 100,000 ohms and the tube plate resistance 7,500 ohms. The amplification factor of the pentode tube used is 36. Calculate the voltage gain of this circuit.
6. In a power amplifier, the plate voltage is 250 volts d-c, plate current 75 ma, and output power 10 w. Calculate the plate efficiency.
7. Describe the biasing of the following amplifiers: (*a*) class-A; (*b*) class-AB; (*c*) class-B; (*d*) class-C.
8. Explain the operation of a simple, self-excited oscillator.
9. Explain the operation of a crystal-controlled oscillator.
10. Calculate the wavelength corresponding to: (*a*) 150 kc; (*b*) 210 megacycles.
11. Describe the operation of a superheterodyne radio receiver.
12. Describe the construction of a phototube.
13. Describe the construction of a photocell.
14. Explain the difference between a photoconductive and a photovoltaic device.

15. Describe the action of an a-c-operated thyratron tube with a phase shifter in its grid circuit.

16. Draw a half-wave thyratron rectifier circuit with grid phase control, and explain its operation.

17. Describe the construction and operation of: (*a*) ignitron tube; (*b*) X-ray tube.

18. Describe the construction of the following transistors: (*a*) *NPN;* (*b*) *PNP.*

19. In a certain common-base transistor circuit, an emitter-current change from 1.5 to 2.5 ma produces a collector-current change from 1.0 to 1.85 ma. Calculate the current amplification factor, α.

20. In a certain common-emitter transistor circuit, a base-current change from 150 to 250 μa produces a collector-current change from 2 to 4 ma. Calculate the current amplification factor, β.

Chapter **18**

Electrical Instruments and Measurements

18.1 CURRENT

Current level is measured with an *ammeter* (contracted form of the term "ampere meter"). Ammeters are available in both a-c and d-c types. For the measurement of very low current values, *milliammeters* and *microammeters* are provided.

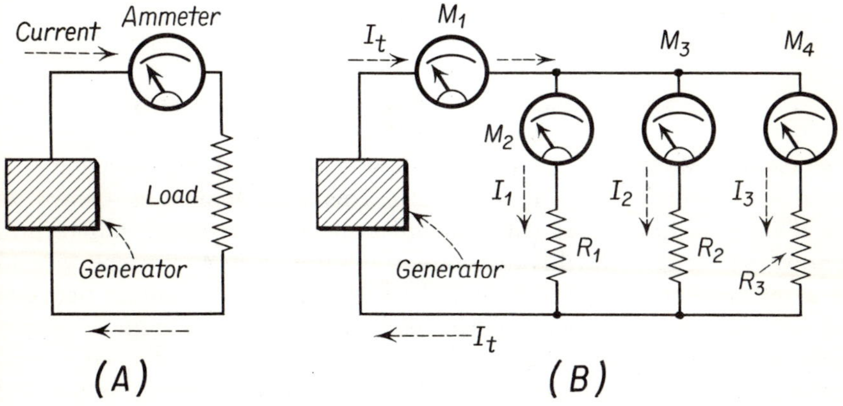

Fig. 18–1 Connection of Ammeter into Circuit

Figure 18–1(A) shows the connection of a current meter into a simple circuit. Current flows from the generator through the ammeter and load in series. This must always be the arrangement when current is to be measured. Because this is a series circuit, the position of meter and load is inconsequential. The meter may be placed either before or after the load. In a complex circuit, the current in each branch may be measured with a separate current meter. Thus in Figure 18–1(B), current from the generator divides into three parts. The first part (I_1) flows through Resistor R_1 and is indicated by Ammeter M_2, the second part (I_2) flows through Resistor R_2 and is indicated by Ammeter M_3, and the third part (I_3) flows through Resistor R_3 and is indicated by Ammeter M_4. Ammeter M_1 indicates the total current (I_t) which is the sum $I_1 + I_2 + I_3$.

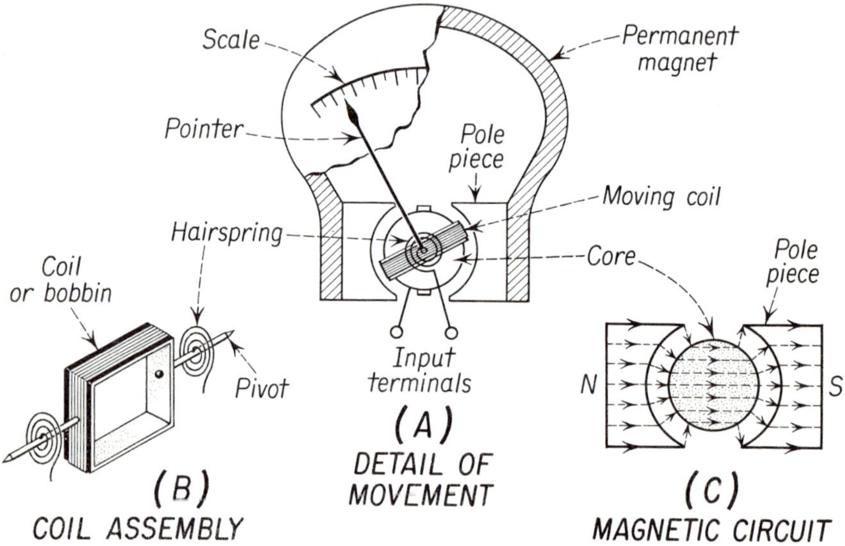

Fig. 18–2 D'Arsonval (Moving-Coil, Permanent-Magnet) Meter

d'Arsonval Meter. Figure 18–2 shows details of the moving-coil, permanent-magnet type of ammeter. This arrangement is known as the *d'Arsonval movement* (after Jacques Arsène d'Arsonval, nineteenth-century physicist). In this instrument [Figure 18–2(A)], a coil wound on a rectangular bobbin is mounted on a shaft so as to rotate between the curved pole pieces of a permanent magnet. The magnetic circuit includes an iron core [Figure 18–2(C)] around which the coil turns and which concentrates the field. The coil-shaft pivots rotate in jeweled bearings.

A lightweight pointer is attached to the bobbin so as to travel over a printed scale. A hairspring on each pivot serves the dual purpose of conducting current to the coil and holding the coil in its zero resting position in the absence of current flow. When current is passed through the coil, a magnetic field is set up about the coil and opposes the permanent magnetic flux. This causes the coil to rotate against the torque of the hairsprings. The amount of movement is proportional to the current, and the pointer is deflected a proportionate distance on the scale which is calibrated to read directly in current units. When the current is stopped, the torque of the springs returns the coil and pointer to the zero position. Figure 18–3 shows a group of d'Arsonval meter movements employing circular magnets.

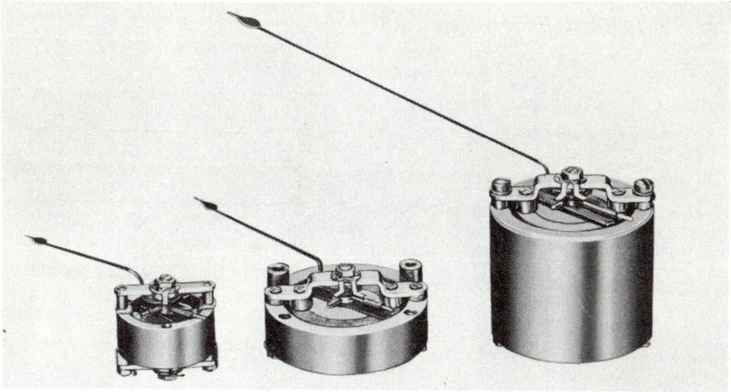

Fig. 18–3 D'Arsonval Movements with Circular Magnets (Courtesy, Simpson Electric Co.)

The moving-coil, permanent-magnet meter is a d-c instrument. It cannot be used *directly* to measure alternating current because the coil will attempt to follow the alternations, although its mechanical inertia prevents it from rising fully to maximum and returning completely to zero along with the a-c cycle. The pointer will buzz or vibrate rapidly, instead of indicating a definite current value. As will be explained later, a rectifier must be used with the d'Arsonval meter when alternating current is to be measured.

Extending Current-Meter Range. The basic d'Arsonval movement requires very little current for full-scale deflection of the pointer. Depending upon the number of turns and size of wire used in the moving coil, the strength of the deflecting current is either milliamperes or microamperes. The coil has a certain amount of resistance, and this determines the *internal resistance* (R_m) of the meter movement. The basic d'Arsonval movement often is regarded as a d-c millivoltmeter, since it is deflected by a small voltage ($E = IR_m$). In order to increase the range of the basic meter to *amperes,* a shunt resistor must be connected externally or internally, as shown in Figure 18–4. The value of this shunt resistance, R_s, is chosen such that the input current divides between R_s and R_m (internal resistance of meter) with the shunt carrying all of the current except that which does not exceed full-scale deflection of the meter.

The required shunt resistance (R_s) for a desired range is determined in terms of the meter resistance (R_m) and the factor (n) by which the basic full-scale deflection of the meter is to be multiplied.

$$R_s = R_m/(n - 1) \quad (18\text{–}1)$$

where R_s = required shunt resistance (in ohms)
R_m = internal resistance of meter (in ohms)
n = scale multiplication factor

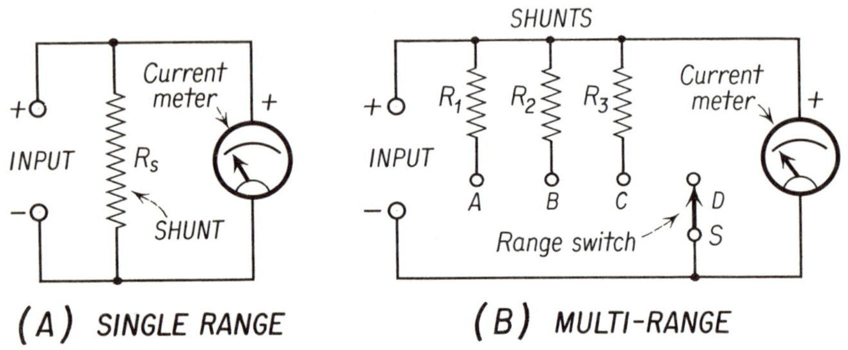

Fig. 18–4 Ammeter Circuits

Example: A 0–1 d-c milliammeter has an internal resistance of 100 ohms. What value of shunt will be needed to convert this instrument into a 0–1 ammeter?

(*a*) Present meter range = 0–1 ma = 0–0.001 amp. Therefore, $n = 1/0.001 = 1{,}000$

(*b*) $R_s = R_m/(n - 1) = 100/(1{,}000 - 1) = 100/999 = 0.1$ ohm

The single shunt in Figure 18–4(A) provides one range for the current meter. Several shunts may be arranged with a selector switch to provide several ranges for the meter. Thus, in Figure 18–4(B), either R_1, R_2, or R_3 may be switched into the circuit in parallel with the meter. When Switch S is at Position *D*, no resistor shunts the meter, and the latter accordingly has its basic, low-current range.

Solenoid-Type Ammeter. Figure 18–5(A) shows a *solenoid-type* or *plunger-type* ammeter. This is a simple movement which does not offer high accuracy and is used only in inexpensive instruments intended for rough, approximate measurements. Magnetic flux due to current flow through the coil draws the curved plunger into the coil against the returning torque of the spring. A pointer attached to the plunger moves over a graduated scale. This type of ammeter may be calibrated for alternating as well as direct current.

Movable-Iron-Type Meter. Figure 18–5(B) shows the mechanism of a movable-iron-type ammeter. Current flowing through the coil causes both the fixed (A) and movable (B) iron vanes to become magnetized. The top of A has the same magnetic polarity as the top of B, and the bottom of A has the same polarity as the bottom of B, although both bot-

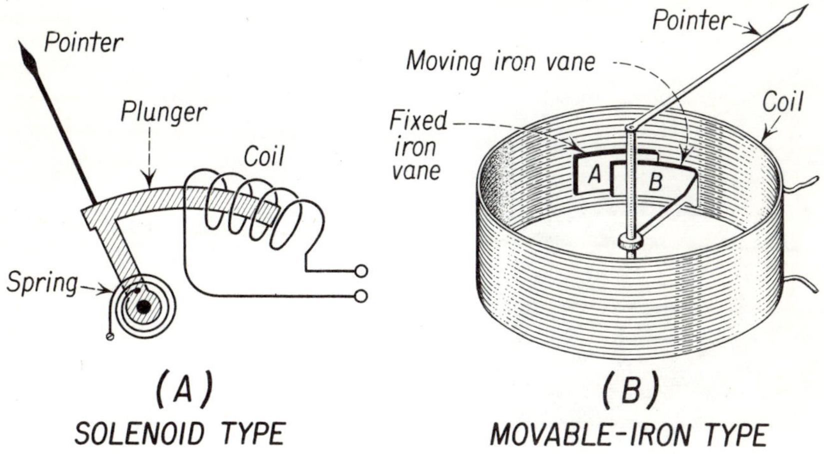

Fig. 18–5 Solenoid and Movable-Iron Type Ammeters

tom polarities are opposite to the tops. This similarity of magnetic poles causes the movable vane to be repelled, moving the shaft and pointer against the returning torque of hairsprings (not shown). The ability of this type of instrument to respond to alternating current without vibrating is due to the fact that the two vanes always have the *same* magnetic polarity, although following the cyclic variations of the alternating current, and hence can only repel, never attract, each other.

Electrodynamometer. The *electrodynamometer-type* ammeter (Figure 18–6) contains a field coil and movable coil connected in series. There is no magnet or core. When current is passed through these coils, the deflection they undergo is proportional to the square of the current, which means that the scale divisions are closer together near the zero end of the scale than at the high end. The electrodynamometer may be used to measure both alternating and direct current on the same scale.

Rectifier-Type Meter. The d'Arsonval movement may be adapted for a-c measurements by means of a small rectifier. Figure 18–7 shows the circuit of such a *rectifier-type* current meter. The semiconductor-bridge rectifier is composed of diodes A, B, C, and D. Tiny, instrument-type rectifiers of this kind usually employ copper oxide as the semiconductor. The circuit operates in the following manner: When a-c input terminal 1 is negative, current flows in the direction of the dotted arrows through Diode *C*, Meter *M*, and Diode *B*. When Terminal 2 is negative, current flows in the direction of the solid arrows through Diode *D*, Meter *M*, and Diode *A*. Notice that the current passes through the meter in the *same* direction during each half-cycle of alternating current.

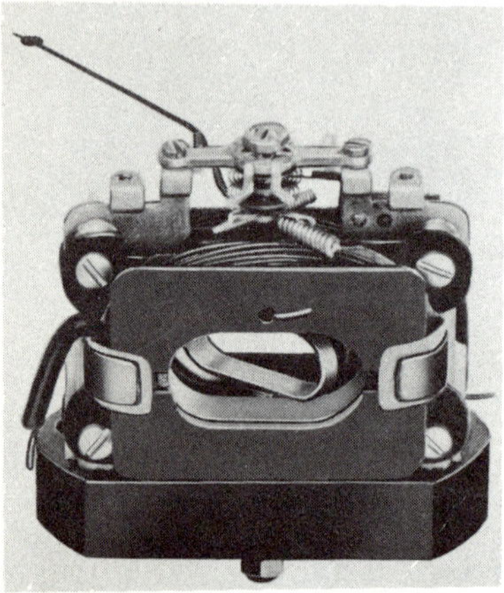

Fig. 18–6 Electrodynamometer-Type Meter Movement (Courtesy, Simpson Electric Co.)

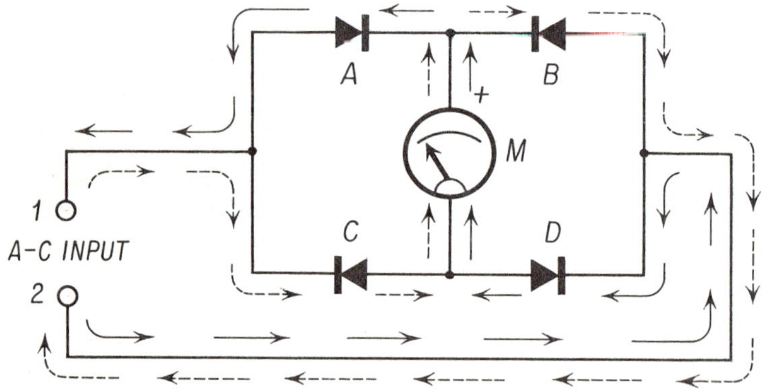

Fig. 18–7 Rectifier-Type Alternating-Current Ammeter

The current readings of the rectifier-type meter are somewhat lower on the scale than the same d-c readings. This is due chiefly to the resistance of the rectifier. Shunt resistors, to increase the meter range, may be connected across Terminals 1 and 2 of this circuit.

Current Transformer. A *current transformer* sometimes is used to change the range of an a-c ammeter. This method is illustrated by Figure 18–8. The current is transformed before presentation to the meter, to the extent determined by the turns ratio. Primary and secondary taps may be provided for several current ranges, as may be seen on the commercial current transformer shown in Figure 18–9.

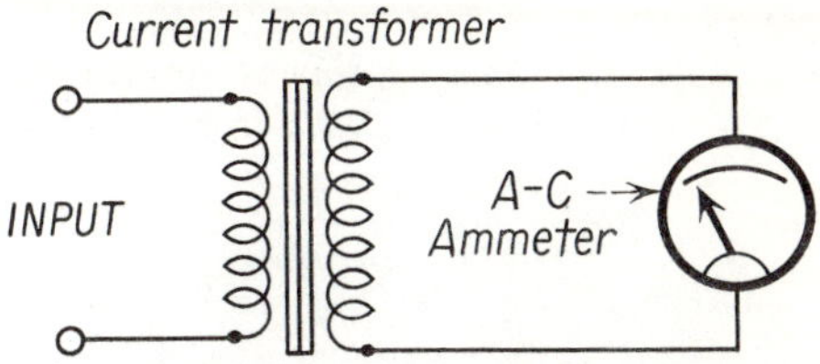

Fig. 18–8 Use of Current Transformer

Fig. 18–9 Commercial Current Transformer (Courtesy, Weston Electrical Instrument Corp.)

Ammeter Resistance. The lowest practicable internal resistance is desired in all current meters. Low resistance is necessary in order to minimize the voltage drop introduced into a circuit by inclusion of the meter. Internal resistance varies between less than 1 ohm for high-current ammeters to several thousand ohms for low-range microammeters.

18.2 VOLTAGE

d'Arsonval Voltmeter. A voltmeter of the d'Arsonval type consists essentially of a current meter movement plus a series resistor. (See Figure 18–10.) The current meter may be of the a-c or d-c type and may be a milliammeter or microammeter.

The series resistor is termed a *multiplier*. Its resistance depends upon the desired voltage range and the full-scale deflection of the current meter.

$$R = E/I \qquad (18\text{–}2)$$

where R = required multiplier resistance (in ohms)
E = desired full-scale voltage (in volts)
I = present full-scale current deflection (in amperes)

Example: What value of multiplier will be required to convert a 0–1 d-c milliammeter into a 0–500 d-c voltmeter?

(*a*) $E = 500$ v

(*b*) $I = 1$ ma $= 0.001$ amp

(*c*) $R = E/I = 500/0.001 = 500{,}000\Omega$

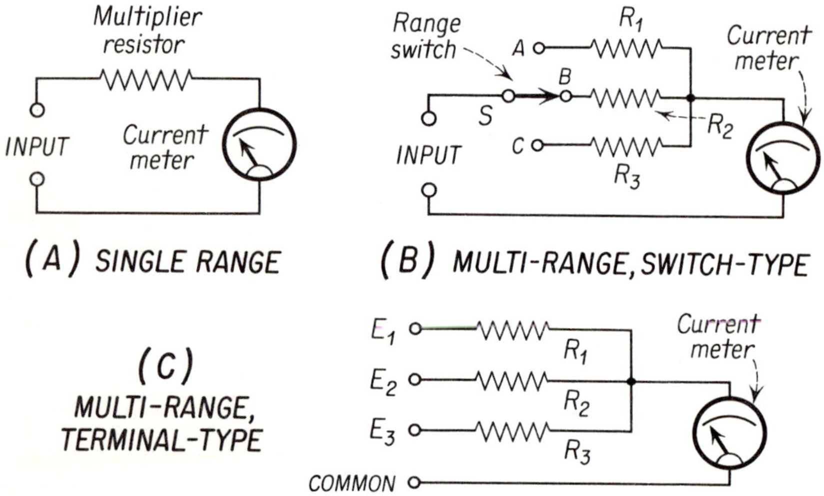

Fig. 18–10 Voltmeter Circuits

The one multiplier resistor in Figure 18–10(A) gives the voltmeter a single range. A range switch [S in Figure 18–10(B)] may be employed to switch different multipliers (R_1, R_2, R_3) into the circuit to change the voltmeter range. In lieu of the switch, binding-post terminals may be used, as shown in Figure 18–10(C). In this instance, the voltage is applied to the COMMON terminal and one of the voltage-range terminals (E_1, E_2, or E_3). Figure 18–12 shows a commercial, portable multirange voltmeter of the terminal type. Figure 18–11 shows a precision, wire-wound resistor of the type employed as a voltmeter multiplier.

Other Types. The movable-iron-type instrument [Figure 18–5(B)] often is employed as a voltmeter as well as an ammeter. When this is done, the coil is wound with many turns of fine wire (as opposed to the few turns of heavier wire employed in the ammeter). The solenoid-type movement [Figure 18–5(A)] also is used as a voltmeter. The electrodynamometer (Figure 18–6) is employed as a standard voltmeter, as well as in general applications, for both a-c and d-c measurements. However,

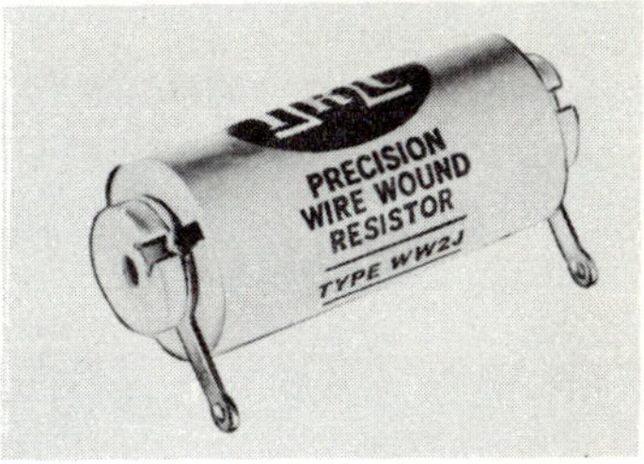

Fig. 18–11 Precision Resistor for Voltmeter Multipliers (Courtesy, International Resistance Co.)

these latter instruments draw relatively high currents, and *potential transformers,* rather than multiplier resistors, usually are employed to

Fig. 18–12 Portable, Multirange Direct-Current Voltmeter (Courtesy, Weston Electrical Instrument Corp.)

change their ranges when making a-c voltage measurements. (The transformer reduces the voltage to the range-value of the instrument and, at the same time, boosts the current.)

Voltmeter Resistance. A voltmeter must be connected *across* a source, load, or circuit element. (See Figure 18–13.) Because of this, the internal resistance of the voltmeter must be very high, in order not to load heavily the circuit or element under test and upset the operation of the latter. When the meter resistance is less than ten times that of the voltage source, the readings will be less than true values. In Figure 18–13, the separate voltmeters measure the voltage drops across the resistors. Meter M_1 indicates the generator voltage.

An important characteristic of a voltmeter is its *ohms-per-volt* rating. This figure is determined by dividing either of the full-scale deflection voltages (E) of the meter into the resistance (R) of the multiplier for that range. Thus,

$$\Omega/V = R/E \tag{18-3}$$

Example: A certain voltmeter has a 1-megohm multiplier resistor for its 0–50-v range. Calculate the ohms per volt.

$$\Omega/V = R/E = 1{,}000{,}000/50 = 20{,}000 \text{ ohms per volt}$$

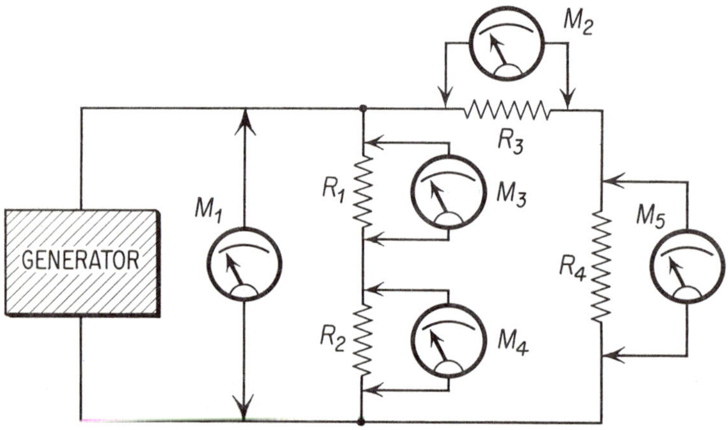

Fig. 18–13 Voltmeter Connections in Circuit

The higher the ohms-per-volt figure, the less the meter will load a circuit under test. A rating of 1,000 ohms per volt is satisfactory for general electrical measurements in low-resistance circuits, but 20,000 ohms per volt or higher will be required for electronic measurements.

Vacuum-Tube Voltmeter. The highest value of instrument resistance (between 10 and 100 megohms in most instances) is provided by the *vacuum-tube voltmeter.* The high resistance of this instrument virtually eliminates loading effects.

Modern v-t voltmeters are available in numerous circuit designs, some of them complicated. Figure 18–14 shows the basic v-t voltmeter circuit. When power is switched on and the tube has heated, plate current flows through the current meter, M, deflecting it. Current obtained from Battery B_2 and adjusted by means of Rheostat R_6 is passed through the meter in the opposite direction to return the pointer to zero against the action of the plate current. Adjustment of R_6 sets the meter initially to zero before the meter is ready for use.

A positive voltage applied to the grid of Tube V in series with the bias voltage developed across cathode resistor R_5 increases the plate current and deflects the meter in proportion to the voltage magnitude. The voltage range of the instrument is changed by means of the RANGE SWITCH, S, which selects various taps along the input voltage divider. The lowest voltage range is E_1 and the highest E_4.

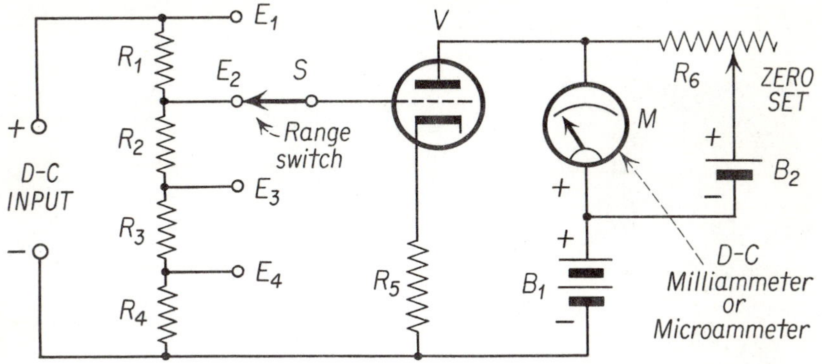

Fig. 18–14 Principle of Vacuum-Tube Voltmeter

The input resistance of the instrument is the sum of the resistances of the voltage-divider sections ($R_1 + R_2 + R_3 + R_4$). Within practical limits, this resistance may be made as high as desired.

A suitable rectifier (connected ahead of the D-C INPUT terminals) may be provided for measuring a-c voltages with the v-t voltmeter.

V-t voltmeters are available for operation from self-contained batteries, as well as being provided with built-in a-c-operated, transformer-rectifier-type power supplies.

Clamp Meter. Figure 18–15 shows a special type of portable voltmeter-ammeter combination. It is called a *clamp-type* meter because this instrument, in effect, is clamped to the insulated line carrying current to be measured. No actual electrical contact is made to the line. The action is inductive.

The heavy loop seen at the top of the instrument is a laminated core. When the lever on the left side of the handle is pressed, the core opens to encircle the current-carrying wire which then acts as a one-turn primary of a transformer. The core provides coupling to a secondary coil inside the meter enclosure to complete the transformer. The magnetic field surrounding the wire induces a current in the secondary via the core, and this current actuates the meter. The ammeter ranges are changed by means of the switch, below the meter, which selects current shunts. The clamp feature is not employed in voltage measurement in the instrument shown in Figure 18–15. The voltage test leads (to the right of the instrument) are plugged into the jacks on the right side of the instrument. This is a terminal-type voltmeter similar to the one illustrated in Figure 18–10(C).

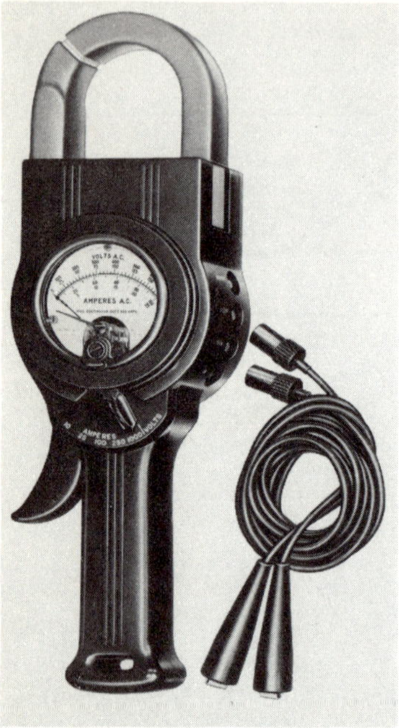

Fig. 18–15 Clamp-Type Volt-Ammeter (Courtesy, Weston Electrical Instrument Corp.)

Precision Measurement of Voltage. In the electrical laboratory, d-c voltage is measured with high accuracy by means of a precision instrument known as a *potentiometer.*

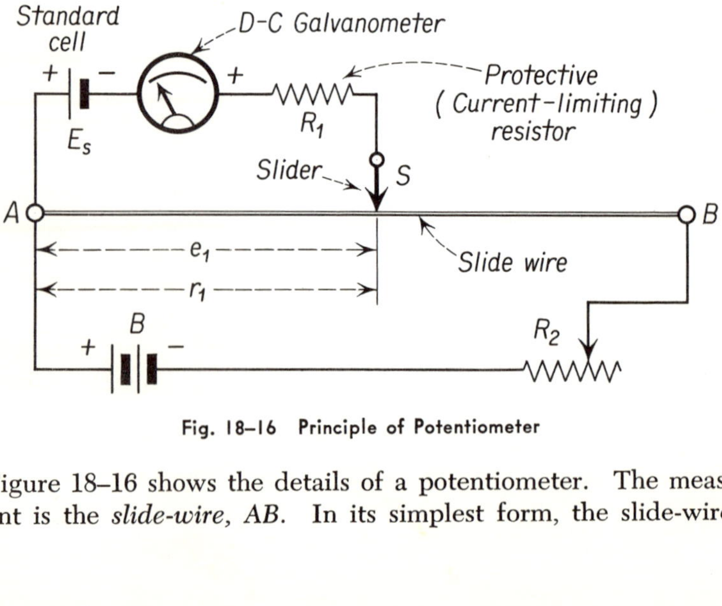

Fig. 18–16 Principle of Potentiometer

Figure 18–16 shows the details of a potentiometer. The measuring element is the *slide-wire, AB.* In its simplest form, the slide-wire is a length of resistance wire, as shown, with a sliding contact, S. In commercial potentiometers, however, the slide-wire usually is a coil or a series of accurately known variable resistors. Two emf circuits are

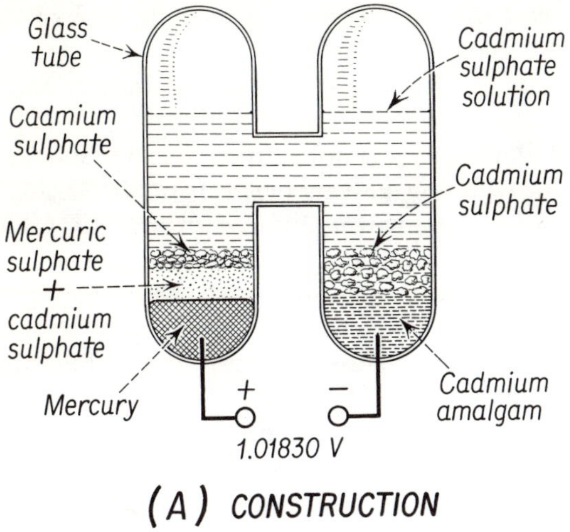

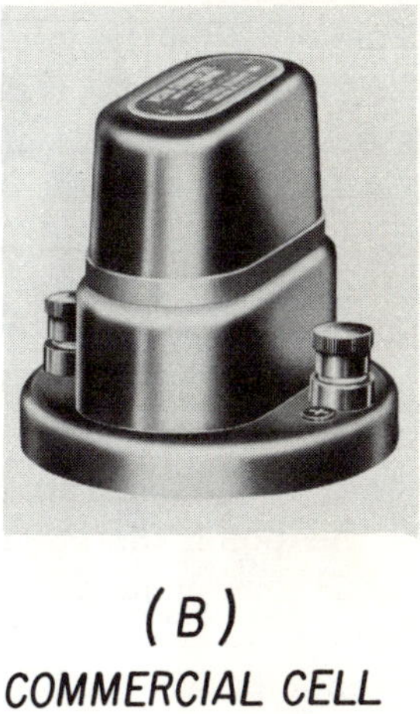

Fig. 18–17 Details of Standard Cell

connected to the slide-wire. One contains Battery B and Rheostat R_2. The other contains a d-c galvanometer [a sensitive current-measuring instrument with zero at the center of its scale (see Figure 18–18)], a

Fig. 18–18 Direct-Current Galvanometer (Courtesy, Weston Electrical Instrument Corp.)

protective, current-limiting resistor (R_1), and a standard cell. The standard cell (shown in detail in Figure 18–17) generates a precise emf of 1.01830 v.

As Slider S is moved along the slide-wire, the resistance between *A* and *S* and *S* and *B* is changed. Initially, the slider is set to Point S, and Rheostat R_2 is set so that current flowing through *AB* from Battery *B* produces a voltage drop e_1 between *A* and *S* equal to the standard cell voltage, E_s. Since E_s and e_1 are equal in magnitude and of the same polarity, the two oppose each other, and no current flows through the galvanometer (condition of *null*). The resistance between *A* and *S* is designated r_1.

If, now, an unknown voltage (E_x) is substituted for the standard cell, E_s, current will flow through the galvanometer (circuit AE_sR_1S) unless $E_x = e_1$, which is unlikely. However, the circuit may be rebalanced (restored to null) by moving the slider along the slide-wire to a new point at which the voltage drops are such that the meter again reads zero. The new resistance (r_2) between *A* and the new position of the slider is determined by reading the slide-wire resistance calibration. At this point, the unknown voltage $E_x = (E_s r_2)/r_1$.

Higher voltages may be measured if they are first stepped down through an accurate voltage divider to values within the range of the potentiometer. Such a laboratory voltage divider is termed a *volt box.*

18.3 POWER

Several methods are available for the measurement of electrical power.

Ammeter Method. D-C and single-phase a-c power (in watts) may be calculated from the measured current (*I*, in amperes) flowing through a load, and the resistance (*R*, in ohms) of the load. Thus,

$$P = I^2R \tag{18–4}$$

This method actually yields volt-amperes and expresses true watts only when the load is a pure resistance, since it does not take into consideration the power factor of the circuit.

Voltmeter Method. Similarly, d-c and single-phase a-c power may be calculated from the measured voltage (*E*, in volts) across a load and the resistance (*R*, in ohms) of the load.

$$P = E^2/R \tag{18–5}$$

Voltmeter-Ammeter Method. Where the load resistance is not known and cannot easily be determined, both current through and voltage across the load may be measured, and the power can be calculated from these values.

$$P = EI \tag{18-6}$$

Wattmeter. The *wattmeter* indicates true watts. This instrument is of the electrodynamometer type and, like the dynamometer ammeter and voltmeter, may be used for either a-c or d-c measurements.

Figure 18–19 shows the basic arrangement of the wattmeter for measuring the power consumed by a load device. The wattmeter contains both a current coil (C) and a potential coil (P). The current coil is connected in series with the load, and the potential coil in parallel with the source. Deflection of the meter is proportional to EI, the product of volts and amperes.

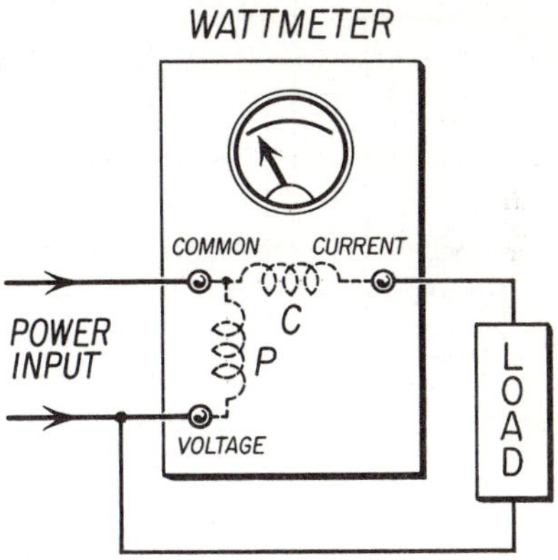

Fig. 18–19 Wattmeter

Three-Phase Power Measurement. The power in a three-phase system may be measured with a three-element polyphase wattmeter. When this instrument is not available, however, three separate single-phase wattmeters may be connected, as shown in Figure 18–20.

Each wattmeter will indicate the power in one phase, and the total power may be determined by adding the three readings.

Watt-Hour Meter. Our electric service bills are computed from readings taken from a *watt-hour meter,* also called a *service meter.*

Figure 18–21 shows the basic mechanism of this instrument. This arrangement essentially is that of the wattmeter, except that a small armature (rotating coil) with commutator and brushes takes the place of the movable coil of the wattmeter. Mounted on the armature shaft is a metal disc which rotates between the poles of two permanent magnets.

Eddy currents are induced in the disc as it spins, and they give rise to a torque which opposes the rotation of the disc. This action causes

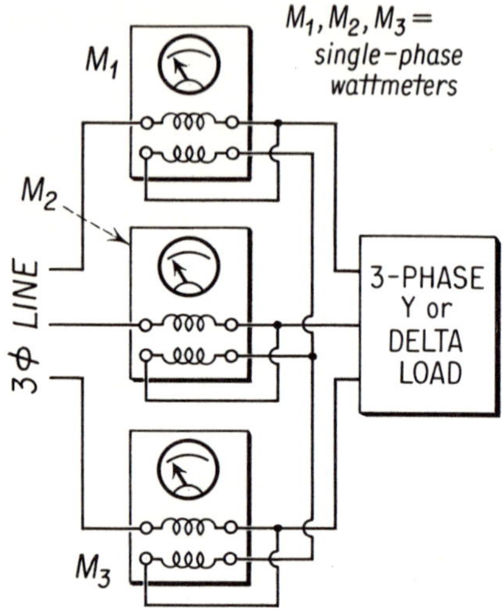

Fig. 18–20 3-Phase Power Measurement

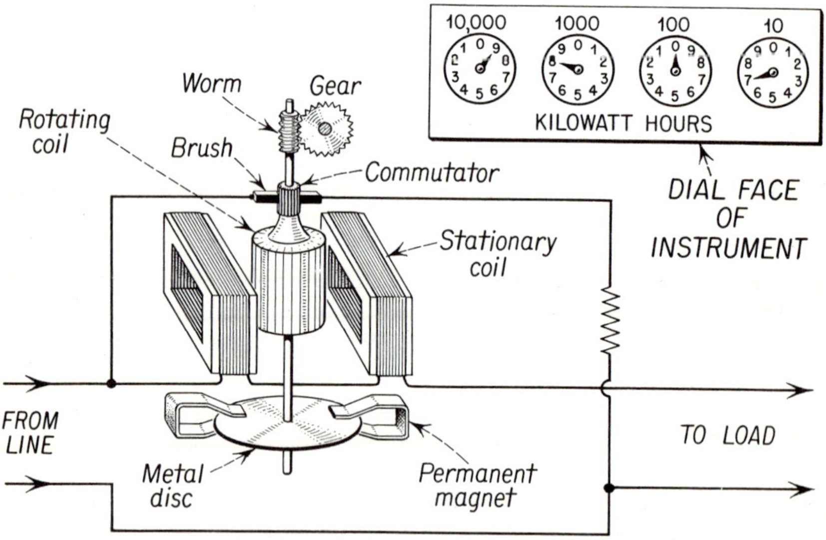

Fig. 18–21 Watt-Hour Meter

the rotational speed of the shaft to be proportional to the watt-hours (or kilowatt-hours) of consumed energy.

A worm on one end of the shaft turns a train of gears which operates the pointers on the dials of the watt-hour meter.

18.4 RESISTANCE

Several methods are available for the measurement of resistance. The principal ones are described in this section.

Current-Meter Method. When no other instruments are available, resistance may be determined from the value of current flowing through the unknown resistance when an accurately known voltage is applied. Figure 18–22 shows the circuit. The instrument may be a microammeter,

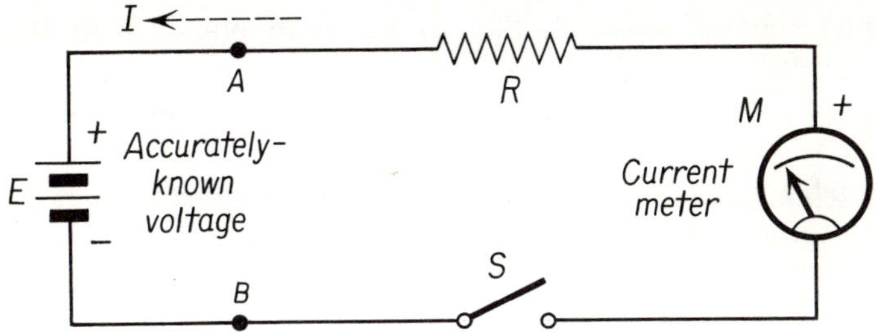

Fig. 18–22 Current-Meter Method

milliammeter, or ammeter, its range depending upon the battery voltage and expected resistance value. The unknown resistance is calculated by means of Ohm's law.

$$R = (E/I) - R_m \qquad (18\text{–}7)$$

where R = unknown resistance (in ohms)
E = applied voltage (in volts)
I = measured current (in amperes)
R_m = internal resistance of meter (in ohms)

Example: An unknown resistance (R) is connected into the circuit of Figure 18–22. The battery supplies 3 v. Meter M is a 0–50 d-c microammeter with an internal resistance of 2,000 ohms. The meter reads 30 μa. What is the value of the unknown resistance?

(*a*) $I = 30\ \mu\text{a} = 0.00003$ amp

(*b*) $R = (E/I) - R_m = (3/0.00003) - 2{,}000$
$= 100{,}000 - 2{,}000 = 98{,}000\Omega$

This method of measurement, while useful in emergencies, requires extraordinary care, since a short circuit in the resistor under test or an actual touching together of the resistor leads can cause the meter to be overloaded and damaged.

The applied voltage, E, must be known quite accurately. A voltmeter often is connected permanently from A to B in the circuit. This latter arrangement therefore has been termed the *voltmeter-ammeter method.*

Voltmeter Method. A d-c voltmeter can be used for resistance measurement without the risk of meter damage which is present in the current-meter method. The unknown resistance (R) is connected in series with a battery and the voltmeter.

Choose a battery voltage (E) as close as possible to (but not exceeding) the full-scale reading of the voltmeter. Read the battery voltage carefully with the meter. Then connect the unknown resistance in series with the battery and meter and take a second voltage reading (e). The unknown resistance is calculated from these two readings and the voltmeter resistance. (The total voltmeter resistance, r, is the sum of the multiplier resistance and the internal resistance of the meter movement. It may be determined by multiplying the ohms-per-volt rating by the full-scale voltage of the meter.) Thus,

$$R = (Er/e) - r \tag{18-8}$$

Example: A test battery gives a reading of 6 v on the 0–10-v scale of a 1,000 ohms-per-volt d-c voltmeter. When an unknown resistance is connected in series with the battery and meter, the reading drops to 2½ v. Calculate the unknown resistance.

(*a*) $E = 6$ v

(*b*) $e = 2.5$ v

(*c*) $r = 1{,}000(10) = 10{,}000\Omega$

(*d*) $R = (Er/e) - r = (6 \times 10{,}000/2.5) - 10{,}000$
$= (60{,}000/2.5) - 10{,}000 = 24{,}000 - 10{,}000$
$= 14{,}000\Omega$

Ohmmeter. The *ohmmeter* gives direct readings in ohms or megohms. No calculations are required. This instrument is a refinement of the voltmeter method of resistance measurement.

Figure 18–23(A) shows the simplest ohmmeter circuit. The voltage of the battery, B, is somewhat higher than that required for full-scale deflection of the meter. When Terminals X–X are short-circuited temporarily, the voltmeter deflection may be set exactly to full scale by adjusting Rheostat R. Full-scale deflection thus corresponds to zero resistance between Terminals X–X. When the terminals are open-circuited, the meter returns to zero volts which is the scale point corresponding to infinite resistance. When some finite value of resistance is connected between the terminals, this resistance is added to the meter multiplier resistance and acts to increase the voltage range. The deflection consequently assumes some point between zero volts and full-scale volts and is proportional to the added resistance.

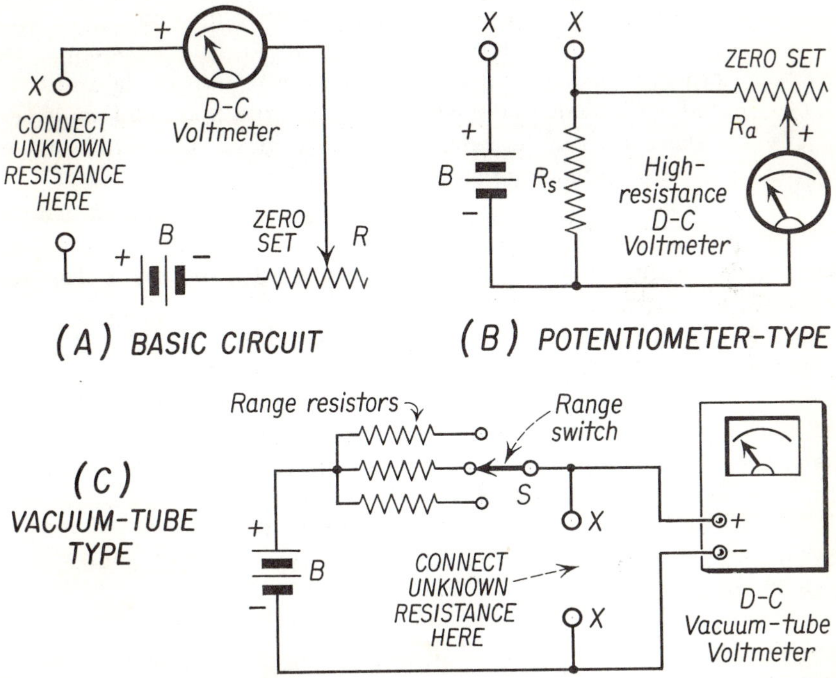

Fig. 18–23 Ohmmeter Circuits

Because of this action, the meter scale may conveniently be graduated in resistance values, with zero ohms at the right (full-scale) and infinite ohms at the left (voltmeter zero). The ohmmeter scale may be multiplied to accommodate higher resistance values than those read on the basic scale, by increasing the battery voltage and the voltmeter range (multiplier resistor).

A troublesome feature of the simple ohmmeter is the over-all crowding of its scale. This condition is relieved by the *potentiometer-type* ohmmeter [Figure 18–23(B)]. In this circuit, the unknown resistance (connected to Terminals *X–X*) is in series with an accurately known standard resistance, R_s. Current from the battery flows through these two resistances in series, setting up a voltage drop across each one. A high-resistance d-c voltmeter is deflected by the voltage drop across the standard resistor. A simple potentiometer or voltage divider thus is formed by R_s and the unknown resistance.

With Terminals *X–X* temporarily short-circuited, the meter is set initially to full scale by adjusting Rheostat R_a. When a resistance then is connected to the terminals, the deflection will drop to some point lower than full scale and will be proportional to the unknown resistance. The meter scale accordingly may be graduated directly in ohms. The scale of this meter is relatively wide-spaced in its low-resistance section, and somewhat crowded at high resistances. In order to change the resistance range, the battery voltage, standard resistance, and voltmeter multiplier usually are switched simultaneously to new values. The same zero-set rheostat is used on all ranges but is readjusted on each.

The *vacuum-tube ohmmeter* circuit is combined with most general-purpose v-t voltmeters. Figure 18–23 shows the circuit arrangement. The resistance range covered by conventional v-t instruments generally extends from 0.1 ohm to 1,000 megohms.

In this circuit, current from the test battery flows through a precision resistor (selected by the RANGE SWITCH, S) and the unknown resistance in series. The d-c v-t voltmeter then is deflected by the resulting voltage drop across the unknown resistance. The higher this resistance, the higher will be the deflection of the meter. The scale of the v-t ohmmeter accordingly reads from left to right.

Before making a resistance measurement, the meter is set initially to full scale, by means of an internal control in the v-t voltmeter circuit, *with Terminals X–X open.*

Wheatstone Bridge. The most accurate measurement of resistance is afforded by a four-resistor network, the *Wheatstone bridge* (named in honor of Charles Wheatstone, 1802–1875). The unknown resistance is one of the four resistances in the network. In the bridge method, the unknown resistance is compared to an accurately known standard resistance by means of a balancing operation.

Figure 18–24 shows the basic circuit of the Wheatstone bridge. A d-c voltage is supplied by the Battery, *B*. Meter *M* is a sensitive d-c galvanometer, the pointer of which normally rests on zero at the center of the scale. (See Figure 18–18 for a typical d-c galvanometer.) Current from the battery divides into two parts—one flowing through the R_1R_2 path and the other through the R_3R_4 path. Current flowing through the upper branch of the circuit produces a voltage drop, e_1, across Resistor R_1, and e_2 across R_2. Similarly, current through the lower branch produces a voltage drop, e_3, across R_3, and e_4 across R_4. These voltages are proportional to the resistances.

Depending upon the relative magnitudes of the resistances and therefore the voltage drops, Point A will be positive with respect to Point B, or it will be negative, or of the same polarity (positive or negative). When A is positive and B negative, the galvanometer will be deflected up-scale. When A is negative and B positive, M is deflected down-scale. When A and B both are positive, or both negative, and both have the same voltage magnitude, the meter will remain at zero (the *null* point) and the bridge is said to be *balanced.*

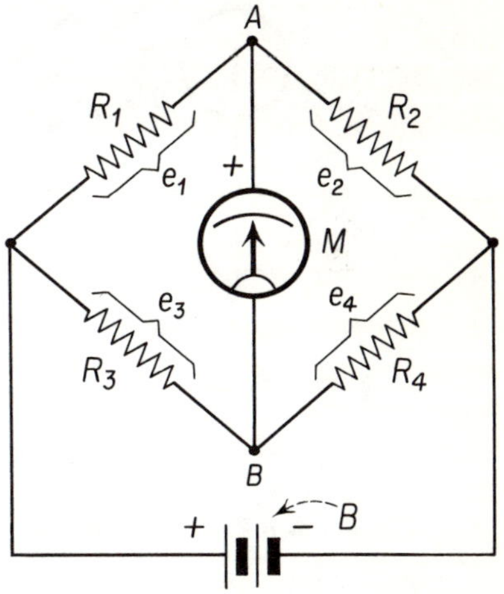

Fig. 18–24 Principle of Wheatstone Bridge

The bridge is balanced only when $e_1/e_2 = e_3/e_4$, and this condition is satisfied only when $R_1/R_2 = R_3/R_4$. In a practical bridge, R_1 is the unknown resistance, and R_2 is a high-precision variable resistance used to adjust the bridge to null. Resistors R_3 and R_4 are termed *ratio arms,* since their values are chosen to have a certain ratio, such as 1:10, 10:1, 1:100, etc. The variable resistor is made direct-reading in ohms. When R_2 is adjusted for null, the unknown resistance (R_1) is to the resistance setting of R_2 as R_3 is to R_4. That is, $R_1/R_2 = R_3/R_4$. From this relationship, the unknown resistance may be calculated. $R_1 = (R_3/R_4)R_2$. From the latter expression, it is seen that the setting of the variable resistor need only to be multiplied by the ratio indicated by the ratio arms to give the unknown resistance.

Figure 18–25 shows the circuit of a practical Wheatstone bridge. The unknown resistance is connected to Terminals X–X. The variable resistance, R_1, may be a precision rheostat or a step-type variable resistor arranged in *decades* (that is, ten 1-ohm sections, ten 10's, ten 100's, and ten 1,000's, all in series). Precision resistors R_2 to R_7, selected by switches S_1 and S_2, adjust the value of the ratio arms (R_3/R_4 in Figure 18–24). With this circuit, R_1 is adjusted for meter null. The resistance reading of R_1 then is multiplied by the ratio indicated by the settings of S_1 and S_2 to obtain the value of the unknown resistance. Thus, when $R_3 = 1{,}000$, and $R_4 = 10$, the ratio is 100:1. If the bridge is balanced when $R_1 = 1{,}500$ ohms, then the unknown resistance = 1,500(100) = 150,000 ohms.

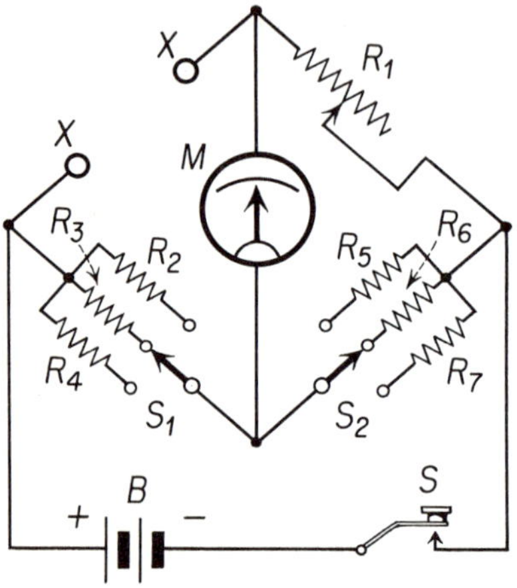

Fig. 18–25 Practical Wheatstone Bridge

18.5 CAPACITANCE

Capacitance may be measured by means of either meters or bridges. The meter method may or may not require calculations, as will be explained later.

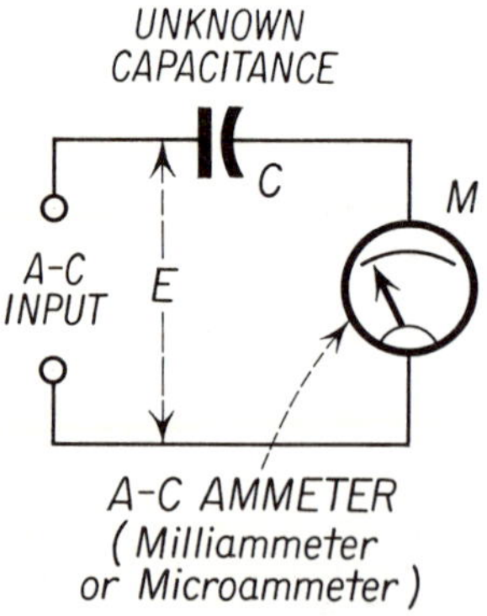

Fig. 18–26 Current-Meter Method of Capacitance Measurement

Current-Meter Method. On the assumption that a capacitor contains reactance only, the current flowing "through" the capacitor, when an alternating voltage of known amplitude (E) and frequency (f) is applied, may be measured in the circuit given in Figure 18–26, and the reactance determined with the aid of Ohm's law for alternating current: $X_c = E/I$. From this value of X_c, the capacitance may be calculated.

$$C = 10^6/(2\pi f X_c) \tag{18-9}$$

where C = unknown capacitance (in microfarads)
π = 3.1416
f = frequency (in cycles per second)
X_c = capacitive reactance (in ohms)
10^6 = 1,000,000

Example: A 60-cps test voltage of 6.3 v is applied to a capacitor in series with a 0–1 a-c milliammeter. A deflection of 0.3 ma is obtained. Calculate the capacitive reactance and the capacitance.

(*a*) $I = 0.3 \text{ ma} = 0.0003 \text{ amp}$

(*b*) $X_c = E/I = 6.3/0.0003 = 21{,}000\Omega$

(*c*) $C = 10^6/(2\pi f X_c) = 1{,}000{,}000/(6.28 \times 60 \times 21{,}000)$
$= 1{,}000{,}000/7{,}912{,}800 = 0.127\ \mu\text{f}$

The current-meter method is convenient when other capacitance-measuring instruments are not available. However, it has the same disadvantages noted earlier for the current-meter method of resistance measurement—namely, low accuracy and the risk of damage to the meter. Moreover, electrolytic capacitors cannot be checked with this circuit unless the applied a-c voltage is low enough to ensure that it will not damage the capacitor.

The accuracy of this method is limited considerably by the incorrect assumption that the capacitor exhibits reactance only. Actually, losses in the capacitor are evidenced by a resistive component, not readily measured in simple meter-type circuits, so that the deflection really is proportional to impedance (Z), not just to reactance.

Voltmeter Method. An alternate method consists of measuring the voltage drop across an unknown capacitor when a known value of alternating current is passed through a reference resistor and the capacitor in series. To ensure minimum loading, an a-c v-t voltmeter is employed. On the assumption that the voltage drop (E) is due entirely to reactance (X_c), the unknown capacitance may be calculated. $C = (10^6 E)/(2\pi f I)\ \mu\text{f}$.

The voltmeter method of capacitance measurement, like the current-meter method, is limited in accuracy, since it offers no way of compensating or correcting for the unknown resistive component of the capacitor.

Microfarad Meter. If the test voltage is known accurately in the current-meter circuit, or the test current in the voltmeter circuit, the meter scale may be graduated to read directly in microfarads, and this will save time by removing the need for calculations. However, the accuracy of this scheme is limited further by the steadiness of the voltage or current.

The ratio-type *microfarad meter* is a more accurate direct-reading instrument. Figure 18–27 shows the connections of this capacitance-measuring meter.

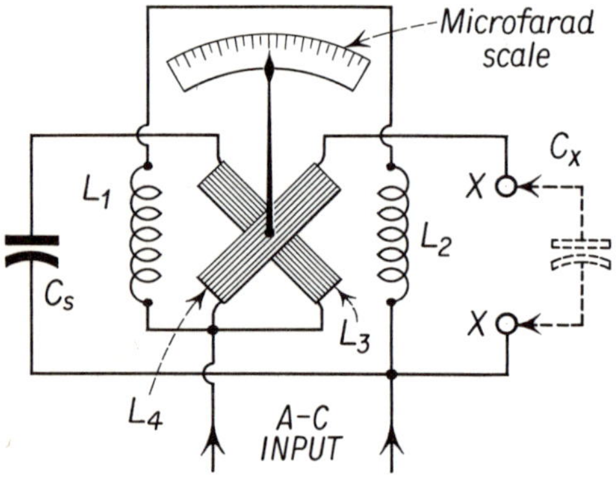

Fig. 18–27 Ratio-Type Microfarad Meter

The moving element is composed of L_3 and L_4, two coils fastened together with perpendicular planes, so as to rotate as a unit. The pointer is attached to the shaft of this moving-coil assembly. The assembly has no return springs. The moving element rotates within the field of two stationary coils, L_1 and L_2.

The unknown capacitance (C_x) is connected to Terminals *X–X*. An a-c input voltage is applied to the circuit which contains an accurate, internal, standard capacitor (C_s), the coils (L_1, L_2, L_3, and L_4 in series), and the unknown capacitor (C_x) in series. When Terminals *X–X* are open, coils L_3 and L_4 set themselves in the plane of the stationary coils (the position of zero on the meter scale). When C_x is connected to the terminals, however, current flowing through the movable coil produces an opposing torque proportional to C_x, and this causes the L_3–L_4 assembly to rotate, moving the pointer over the scale.

The ratio-type meter is available for operation at 100 to 130 and 200 to 250 volts, and at 50, 60, and 500 cps. Like the electrodynamometer ammeter and voltmeter, this type of microfarad meter draws relatively high currents for its operation—150 to 300 ma, depending upon capacitance range and frequency as well as operating voltage.

Capacitance Bridge. As in the case of resistance, the most accurate measurements of capacitance are afforded by bridge methods.

Figure 18–28 shows a rudimentary *capacitance bridge.* With this circuit, the unknown capacitance (C_x), connected to Terminals *X–X*, is compared to an accurately known standard capacitor (C_s). An a-c input voltage must be used. The bridge is balanced (set to null) by adjustment of the variable resistor, R_b, which then establishes a ratio with the precision resistor, R_1, to express the relationship of C_x to C_s. Thus, at null, C_x is to C_s as R_b is to R_1. While, for simplicity, headphones are shown as the null indicator (bridge detector), this might be a sensitive v-t voltmeter.

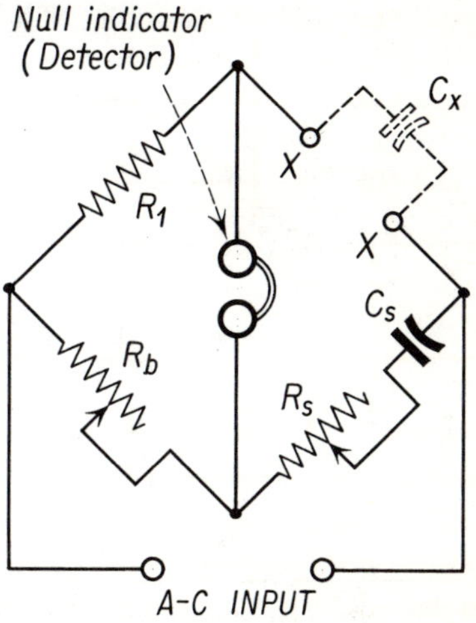

Fig. 18–28 Capacitance Bridge

The bridge circuit has two balances, each of which must be adjusted separately for null—R_b for the reactive component of the unknown capacitor, and R_s for the resistive component (power factor) of the capacitor. R_b and R_s are adjusted in that order. At null

$$C_x = (R_1/R_b)C_s \quad (18\text{–}10)$$

where C_x and C_s are in microfarads, and R_1 and R_b in ohms.

$$\text{Power factor, pf} = (2\pi f R_s C_x)10^{-7} \quad (18\text{–}11)$$

where C_x is in microfarads, f in cycles per second, pf in per cent, and R_s in ohms.

The capacitance range of the bridge may be changed by switching in different values of R_1 (range resistor) and/or C_s. The dial attached to R_b often is graduated directly in microfarads, and its settings at null are multiplied by a factor (0.1, 1, 10, 100, 1,000, etc.) depending upon the value of R_1.

18.6 INDUCTANCE

Like resistance and capacitance, inductance may be measured by the meter method and the bridge method. Here also, the bridge method is the most accurate.

Meter Method. The meter method of inductance measurement is satisfactory only for checking high-inductance coils, which may be measured satisfactorily at low frequencies. Figure 18–29 shows the simple circuit. The current (I) is proportional to the applied voltage (E) and the impedance (Z) of the coil. The inductance, in turn, is proportional to the impedance and frequency.

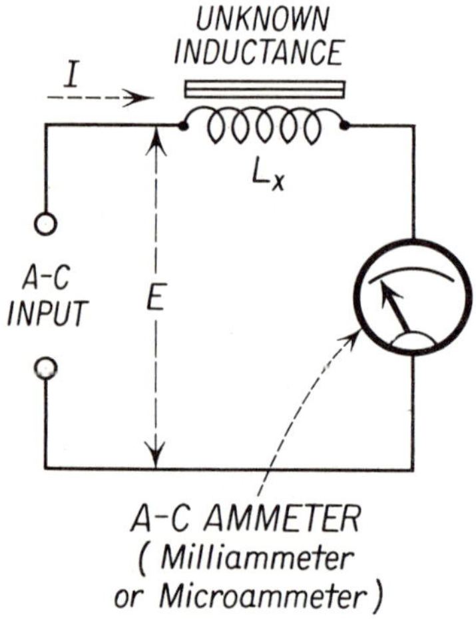

Fig. 18–29 Meter Method of Inductance Measurement

Since the resistive component of high-inductance coils usually is large, it cannot be neglected. The measurement embraces three steps: (1) resistance, (2) current, and (3) inductance calculation. The resistance (R) of the coil is measured with an ohmmeter or Wheatstone bridge before the coil is connected into the circuit. At low test frequencies, the true resistive component will not differ greatly from this measured R value. With the coil in the circuit, the input voltage is adjusted for a deflection (I) in the upper quarter of the current-meter scale. The unknown inductance (L_x) then is calculated with the aid of the following equation which is based on the relationships $Z = E/I = \sqrt{R^2 + X_L{}^2}$ and $X_L = 2\pi f L_x$.

$$L_x = \frac{\sqrt{(E/I)^2 - R^2}}{2\pi f} \quad (18\text{–}12)$$

where L_x = unknown inductance (in henrys)
E = applied voltage (in volts)
I = current deflection (in amperes)
R = resistance (in ohms)
2π = 6.28
f = test frequency (in cycles per second)

Example: The resistance of a certain inductor is checked as 100 ohms. This inductor is tested at 10 v, 60 cps, in the circuit shown in Figure 18–29. The current deflection is 20 ma. Calculate the inductance with the aid of Equation (18–12).

Here $E = 10$ v, $I = 20$ ma $= 0.02$ amp, $R = 100\Omega$, and $2\pi f = 376.8$

From Equation (18–12),

$$L_x = \frac{\sqrt{(10/0.02)^2 - 100^2}}{376.8} = \frac{\sqrt{500^2 - 100^2}}{376.8}$$

$$= \frac{\sqrt{250{,}000 - 10{,}000}}{376.8} = \frac{\sqrt{40{,}000}}{376.8}$$

$$= \frac{489.9}{376.8} = 1.29 \text{ h}$$

The current-meter method is not used for the measurement of small inductances, since the test frequency for these inductances must be made so high for a suitable current deflection that distributed capacitance in the coil and circuit becomes significant. Moreover, the measured resistance (R) does not indicate the true high-frequency resistance component.

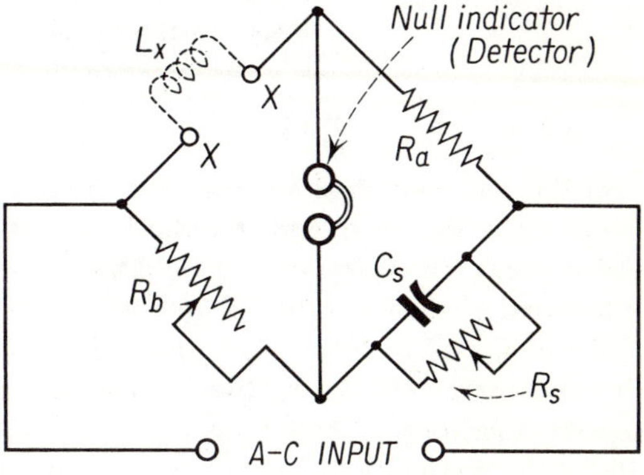

Fig. 18–30 Inductance Bridge

Inductance Bridge. Figure 18–30 shows one type of inductance-bridge circuit. This type is known as the *Maxwell bridge.* In this circuit, the unknown inductance (L_x), connected to Terminals X–X, is compared to an accurately known standard capacitance, C_s. The bridge is balanced for the reactive component of the inductor by means of the variable resistor (R_b) and for the resistive component by means of R_s.

At null, the unknown inductance $L_x = R_a R_b C_s$. Here, L_x is in henrys, R_a and R_b in ohms, and C_s is in farads.

In practical bridges, the dial attached to variable resistor R_b often is graduated to read directly in henrys. Its readings are multiplied by a factor (0.001, 0.01, 0.1, 1, 10, 100, 1,000, etc.) determined by the resistance of R_a. The inductance range is changed by switching the value of R_a and/or C_s.

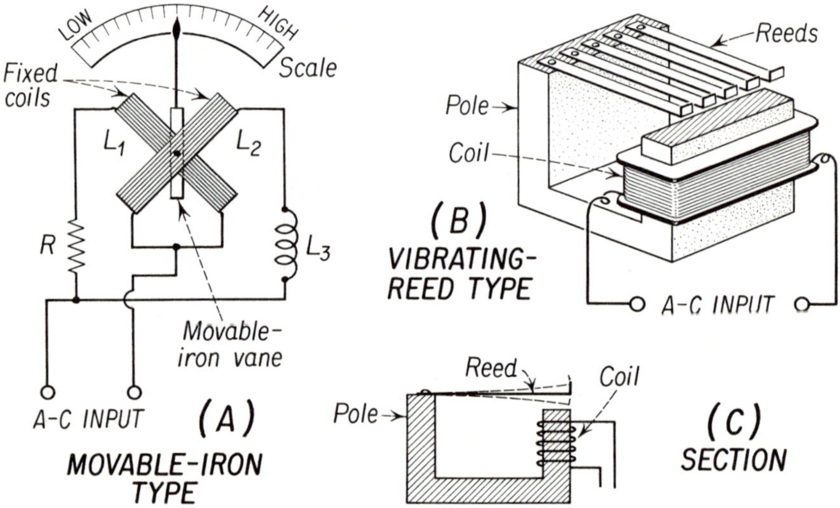

Fig. 18–31 Frequency Meters

18.7 FREQUENCY

Various electronic instruments are available for the measurement of frequencies up to several thousand megacycles. However, these devices are beyond the scope of this book and, for their detailed treatment, the interested reader is referred to the author's book *Basic Electronic Test Instruments* (Rinehart & Company, Inc.).

Figure 18–31 shows two of the frequency meters commonly used for the measurement of power-line and low audio frequencies. Figure 18–31(A) resembles somewhat the ratio-type microfarad meter (Figure 18–27). Coils L_1 and L_2 are attached together at right angles but are stationary. A soft-iron vane, to which the pointer is fastened, rotates within this coil assembly.

The vane rotates to indicate the direction of the resultant field of L_1 and L_2. This direction varies with frequency because of the phase difference introduced by Resistor R and Inductor L_3.

The scale is graduated directly in cycles per second, with some common value (such as 60, 120, 400, or 500 cps) at the center. The complete scale range usually extends from several cycles below to several cycles above the center-scale frequency.

The mechanism shown in Figure 18–31(B) uses tuned vibrating reeds to indicate unknown frequencies. The reed is of magnetic material (iron or steel) and is made a part of a magnetic circuit. One end is attached to one pole of the magnet assembly, while the tip rests a short distance above the other pole. [See Figure 18–31(C).] The reeds may be tuned by grinding each one carefully to a different thickness, although instead they usually are cut carefully to different lengths.

When current is passed through the coil, the alternating magnetic field affects several adjacent reeds, but the one vibrating most vigorously (its tip usually disappearing from view) is the one which is resonant at the applied frequency. In the completed reed-type frequency meter, a dial is mounted over the front of the assembly with a number placed above each reed to indicate its resonant frequency. This allows the frequency to be read directly as the number identifying the reed vibrating most strongly.

18.8 POWER FACTOR

Figure 18–32 shows the circuit of a single-phase *power-factor meter.* This electrodynamometer-type instrument is similar to the microfarad meter (Figure 18–27) and to the dynamometer-type ammeter and voltmeter. The rotating-coil assembly consists of two coils (L_1 and L_2) attached together perpendicularly and rotating inside the stationary coil, L_3.

Current flowing in Coil L_1 is in phase with the input voltage, but the current in L_2 lags the voltage by 90 deg because of the inductance of L_4. The stationary coil (L_3) is in series with one input line and the load.

When the power factor is unity (1.0), the current in L_1 is in phase with that in L_3. The rotating-coil assembly accordingly turns to align the flux of L_1 with that of L_3, the position shown in Figure 18–32. When the load current is 90 deg out of phase with the input voltage, the flux of L_2 is in phase with that of L_3. Coil L_2 therefore turns the rotating-coil assembly 90 deg away from the unity-power-factor (center-scale) position. Rotation is to the left for lagging power factor and to the right for leading. Phase angles between zero and 90 deg cause rotations between the maximum limits just mentioned, in accordance with the interaction between the fields of L_1, L_2, and L_3.

The rotating-coil assembly in a *three-phase power-factor meter* (Figure 18–33) consists of three coils (L_2, L_3, and L_4) fastened together and rotating as a unit within the stationary coil, L_1. Currents in L_2, L_3, and L_4 have a phase difference of 120 deg which produces a rotating magnetic flux. The effect of the latter, in conjunction with the alternating flux of L_1, causes the L_2–L_3–L_4 assembly to rotate to a position proportional to the average three-phase power factor.

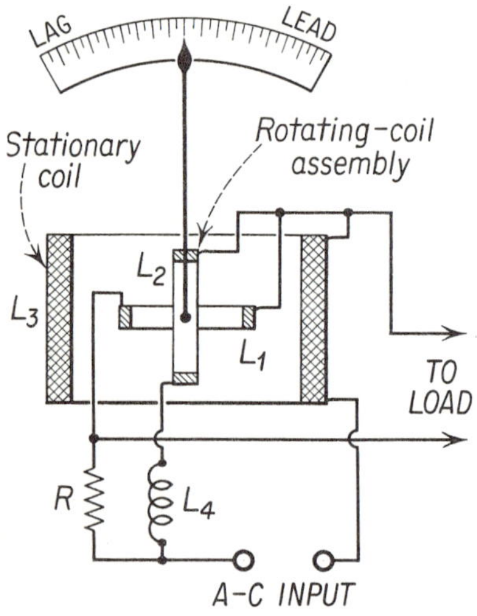

Fig. 18–32 Power-Factor Meter

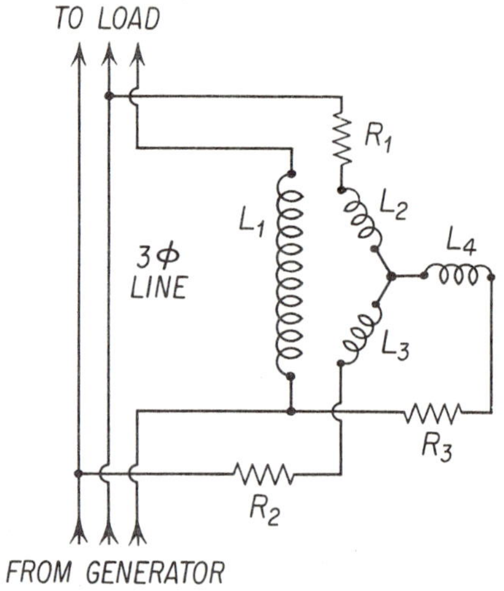

Fig. 18–33 Three-Phase Power-Factor Meter

18.9 GENERAL-PURPOSE TEST SET

Several meters covering the most important test functions may be grouped together in a single, portable instrument. Such a *general-purpose test set* is the industrial analyzer shown in Figure 18–34.

This instrument contains separate meters for a-c volts, amperes, kilowatts, and power factor—a combination which is invaluable for a-c measurements. Other meter combinations may be put together in test sets for other types of a-c and d-c measurements.

Assembled test sets offer the advantage of a wide range of test functions without loose instruments. The instruments normally employed in a given type of measurement activity are available in a compact assembly.

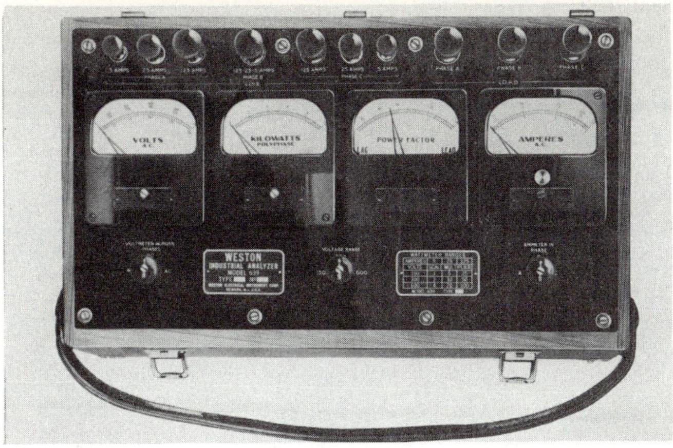

Fig. 18–34 A General-Purpose Test Set (Courtesy, Weston Electrical Instrument Corp.)

18.10 OSCILLOSCOPE

The *oscilloscope* enables wave forms to be observed visually. Although a voltage may be alternating many thousand times per second, its pattern appears stationary on the oscilloscope screen just as if it had been drawn on graph paper. With this instrument, it becomes an easy matter to measure directly peak, rms, average, and instantaneous values and to determine whether a wave is sinusoidal or nonsinusoidal.

The heart of the oscilloscope is the *cathode-ray tube*. This is a special type of vacuum tube in which an electron beam is caused to trace a luminous pattern on a fluorescent screen. This tube is an elaboration upon the simple cathode-ray tube shown in Figure 4–4, Chapter 4.

Figure 18–35 shows details of the cathode-ray tube. The narrow electron beam needed to trace a fine, sharp line on the screen is produced by the *electron gun* mounted in the back end of the evacuated glass tube. The gun [Figure 18–35(C)] consists of an indirectly heated cathode which produces the electrons, and several other electrodes for speeding up the flow of the electrons and focusing the beam. A small beam (pencil) of electrons passes through a hole in a positively charged, cylindrical *control electrode* surrounding the cathode. The beam then continues through the rear and front holes in the first accelerating electrode (*Anode No. 1*) which has a high positive voltage applied to it. Next, it passes through a hole in the focusing electrode. This electrode narrows the beam to an extent proportional to the positive voltage impressed upon the electrode. Finally, the beam passes through a second, highly positive, accelerating electrode (*Anode No. 2*).

After emerging from the gun, the electron beam passes between two pairs of metal plates on its way to the fluorescent screen. One pair, the *horizontal deflecting plates,* is arranged on each side of the beam; the other pair, the *vertical deflecting plates,* is arranged above and below the beam. [See Figure 18–35(A).] Since the beam consists of electrons which are negative particles, it will be repelled when a negative voltage is applied to a deflecting plate and attracted when a positive voltage is

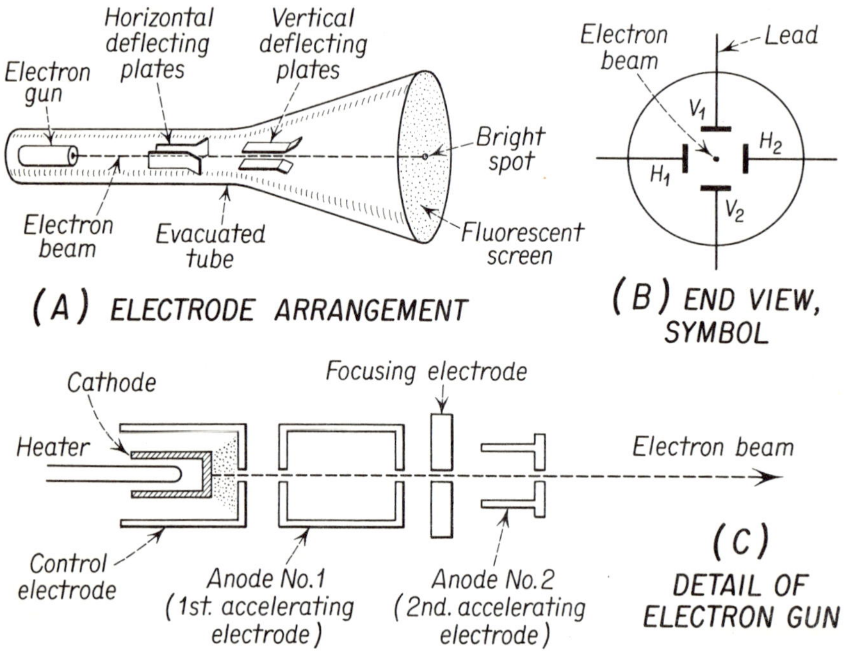

Fig. 18–35 Cathode-Ray Tube

applied. In this way, the beam may be moved from side to side or up and down, and the point at which it strikes the screen thus can be dictated.

The inner surface of the front end of the glass tube is coated with a phosphor, such as willemite, which glows when it is struck by an electron beam. Thus, in Figure 18–35(A), the slender beam produces a bright dot, or point, where it strikes the fluorescent screen.

Figure 18–35(B) shows the arrangement of the deflecting plates, as viewed from the screen end of the cathode-ray tube. The beam is shown passing through the center of the space bounded by the plates. This condition occurs when neither plate has a voltage impressed upon it and there consequently is no deflecting action. The top vertical deflecting plate is labelled V_1, the bottom vertical V_2, left horizontal H_1, and right horizontal H_2.

Figure 18–36 shows the various basic images appearing on the oscilloscope screen. Figure 18–36(A) is the condition referred to in the preceding paragraph, in which neither deflecting plate has a voltage

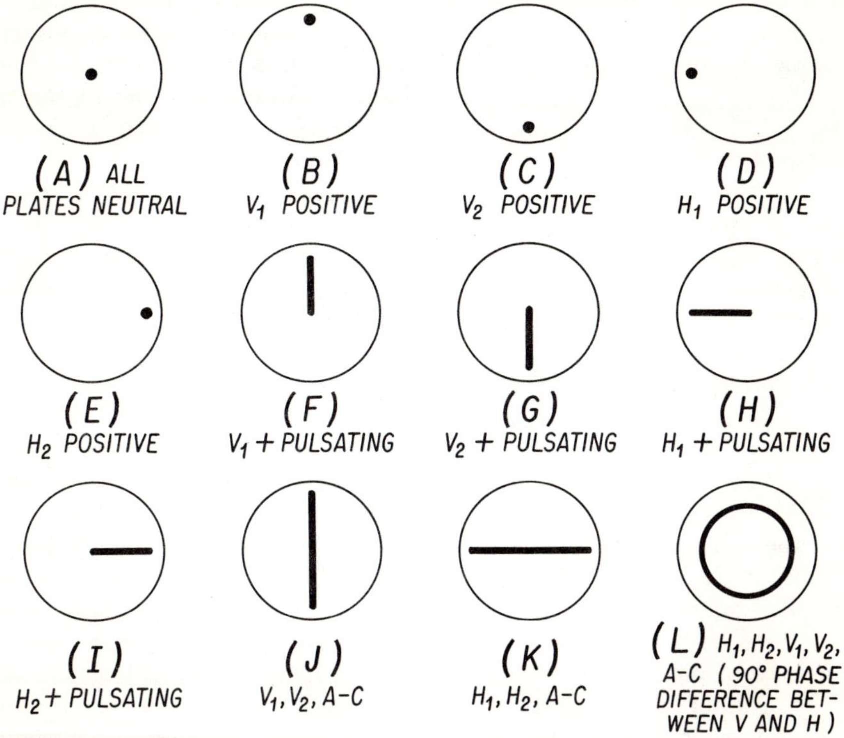

Fig. 18–36 Basic Oscilloscope Deflections

and the beam passes unaffected through the plates to produce a single dot in the center of the screen.

When the top vertical plate (V_1) is made positive with respect to the electron gun, and the horizontal plates and V_2 are left neutral, the beam is attracted upward, and the dot moves vertically to a spot above center, as shown in Figure 18–36(B). The distance through which the spot moves is proportional to the deflecting voltage. When V_2 is made positive, the beam is attracted downward, and the spot moves to a point such as is shown in Figure 18–36(C). Making the left horizontal plate (H_1) positive deflects the spot to the left [Figure 18–36(D)], while making H_2 positive deflects the spot to the right [Figure 18–36(E)].

If a pulsating voltage (positive half-cycles only) is applied to V_1, the beam will travel up from the center of the screen as the positive half-cycle rises from zero to maximum, and will return to center as the half-cycle returns to zero. When the pulsations are very rapid, as they usually are in practice, the eye cannot follow the movement of the spot on the screen. Persistence of vision accordingly makes it appear that the beam has traced a straight-line vertical pattern from center-screen upward, as shown in Figure 18–36(F). Similarly, when a positive-pulsating voltage is applied to V_2, the straight-line trace extends down from center. When the positive-pulsating voltage is applied to H_1, the trace is from center screen to the left [Figure 18–36(H)]; when it is applied to H_2, the trace is to the right [Figure 18–36(I)].

When an a-c voltage is applied to both vertical plates (V_1 and V_2), the spot traces a vertical line upward [similar to Figure 18–36(F)] when V_1 is positive, and downward [similar to Figure 18–36(G)] when V_2 is positive. The result is a vertical straight-line trace extending both above and below center [Figure 18–36(J)]. Similarly, when an a-c voltage is applied to both horizontal plates (H_1 and H_2), a horizontal straight-line trace [Figure 18–36(K)] is obtained.

When voltages are applied simultaneously to both pairs of plates, the spot displacement becomes the resultant of the horizontal and vertical movements. The instantaneous position of the spot therefore is intermediate between the horizontal and vertical axes—a position determined by the relative, instantaneous horizontal and vertical deflecting voltages. Thus, if a-c voltages of the same amplitude, frequency, and phase are applied to horizontal and vertical plates simultaneously, a straight 45-deg line will be traced across the screen. When the two voltages are 90 deg out of phase, a circle will be traced [Figure 18–36(L)].

In order for the oscilloscope to be useful for observing wave forms accurately, it is necessary that the spot be moved across the screen horizontally at a uniform rate, in order to establish a time base. Furthermore, the spot always must move from left to right so as not to "back trace." The voltage to be studied then is applied to the vertical deflecting plates.

A satisfactory method of *sweeping* the spot horizontally is to apply saw-tooth-wave voltage to the horizontal plates. Such a wave is shown in Figure 18–37. At the negative value, $-E$, the spot is at the extreme left end of the screen. The voltage then rises linearly from A to B, moving the spot at an even rate across the screen to the right end. The extreme right end of the screen corresponds to voltage $+E$. The voltage now changes very rapidly from B to C (back to $-E$), and this action snaps the spot quickly back to the starting point. The voltage then rises once again linearly to D, sweeping the spot from left to right uniformly to repeat the cycle. The saw-tooth voltage is generated by a *sweep oscillator.*

Figure 18–38 illustrates the condition when a saw-tooth sweep voltage (*timing wave*) is applied to the horizontal plates of a cathode-ray tube and a sine-wave voltage to be studied is applied to the vertical

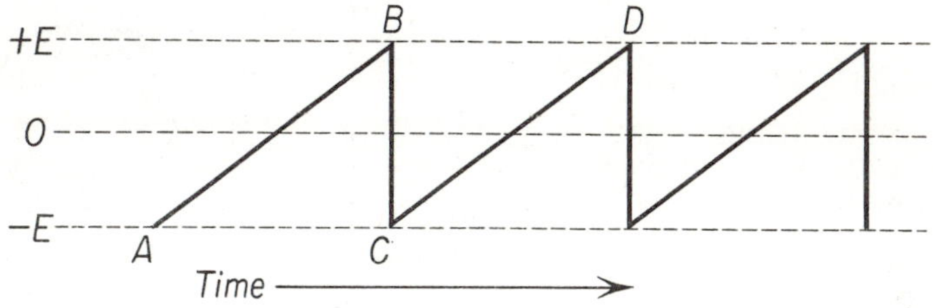

Fig. 18–37 Sawtooth Sweep Voltage

plates. The resulting pattern on the screen is a reproduction of the sine wave. Waves having other shapes than sinusoidal may be reproduced in the same way. Thus, the oscilloscope is employed to observe the wave forms of speech, music, radio and television signals, heartbeats, nerve currents, switching transients, and many other phenomena other than simple alternating voltages.

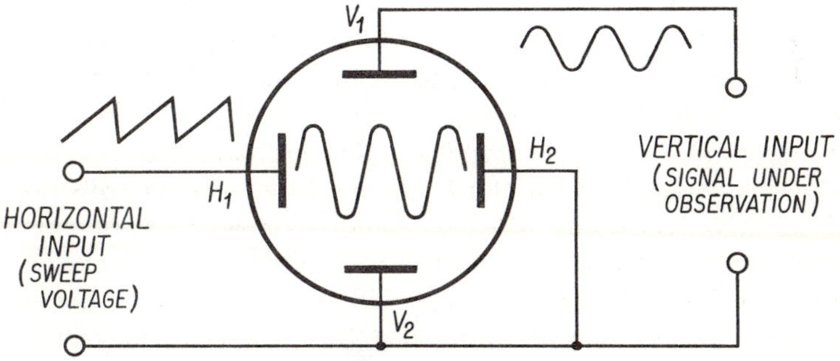

Fig. 18–38 Reproduction of Sine-Wave Signal

Figure 18–39 shows the simplest configuration of a complete oscilloscope. Separate voltage amplifiers are employed for the horizontal and vertical deflecting plates. This allows full deflection to be obtained with weak signal voltages. Switch S permits the use of either an external signal voltage (applied to input terminals 3 and 4) or a saw-tooth voltage (generated internally by the sweep oscillator) to be used to sweep the trace horizontally. When S is set to its Position *A*, the external signal is selected; at Position *B*, the saw-tooth sweep voltage is selected.

For additional information regarding use of the oscilloscope in electronic measurements, see the author's *Basic Electronic Test Instruments,* and Jacob H. Ruiter's *Modern Oscilloscopes and Their Uses* (both Rinehart & Company, Inc.).

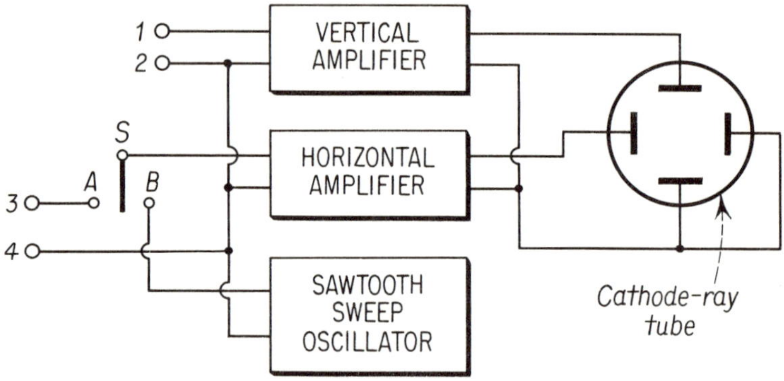

Fig. 18–39 Functional Oscilloscope

18.11 INDUSTRIAL INSTRUMENTATION

Electrical measurements provide the engineer, electrician, and researcher with a means for monitoring and measuring industrial processes, a number of which are nonelectrical in nature. An everyday example is the fuel gauge in an automobile. Although this is an electric meter (usually a d-c milliammeter), it gives an indication of the amount of gasoline in the tank. It is a good example also of a simple remote-type of indicator, since it displays its vital information at the dashboard and saves the driver the inconvenience of going back to the tank to take a reading.

Instrumentation systems almost always consist of a *transducer* and *indicator.* The transducer is a device which converts some nonelectrical phenomenon (such as temperature, humidity, pressure, sound, etc.) into electrical impulses and transmits the latter. The indicator is an electric meter or other instrument the deflection of which may be calibrated in applicable units (such as degrees, pounds per square inch, gallons, etc.). Some transducers generate voltages directly, while others vary the value of a current flowing through them to the indicator. Some transducers supply enough energy to operate an indicator directly [Figure 18–40(A)], while the output of others is so low that an amplifier must be inserted between the transducer and indicator [Figure 18–40(B)].

Figure 18–41 shows a simple, liquid-level gauge. Here, the transducer is a rheostat (R_1) operated by a float riding on the surface of the liquid inside a tank. As the level rises, the float ascends vertically and, through lever action, reduces the resistance setting of R_1. This increases the current flowing in the circuit from Battery B. The increased current

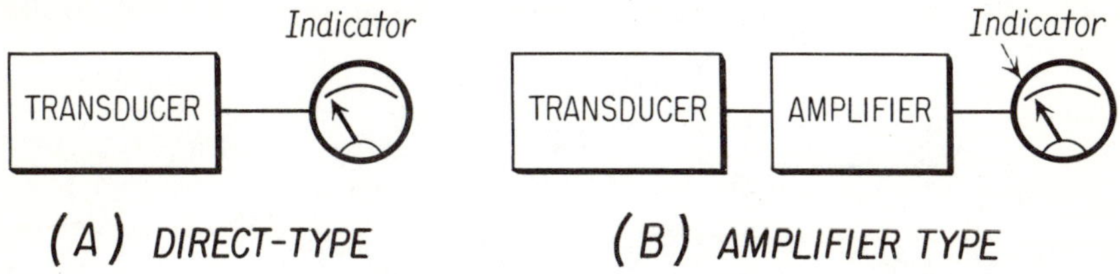

Fig. 18–40 Transducer-Indicator Systems

deflects the meter upscale to indicate a full tank. The meter may be calibrated to read gallons, feet, or some comparative units (such as FULL, HALF, and EMPTY). Resistor R_2 serves to limit the maximum current to the full-scale deflection of the meter.

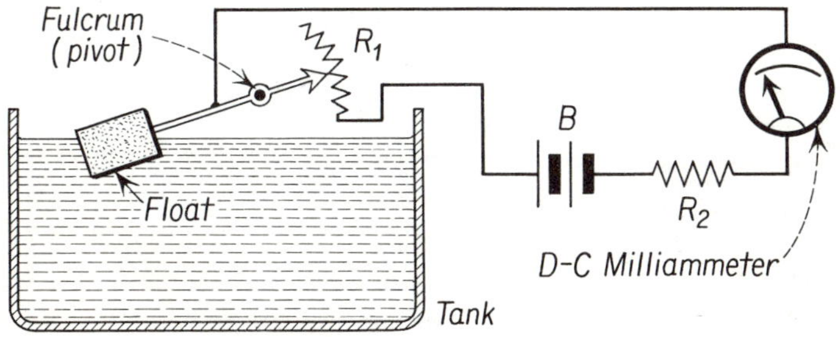

Fig. 18–41 Liquid-Level Gauge

Many hundreds of industrial instrumentation systems have been developed, and new ones constantly are being invented. Entire books are devoted to this one subject. This section can describe only a few examples which have been chosen to illustrate typical applications of electrical measurements to industrial processes.

Temperature Measurement. Figure 18–42 shows several circuits for temperature measurement. In Figure 18–42(A), the thermocouple generates a d-c voltage proportional to the temperature to which it is exposed. This voltage deflects the sensitive d-c millivoltmeter. The scale of the latter may be calibrated to read in degrees.

In Figure 18–42(B), the thermocouple has been replaced with a *thermistor,* which is a heat-sensitive resistor. Current from Battery B flows through the thermistor, and its value is set by the resistance of the latter. Rheostat R is set to limit the maximum amount of current which may flow through the meter to the safe full-scale value.

A more sensitive arrangement is shown in Figure 18–42(C) in which the thermistor is one arm of a resistance bridge. (See Section 18.4 for an explanation of bridge action.) The bridge is nulled, by adjustment

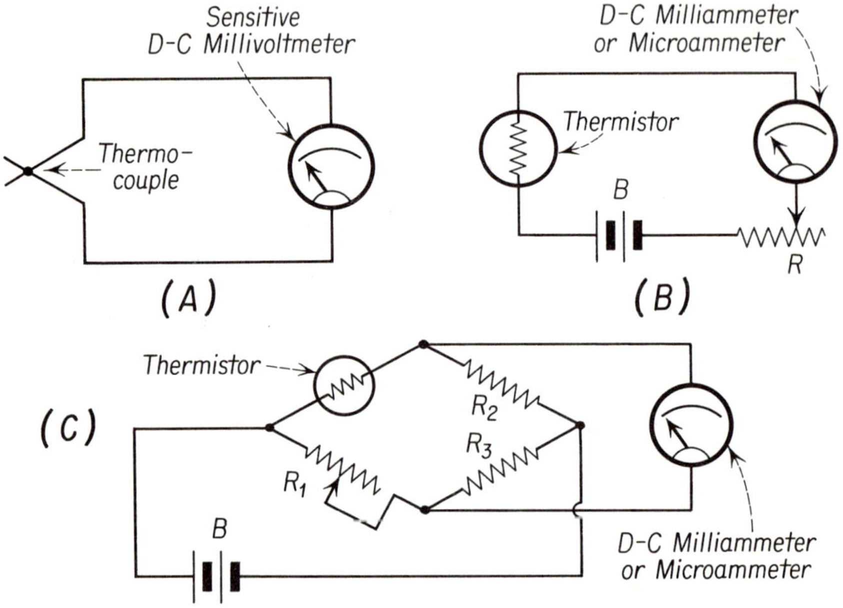

Fig. 18–42 Temperature Measurement Circuits

of R_1, with the thermistor cooled to zero degrees centigrade (or to some other desired reference temperature). As the thermistor subsequently is heated, its changing resistance unbalances the bridge and current proportional to the temperature passes through the meter.

Humidity Measurement. In one type of electrical hygrometer, or moisture meter, the transducer is a *sensor unit* the resistance of which decreases as the moisture content of the surrounding air increases. These units are used with alternating current only.

When the sensor is connected in the circuit shown in Figure 18–43, the deflection of the a-c milliammeter will be proportional to the humidity. Rheostat R permits the maximum possible current passed by the sensor to be limited to the safe full-scale deflection of the meter.

Speed Measurement. Speed is measured directly by the use of a *tachometer*. Figure 18–44 shows the arrangement of a simple electrical tachometer. The transducer here is a small, magneto-type a-c generator. The indicator is an a-c voltmeter. The generator shaft is coupled mechanically to the rotating machine under observation.

The faster the generator shaft turns, the higher is the a-c output. The voltmeter scale therefore may be calibrated to read directly in

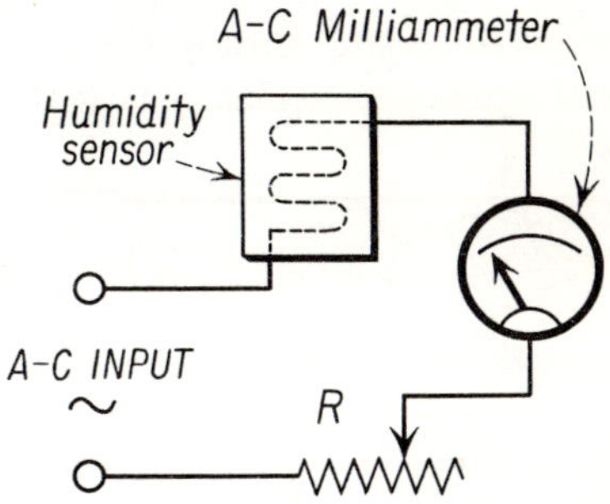

Fig. 18–43 Humidity Measurement

revolutions per minute or revolutions per second. A variety of speed ranges may be obtained by means of multiplier resistors for the voltmeter.

The speed of a rotating machine or the rate of movement of a recurrently moving device (such as a vibrator) may be measured

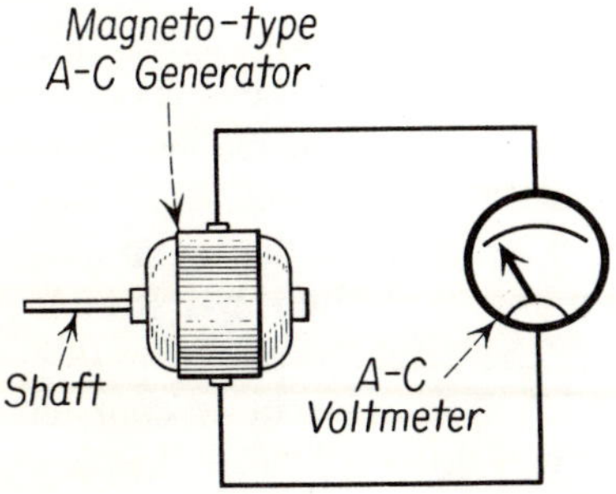

Fig. 18–44 Simple Tachometer

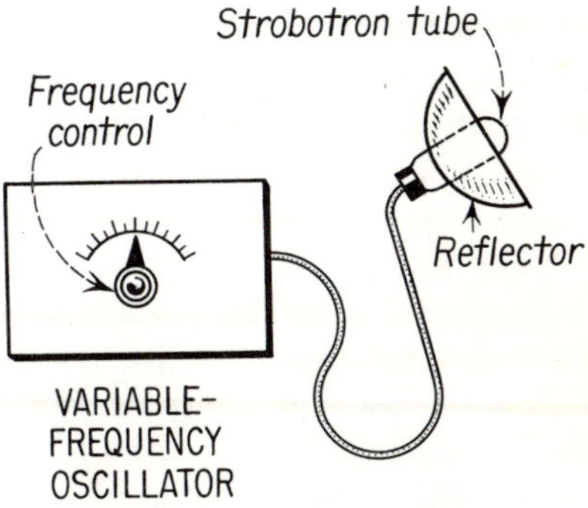

Fig. 18–45 Stroboscope

indirectly by means of a *stroboscope.* This instrument emits a series of bright, regularly occurring, short light flashes which illuminate the moving object. When the flashing rate equals the rotational speed or vibration rate of the object, the latter appears to stand still. If the flashing mechanism is calibrated, the speed or rate then may be read directly from the dial of the flashing-rate control.

Figure 18–45 shows the arrangement of an electronic type of stroboscope. The flashes are emitted by a strobotron tube mounted in a suitable reflector for intensification of the light. The tube is switched on and off by short-timed voltage pulses delivered by a special oscillator and power supply. The oscillator frequency control is graduated in flashes per minute or flashes per second.

Another important industrial use of the stroboscope is the "stopping of machines in motion." Since the machine appears to stand still when the flashing rate is adjusted properly, the various parts of the machine may be observed closely for bending, whipping, hunting, etc., while in actual operation. This makes the stroboscope a powerful dynamic tool.

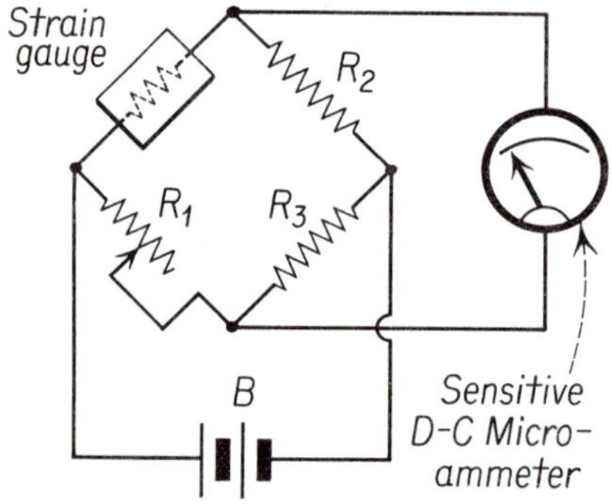

Fig. 18–46 Strain Measurement

Strain Measurement. The measurement of the effects of stress and strain in machinery of all kinds is of the greatest importance in the design of such equipment and in its preventive maintenance. The strain factor is studied minutely in the development of airplanes, guided missiles, automobiles, railway rolling stock, and building structures.

The heart of practical measurements of this type is the *strain gauge,* a tiny wire device the resistance of which changes when it is stretched. The strain gauge is cemented to the part of the machine under investigation and in one application of this type of measurement is connected into a resistance bridge circuit, as shown in Figure 18–46. The bridge is nulled by adjustment of R_1. When the machine part subsequently is stressed, the strain gauge becomes stretched, and its change in resistance unbalances the bridge, causing the meter to be deflected by an amount proportional to the stress.

The output of the bridge often is so low, because of the small resistance changes in the strain gauge, that an amplifier must be inserted into the measurement setup between the bridge and indicator.

Thickness Measurement. Figure 18–47 shows an electronic setup for measuring the thickness of a continuously moving strip of nonconducting material. The beat-note method is employed.

The two rigidly supported metal plugs, *A* and *B*, form the plates of a capacitor. The thickness of the material passing between the faces of these plugs largely determines the capacitance. The plugs are connected in parallel with the tank circuit of the variable-frequency r-f oscillator which usually is tuned to a low frequency such as 100 kc. If the capacitance of the plug assembly (*A–B*) changes, the frequency

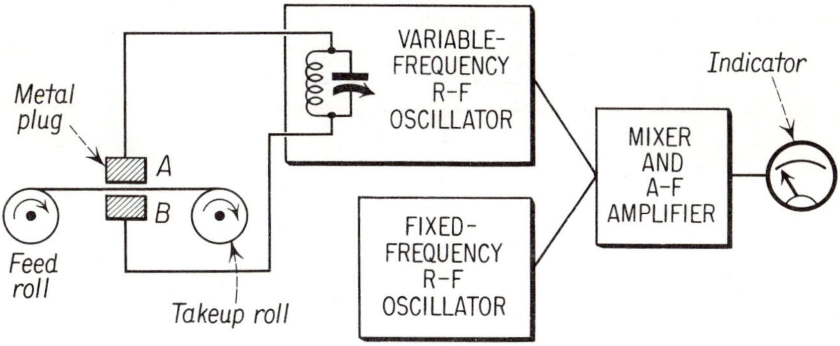

Fig. 18–47 Setup for Thickness Measurement

of this oscillator will be changed accordingly. The output of this oscillator is fed into a mixer and a-f amplifier simultaneously with the output of a fixed-frequency r-f oscillator.

In operation, the system is set to zero beat by tuning the variable-frequency oscillator, with the correct thickness of material between plugs *A* and *B*. There now is no beat note, and the indicator meter is not deflected. When the thickness of the material changes, the variable-frequency oscillator is detuned proportionately, the frequency increasing when the thickness increases, and vice versa. An audio-frequency beat note now occurs between the two oscillators, and the note is amplified by the a-f amplifier and causes the meter to be deflected. The meter scale may be graduated in thickness units such as inches, mils, centimeters, etc.

Vibration. Studies of vibration in machinery, vehicles, and other equipment are made with the aid of amplifying and indicating instruments typified by the arrangement shown in Figure 18–48. Here, the transducer is a piezoelectric crystal pickup (similar to a microphone) which is placed in contact with the vibrating object. The resulting electrical impulses generated by the pickup as a result of the vibration are amplified by a high-gain amplifier and caused to deflect the indicating meter. The scale of the meter is graduated to indicate vibration amplitude.

A complete, professional vibration meter often has a circuit also for measuring the various vibration frequencies, as well as provision for connecting headphones to determine the aural nature of the impulses.

The electrical maintenance departments of many large, modern factories make frequent vibration studies in connection with the installation and operation of production machinery and other equipment such as air conditioners, heaters, elevators, and the like. Such studies often are a part of noise-abatement programs.

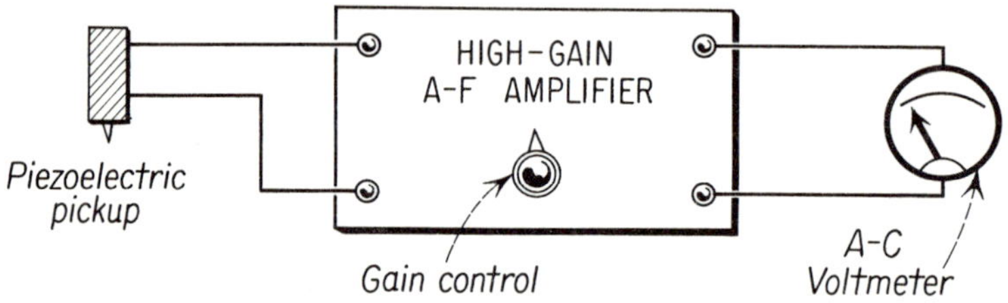

Fig. 18–48 Vibration Measurement

In the aircraft and automotive industries, vibration studies using the general type of equipment illustrated by Figure 18–48 are made in the interest of quieter interiors. A different type of equipment is employed in the study of vibration in functional aircraft and automotive structures.

Telemetering. The word "telemetering" means literally "measuring at a distance." But the art of telemetering involves more than the simple procedure which the reader might guess of locating a meter some distance from an electrical system to which it is connected. Radio links often are involved in the process. For example, temperature, velocity, strain, voltages, currents, and other parameters need to be known on a guided missile or rocket in flight, and we have no access except by means of a radio link with the ground station. Similarly, we might want to know the temperature or barometric pressure at an unmanned weather observation post atop a distant mountain, or to check various phenomena in the no-man's land of the immediate zone of an exploding atomic bomb.

Like other fields of electrical instrumentation, telemetry is wide in scope. A great many systems have been invented. Many of them are extremely complicated. As simple examples, changes in temperature, humidity, pressure, and similar phenomena may be converted by means of transducers (such as two-plate variable capacitors of suitable construction) into capacitance changes. These capacitance changes then may be employed to control the frequency of a radio transmitter. At the receiving station, the resultant variations in signal frequency may be used to obtain corresponding deflections of a meter.

A somewhat more stable method uses the phenomenon changes at the transmitting end to change the modulation frequency of an a-m radio transmitter. Various modulating tones obtained by this process then are interpreted at the receiving station (where they actuate indicators) as various values of the phenomena.

SUMMARY OF CHAPTER 18

Current Meters are classified, according to their ranges, as *ammeters, milliammeters,* and *microammeters.* The position of a current meter in an electric circuit is in series with the current-carrying branch. The meter, accordingly, must have low internal resistance.

The *d'Arsonval* meter contains a moving coil rotating between the poles of a permanent magnet. The pointer is attached to the coil.

Some other types of current meters include the *solenoid type, movable-iron type,* and *electrodynamometer type.* These types may be used for either a-c or d-c measurements. The d'Arsonval movement is adapted for alternating current by means of a rectifier.

A **Shunt Resistor** increases the range of a current meter. The formula for calculating the shunt resistance value is $R_s = R_m/(n - 1)$.

Voltmeters may be of the following types: *d'Arsonval, movable-iron, solenoid,* and *electrodynamometer.* Special voltmeters include the *vacuum-tube* and *clamp types.* The position of a voltmeter in an electric circuit is in parallel with the voltage source (or voltage-dropping branch). In order not to load the circuit, the voltmeter, accordingly, must have a higher internal resistance than the circuit branch across which it is connected.

A series **Multiplier Resistor** converts a d-c meter into a voltmeter. The formula for multiplier resistance calculation is $R = E/I$.

Voltmeter Resistance (sometimes called *sensitivity*) is expressed in *ohms per volt.* $\Omega/V = R/E$.

The **Potentiometer** permits precision measurements of voltage. In this instrument, the unknown voltage is compared to the emf of a standard cell.

Power is measured by means of: (1) *ammeter method* ($P = I^2R$), (2) *voltmeter method* ($P = E^2/R$), (3) *voltmeter-ammeter method* ($P = EI$), and (4) *wattmeter* (direct reading in watts).

Resistance may be measured by means of: (1) *current-meter method* $[R = (E/I) - R_m]$, (2) *voltmeter method* $[R = (Er/e) - r]$, (3) *ohmmeter* (direct reading in ohms), or *Wheatstone bridge* $[R_1 = (R_3/R_4)R_2]$.

Capacitance may be measured by means of: (1) *current-meter method*

$[C = 10^6/(2\pi f X_c)]$, (2) *voltmeter method* $[C = (10^6 E)/(2\pi f I)]$, (3) *microfarad meter* (direct reading in microfarads), or (4) *capacitance bridge* $[C_x = (R_1/R_b)C_s]$.

Inductance may be measured by means of *meter methods* or *bridge methods.* Meters are employed to measure the a-c voltage across and a-c current through a coil, and the unknown inductance determined from the equation $L_x = \frac{\sqrt{(E/I)^2 - R^2}}{2\pi f}$

The **Maxwell Bridge** is one type of inductance bridge. When the bridge is set for null, the unknown inductance $L_x = R_a R_b C_s$.

Frequency may be measured by means of a number of electronic and nonelectronic instruments, depending upon the magnitude of the frequency. Two types of *frequency meter* widely used at low frequencies are the *iron-vane* and *vibrating-reed* types.

Power Factor at power frequencies is measured with an electrodynamometer-type *power-factor meter.*

The **Oscilloscope** is a convenient instrument for the visual examination of wave forms. Its basis is the *cathode-ray tube* in which an electron beam is deflected both horizontally and vertically by the electrostatic action of voltages applied to pairs of deflecting plates. The deflected beam strikes a fluorescent viewing screen situated in the end of the tube, and produces a visible-light trace of the beam movement. The deflecting voltages usually are amplified before they are applied to the deflecting plates.

QUESTIONS AND EXERCISES

1. Describe three types of current meters.

2. Calculate the value of shunt resistance required to convert a 0–1 milliammeter into a 0–1-amp instrument. $(R_m = 100\Omega)$.

3. Calculate the value of multiplier resistance required to convert a 0–50 d-c microammeter into a 0–1 d-c voltmeter.

4. A certain voltmeter has a 500,000-ohm multiplier resistor for its 0–500-v range. Calculate the ohms-per-volt rating of this instrument.

5. Describe the operation of a simple v-t voltmeter.

6. Explain the operation of a voltage-measuring potentiometer.

7. A certain electric heater draws 5 amp from the 115-v power line. Calculate its power drain and explain how this power might be determined by: (*a*) ammeter method; (*b*) voltmeter method; (*c*) voltmeter-ammeter method.

8. Describe the construction of a watt-hour meter.

9. A potential of 6 v is applied to an unknown resistance connected in series with a 0–10-v voltmeter (resistance rating of meter = 20,000 ohms per volt). The meter deflection is 1½ v. What is the value of the unknown resistance?

10. Explain the measurement of resistance with a v-t voltmeter.

11. Explain the operation of a Wheatstone bridge in resistance measurement.

12. Explain three methods of measuring capacitance.

13. Using the voltmeter method for capacitance measurement, the 60-cps voltage drop across the capacitor under test is 2 v for a current of 1 ma. Calculate the value of the unknown capacitance.

14. Explain the operation of the Maxwell bridge.

15. Using the meter method for inductance measurement, the following values were checked: $R = 100\Omega$, $E = 10$ v at 400 cps, and $I = 10$ ma. Calculate the value of the unknown inductance.

16. At null, the adjustable arms, R_a and R_b, of a Maxwell bridge are set to 1,500 and 10,000 ohms, respectively. The standard capacitor in the bridge is 0.1 μf. Calculate the value of the unknown inductance.

17. Describe the operation of a vibrating-reed frequency meter.

18. Describe the operation of a power-factor meter.

19. Describe the construction and operation of the electron gun in a cathode-ray tube such as is used in an oscilloscope.

20. Explain how a sine-wave alternating-voltage wave form becomes displayed on the screen of an oscilloscope.

INDEX

Index

B

C

D

E

J

K

L

M

N

O

P

S

T

U

V

W

X